"An engaging and accessible book which weaves together theoretical arguments and diverse case studies to enhance student understanding. It draws on voices and debates from across the region to provide insights on geographical concepts which are often ignored in English-language scholarship."

Professor Katie Willis, *Royal Holloway, University of London*

"*Latin American Geographies* is setting its subfield free from an Anglophone echo chamber. By creating genuine dialogues across continents and languages, this volume breathes new life into a Latin American geography as a discipline and a practice. Lively and persuasive, it is the new essential text for students and scholars alike."

Dr. Andrea Marston, *Rutgers University, New Brunswick*

W0234687

Latin American Geographies

Latin American Geographies introduces student readers to cutting-edge scholarship on a range of topics from Indigenous geographies to sustainable development and dependency theory. The book is written primarily by a Latin American-based authorship and blends complex theory with in-depth case studies in an accessible way for students with little prior knowledge.

Each chapter contains a general overview of the topic and includes summary boxes, review questions and annotated further readings. The book is divided into three sections. Section 1, "Core Themes," gives the reader the necessary historical, conceptual and theoretical tools to make sense of and engage in contemporary geographical debates in Latin American geographies. It is divided into four areas, covering major sub-themes (historical and colonial geographies, political geographies, economic and urban geographies, and development and environmental geographies). Section 2, "Key Perspectives," outlines key identities and positionalities that have had a profound impact on Latin American geographies, exemplifying their significance through a range of case studies. Section 3, "Uneven Processes," provides an in-depth analysis into core geographical trends across three main themes: ecologies, urbanisation and resistance.

The book is unique in providing an introduction to Latin American geography that showcases the ideas of some of the region's leading geographical thinkers. Aimed at undergraduate students, chapters will also be of relevance to advanced researchers looking for introductions to specific areas. The book is designed for use in the classroom as well as for independent learning.

Sam Halvorsen is Reader in the School of Geography at Queen Mary University of London. His research examines the role of territory in grassroots politics and, more broadly, political participation and democracy in Latin American cities. He has published widely in journals such as *Progress in Human Geography*, *Transactions of the Institute of British Geographies* and the *Annals of the American Association of Geographers*. He is founder and chair of the Latin American Geographies Research Group of the Royal Geographical Society. He is currently editorial board member of *Transactions of the Institute of British Geographers* and *Punto Sur* and sits on the international editorial boards of *Antipode: A Radical Journal of Geography*, *Journal of Latin American Geography* and *Third World Quarterly*.

Latin American Geographies

Edited by Sam Halvorsen

Routledge
Taylor & Francis Group

LONDON AND NEW YORK

Designed cover image: Getty Images

First published 2025
by Routledge
4 Park Square, Milton Park, Abingdon, Oxon OX14 4RN

and by Routledge
605 Third Avenue, New York, NY 10158

Routledge is an imprint of the Taylor & Francis Group, an informa business

British Library Cataloguing-in-Publication Data
A catalogue record for this book is available from the British Library

Library of Congress Cataloging-in-Publication Data
Names: Halvorsen, Sam, editor.
Title: Latin American geographies / edited by Sam Halvorsen.
Description: First edition. | New York : Routledge, 2025. | Includes bibliographical references and index.
Identifiers: LCCN 2024033824 (print) | LCCN 2024033825 (ebook) | ISBN 9781032554839 (hbk) | ISBN 9781032554815 (pbk) | ISBN 9781003430926 (ebk)
Subjects: LCSH: Geography—Study and teaching—Latin America. | Latin America—Geography.
Classification: LCC G76.5.L29 L38 2025 (print) | LCC G76.5.L29 (ebook) | DDC 918.0071—dc23/eng/20241209
LC record available at https://lccn.loc.gov/2024033824
LC ebook record available at https://lccn.loc.gov/2024033825

ISBN: 978-1-032-55483-9 (hbk)
ISBN: 978-1-032-55481-5 (pbk)
ISBN: 978-1-003-43092-6 (ebk)

DOI: 10.4324/9781003430926

Typeset in Sabon
by Apex CoVantage, LLC

Contents

CONTENTS

Figures

Tables

Contributors

Mariana Arzeno holds a degree and a PhD in Geography from University of Buenos Aires (UBA). She is an independent researcher at CONICET, assistant professor at Department of Geography (UBA) and postgraduate professor at various universities in Argentina. She coordinates the Emerging Geographies Studies Group (Institute of Geography-UBA). Her research lies at the intersection of political and rural geography with an interest in processes of resistance and disputes around socio-spatial order(ing) among state and social organizations; resistances and politics around revalorization of places and traditional food; and alternative food production/consumption networks. She has published several articles in scientific journals and books. She recently co-edited the book: *Order, Regulate, Resist. Political Disputes Over Space* (2021).

Matthew C. Benwell (School of Geography, Politics and Sociology, Newcastle University) is a political geographer at Newcastle University (UK) with a particular interest in people's everyday engagements with geopolitical events past and present. His research contributes to debates in feminist geopolitics, focusing specifically on the perspectives and experiences of children, young people, and families, and has engaged the concepts of intergenerational memory (of, for example, the Falklands/Malvinas War), alter-geopolitics, and everyday nationalism, as well as the diplomatic practices and performances of citizens and diplomats in the context of territorial sovereignty disputes. His regional specialism focuses on the Southern Cone, encompassing Argentina, Chile, British Overseas Territories in the South Atlantic (including the Falkland Islands/Islas Malvinas), and Antarctica. He was formerly a Leverhulme Early Career Fellow working on a project exploring 'The making of the geopolitical citizen: the case of the Falklands/Malvinas'. More recently, he has been involved in projects (funded by HERA and UKRI) exploring the everyday experiences of young refugees and asylum seekers in public spaces and during the Covid-19 pandemic.

Eric D. Carter is the Edens Professor of Geography and Global Health at Macalester College in Saint Paul, Minnesota. His interdisciplinary research combines geography, environmental history, and the history of public health, with a regional focus on Latin America. His main areas of research interest include the political ecology of infectious and vector-borne diseases; environmental and social history of disease control; social medicine and public health in Latin America; and the biopolitics of public health interventions. He has published two books, *Enemy in the Blood: Malaria, Environment, and Development in Argentina* (2012) and *In Pursuit of Health Equity: A History of Latin American Social Medicine* (2023), and articles in such journals as *Social Science and Medicine*, *Local Environment*, *Journal of Historical Geography*, and *Health and Place*.

Veronica Crossa is Associate Professor in Urban Studies at the Centro de Estudios Demográficos, Urbanos y Ambientales at El Colegio de Mexico. Her research lies at the intersection of urban political and cultural geography with an interest in the multiplicity of ways in which 'order' is (re)produced and performed at the level of everyday life in the streets of contemporary cities. Most of her work centres around changing notions of urban order resulting from the implementation of revitalization policies in Mexico City's public spaces. Her recent book, entitled *Luchando por un Espacio en la Ciudad de México: Espacio público urbano y el comercio ambulante* examines the ways in which street vendors in Mexico City negotiate and struggle over changing power structures in their everyday lives.

Mónica Farías holds a PhD in Geography from the University of Washington. She is a faculty member at the Department of Geography at the University of Buenos Aires (UBA), a researcher at the Institute of Geography (UBA), and a grassroots political activist. Her research focuses on the modalities of political change and the geographies of poverty and inequality, specifically in the city of Buenos Aires. She has published both in English and Spanish and has participated in several national and international projects and collaborations with colleagues from the Global South and North. She has taught graduate seminars as guest lecturer in Universidad Nacional de La Plata and Universidad Nacional del Litoral.

Bernardo Mançano Fernandes is a geographer at São Paulo State University, holder of the UNESCO Chair in Education for the Countryside and Territorial Development, and professor at the Postgraduate Program in Territorial Development in Latin America and the Caribbean – TerritoriAL. He graduated in Geography (1988), MA in Geography (1994), and Ph.D. in Geography (1999) from the University of São Paulo. He completed a postdoctorate at the Institute for the Study of Latin American and Caribbean – University of South Florida (2008) and became Full Professor at UNESP (2013). He is professor of undergraduate and postgraduate studies in Geography at the Universidade Estadual Paulista (Unesp), Presidente Prudente campus. Professor of the Post – Graduate Program in Territorial Development in Latin America and the Caribbean – TERRITORIAL – the Institute for Public Policy and International Relations – IPPRI/UNESP, São Paulo campus, and Vice President of the International Geographical Union – IGU.

Federico Ferretti received his PhD from the Universities of Bologna and Paris 1 Panthéon-Sorbonne in 2011. After conducting research and teaching experiences in Italy, France, Switzerland, Brazil, and Ireland, he is now Full Professor of Geography at the University of Bologna. His research and teaching interests lie in the philosophy and history of geography, cultural and historical geography, and in the international circulation of geographical knowledge through critical, feminist, decolonial, and anarchist approaches, with a special focus on Latin America. He authored, co-authored, or edited 15 books in Italian, French, and English, and published research articles in the major international peer-reviewed journals in his field of study, in English, French, Italian, Spanish, and Portuguese. He is Secretary/Treasurer for the Commission History of Geography of the International Geographical Union, Article Forum Editor for *Dialogues in Human Geography*, and Associate Editor for the *Journal of Historical Geography*.

Jean Grugel is Professor of Development Politics at the University of York, UK and Research Professor at the Institut Barcelona Estudis Internacionals (IBEI), Spain. Her research interests centre on the political economy of development, regionalism, neoliberalism and (post)neoliberalism, and human rights, with a focus on Latin America. She also works on health governance in development, with a particular focus on gender and care. Recent publications include *The Gendered Face of Covid-19 in the Global South* (Bristol University Press, 2022 with Matt Barlow, Tallulah Lines, Maria Eugenia Giraudo, and Jessica Omukuti), 'Visual narratives of care and reproduction in forced migration: women displaced from Venezuela to Brazil' *Bulletin of Latin American Research* (2024, with Tallulah Lines, Bruno Curcio, Pia Riggirozzi, and Natalia Cintra), 'Regional governance, gender and the Covid-19 pandemic in the global South' *Globalizations*, (2024, with Matt Barlow), 'Imaginaries of Soy and the Costs of Commodity-Led Development: Reflections from Argentina' *Development and Change* (2022, with Maria Eugenia Giraudo), and 'Depletion, intersectionality and the limits of social policy: child carers in Mexico City' *European Journal of Politics and Gender* (2020, with Susana Macias and Shirin Rai).

Ana Laura Zavala Guillen is a Lecturer in Human Geography at the University of Northumbria, having previously held a British Academy Postdoctoral Fellowship in the School of Geography, Queen Mary University of London. Her research combines participatory mapping, participatory use of archival records, and oral tradition to delve into Maroon spatialities and territories from colonial times to the present day. Her latest research project, 'Blackness in Resistance: Territory and Regime Violence in Uruguay', was sponsored by the British Academy Postdoctoral Fellowship. She is also interested in care practices in space production and reclaiming territories by Afro-descendant women. Ana Laura is also the co-founder of the Network of Women Doing Fieldwork, where she advocates about embodiment, emotions, and risks during data collection from a decolonial feminist perspective. Ana Laura holds a doctorate in human geography from the University of Sheffield, a Masters in Conflict Resolution from Bradford University, a Masters in Fundamental Rights from the Universidad Carlos III de Madrid, and Bachelor of Laws from the Universidad Nacional de La Plata in Argentina. Her most recent publications, 'Geographies of Slavery in the

Les Malouines/Las Malvinas/Falklands Islands: The Maroon Connection', 'Maroon-Socioterritorial Movements', and 'Afro-Latin American geographies of in-betweenness: Colonial marronage in Colombia', were published in *Transactions of the Institute of British Geographers*, *Annals of the American Association of Geographers*, and the *Journal of Historical Geography*, respectively.

Rógerio Haesbaert is Full Professor of the Postgraduate Program in Geography at the Fluminense Federal University (UFF), Niterói, Rio de Janeiro, Brazil. He holds a PhD in Human Geography from the University of São Paulo and a post-doctorate/visiting research fellowship from the Open University. He was also a professor at the University of Buenos Aires and a visiting professor at the Universities of Paris VIII, Toulouse, Antioquia (Colombia), and Mexico. He is author of many books such as *The Myth of Deterritorialization*, *Living on the Edge* and *Territory and Decoloniality*.

Sam Halvorsen is a Reader in the School of Geography at Queen Mary University of London. His research examines the role of territory in grassroots politics and, more broadly, political participation and democracy in Latin American cities. He has published widely in journals such as *Progress in Human Geography*, *Transactions of the Institute of British Geographies*, and the *Annals of the American Association of Geographers*. He is founder and chair of the Latin American Geographies Research Group of the Royal Geographical Society. He is currently editorial board member of *Transactions of the Institute of British Geographers and Punto Sur* and sits on the international editorial boards of *Antipode: A Radical Journal of Geography*, *Journal of Latin American Geography*, and *Third World Quarterly*.

Chris Hesketh is a Lecturer in International Relations at the University of Sussex. He received his BA, MA, and PhD all from the University of Nottingham. Before joining Sussex in 2023, he taught at Oxford Brookes University, Birkbeck College, and the University of Nottingham. He has an inter-disciplinary research agenda that combines political economy, the historical sociology of international relations, political geography, political theory, and Latin American studies. These interests are captured in his monograph, *Spaces of Capital/Spaces of Resistance: Mexico and the Global Political Economy* (University of Georgia Press, 2017). His recent work focuses on issues of Indigenous resistance to extractivist development in the cases of Bolivia and Mexico, as well as applying the insights of Antonio Gramsci to state theory more broadly.

Jessica Hope is a Lecturer in Sustainable Development at the University of St Andrews, Scotland. Her PhD is from the Global Development Institute, University of Manchester, and she has held posts at UCL, Cambridge and Bristol. She is a political ecologist, interested in the human/more-than-human relations needed to navigate the climate emergency. Her current ERC funded project researches how road infrastructure, and its alternatives, influences environmental knowledges and politics in the Western Amazon. Previous projects investigated the anti-politics of the SDGs and post-neoliberal conservation in Bolivia. She is currently an editor at *Geoforum Journal* and on the editorial board at *Transactions of the Institute of British Geographers*.

Juan Miguel Kanai is a Senior Lecturer in the Department of Geography, University of Sheffield. With interests in global urban development, his research agenda focuses on the economic drivers and socio-environmental consequences of infrastructure provision in Latin America.

Vania Reyes Muñoz is a Geographer, Master in Residential Habitat from the Universidad de Chile, and PhD in Architecture and Urban Studies at Pontificia Universidad Católica de Chile. Vania's research interests are in urban studies and cultural geography, migrations, racialization, and gendered spatialities. Her doctoral thesis addresses the configuration of racialized representations of urban space in the metropolitan area of Santiago de Chile. Vania has worked as a technical advisor for social and interdisciplinary research organizations, as well as an analyst in the Chilean Public Service on housing and social development. Vania is a member of the research group Comparative Analysis in International Migration and Displacement in the Americas (CAMINAR), and member of the Association of Feminist Geographers of Chile.

Nadia Mosquera Muriel is Assistant Professor in the Department of African, African American and Diaspora studies at the University of North Carolina at Chapel Hill. She examines the intersections of race, class, gender, culture, space, and Black political mobilizations under Venezuela's Bolivarian Revolution (1999-present). Her work focuses on the struggles for spatial justice among Black grassroots cultural producers, artists, and political activists as they challenge structural forms of anti-Blackness and spatial inequalities in Venezuela. She was a Postdoctoral Provost Fellow at the University of Texas at Austin, and a Stipendiary Fellow at the Institute of Latin American Studies (ILAS), School of Advanced Study, University of London. Nadia earned a Bachelor's degree in International Studies at the Central University of Venezuela, a Master of Arts in Development Studies at the Institute of Development Studies at the University of Sussex, and a PhD in Development Studies at the University of Sussex. She is an editorial board member of *Gender, Place and Culture*.

Andrés Núñez is a cultural geographer at the Pontificia Universidad Católica de Chile. He has a PhD in History and a post-doctorate in Geography from the same institution. His research area is framed in the line of social, cultural, and historical geography in dialogue with critical theory. His research engages with the concepts of frontiers, territorial representations, geographical imaginaries, and processes associated with the social production of space. His latest work focuses on the culture-nature relationship, geographical images, decolonial geographies, posthuman geographies or more-than-human-geography, and what is known as non-representational theories or relational ontologies. He has published numerous articles and books and has led multiple research projects (ANID) related to the territory of Patagonia in Chile. His current project is entitled 'Posthuman geographies in Patagonia: interactions between nature, capital and desire', which reflects his current Deleuzean-Guattarian influence.

Matthew A. Richmond is Lecturer in Political Geography at Newcastle University. He holds a PhD in Human Geography from King's College London. He was Postdoctoral Researcher at the Centro de Estudos da Metrópole (CEM) and the Universidade

Estadual Paulista (UNESP), São Paulo, and Leverhulme Trust Early Career Fellowship at the London School of Economics (LSE). Matthew's research explores themes of urban development, governance, spatial practices, and subjectivity in Brazilian cities. He is Secretary of the Latin American Geographies Research Group and associate editor of the *Revista Brasileira de Estudos Urbanos e Regionais*.

Pia Riggirozzi is Professor of Global Politics at the University of Southampton, United Kingdom (UK). Her research focuses on the political economy of development, human rights and regional governance in Latin America. She is currently working on research projects regarding gendered health inequalities, poverty and the challenges of inclusive development in Latin America. Pia is Principal Investigator in the project 'Redressing Gendered Health Inequalities of Displaced Women and Girls in contexts of Protracted Crisis in Central and South America' (ReGHID). Recent publications include *Displacement, Human Rights and Sexual and Reproductive Health: Conceptualising Gender Protection Gaps in Latin America* (Bristol University Press, 2023 with Natalia Cintra and David Owen); Securitisation, Humanitarian Responses and the Erosion of Everyday Rights of Displaced Venezuelan Women in Brazil, *Journal of Ethnic and Migration Studies* (w/Jean Grugel, Natalia Cintra, Zeni Lamy, and Gabriela Garcia, 2023), 'The credibility of regional policymaking: Insights from South America', *Globalizations* (2022, with Matt Ryan); and 'Everyday political economy of human rights to health: dignity and respect as an approach to gendered inequalities and accountability', *New Political Economy* (2021).

Renato Emerson dos Santos holds a PhD in Geography, and is Professor at the Urban and Regional Planning Research Institute (IPPUR/UFRJ) at Federal University of Rio de Janeiro (Brazil). He has published (as author or editor) books in Brazil like *Diversity, Space and Ethnic/Race Relations: Blacks in Brazilian Geography* (2007), *Social Movements and Geography: the Spatialities of Action* (2011), *Racism and Urban Questions* (2012), and *Black Territories: Heritage and Education in Little Africa in Rio de Janeiro* (2022), dozens of book chapters, and papers in scientific journals. He was president of the Brazilian Geographers Association (2012–2014). His research is focused on spatialities of social movements, cartographical activisms, and the Brazilian Black Movement's struggles. He is coordinator of the Laboratory of Studies and Research in Geography, Racial Relations and Social Movements (NEGRAM/IPPUR/UFRJ).

Laura Sariego-Kluge earned a Ph.D. in Economic and Political Geography at CURDS, Newcastle University, in the UK; a M.Sc. in Local Economic Development from the London School of Economics and Political Science, UK; and undergraduate honours degree in Customs Administration and International Trade (UCR). She is the deputy director of the Escuela de Administración Pública (EAP), lecturer, and a researcher at the Instituto de Investigaciones en Ciencias Económicas (IICE), both departments from the Universidad de Costa Rica (UCR). She also directs the new journal *Administrar lo Público* from the Centro de Investigación y Capacitación en Administración Pública, UCR. She has experience conducting research with mixed methods and comparative case studies, and subscribes to a philosophy of science framed in critical realism. In the past she has published on topics around trade and development, Central American integration, cooperatives, fair trade, and public-private partnerships for development. Currently, her research interests revolve around territories and sustainable development in Latin America, economic geography, sustainability transitions, institutions, learning, policy, and public sector innovation, as well as teaching and research methods.

Marcelo Lopes de Souza is a professor of environmental geography and political ecology at the Department of Geography of the Federal University of Rio de Janeiro (UFRJ), Brazil. He holds a degree in geography from the Federal University of Rio de Janeiro/UFRJ (1985), a specialisation in urban sociology from the State University of Rio de Janeiro/UERJ (1987), a Master's degree in geography from the Federal University of Rio de Janeiro/UFRJ (1988) and a Doctorate (Dr. Phil.) in geography from the Eberhard-Karls Universität Tübingen (Germany) (1993). He received the first prize of the German Society of Research on Latin America/ADLAF in 1994 for his doctoral thesis (which was published in Germany) about the urban question in Brazil. He further received the Jabuti Award of the Brazilian Book Chamber for his book *O desafio metropolitano* (The Metropolitan Challenge) in 2001. His book *Fobópole: O medo generalizado e a militarização da questão urbana* (Phobopolis: Generalized Fear and the Militarization of the Urban Question), published in 2008, was nominated for the Jabuti Award in 2009. He acted as an academic visitor or visiting professor at several universities in Europe (Germany, United Kingdom, and Spain) and Latin America (Mexico). Since the mid-1990s, he has acted as a coordinator of a number of research projects, funded by the Brazilian Research Council/CNPq, the German Academic Exchange Service/DAAD, the British Academy or the European Union. He has published 17 books (13 monographs and four edited volumes) and about 150 papers and book chapters in several languages, covering subjects such as the spatial dimension of social movements, urban political ecology (focusing especially on environmental justice), and the epistemology of geography.

Maristella Svampa, Universidad Nacional de La Plata (UNLP) and CONICET, is a researcher, sociologist, activist, and writer. She has an undergraduate degree in Philosophy from the Universidad Nacional de Córdoba and a Ph.D. in Sociology from Paris, at Ecole des Hautes Etudes en Sciences Sociales. She lives in Argentina and is a Researcher at the Conicet (National Center for Scientific and Technical Research), Argentina, and Professor at the Universidad Nacional de la Plata (province of Buenos Aires). She is the coordinator of the Group of Critical and Interdisciplinary Studies on the Energy Problem (www.gecipe.org), and participates in the Group of Alternative Development, since 2011. Her fields of research are Political Ecology, Social Theory, and Political Sociology. She researches about Socio-ecological Crisis, Socio-environmental Conflicts and Resistance in Latin America, Neo-extractivism, and the challenges of the Eco-social Transition, from the South. She received the Guggenheim Fellowship (2007) and the Kónex award in Sociology, and Political Essays (Argentina) in 2006 and 2016, and the Platinum Kónex Award in Sociology (2016). In 2018, she received the National Award in Sociology in Argentina.

Fernanda Valeria Torres is a researcher at CONICET and professor at the Universidad Nacional de La Plata (UNLP). She holds a degree in Sociology (UNLP), a Master's degree in Social Sciences and Humanities (UNQ), and a PhD in Social Sciences (UNLP). She works at the Institute for Research in Humanities and Social Sciences IdIHCS-UNLP/CONICET and the Department of Sociology of the UNLP. She directs and/or is a member of several research projects based in Argentina, Brazil and Spain. She has taught seminars and undergraduate and graduate courses at several universities in Argentina, Brazil and Spain. Her topics of study and research are located in the field of political sociology and social movement studies. She proposes a spatial perspective of analysis to understand urban social movements, problematizing the category of socio-territorial movements to understand certain processes of production of territory in pursuit of collective political projects. She has published several books and articles on these topics in scientific journals in Argentina, Mexico, Brazil, Spain, and USA.

Astrid Ulloa is a full professor in the Department of Geography at the Universidad Nacional de Colombia. She is an anthropologist, and received her Ph.D. from the University of California, Irvine. Her main research interests include Indigenous movements, Indigenous autonomy, Indigenous feminisms, gender, climate change, water, territoriality, extractivisms, and feminist political ecology. She has published books, chapters, and articles on Indigenous peoples and women, as well as on environmental issues and extractivism. Her current research topics are gender and extractivism, indigenous women, territorial-environmental demands and medialities, and just energy transition and indigenous territories.

Marcia A. Vera Espinoza is a Reader and Deputy Director of the Institute for Global Health and Development (IGHD) at Queen Margaret University, in Edinburgh. Her research focuses on the study of processes of inclusion and migration governance in Latin America, Scotland, and beyond. Marcia is a co-founding member of the research group Comparative Analysis in International Migration and Displacement in the Americas (CAMINAR). She is Associate Editor of *Migration Studies* and member of the Advisory Board of *Forced Migration Review*. Marcia has recently published in *International Migration Review, Journal of Immigrant & Refugee Studies, Comparative Migration Studies, Frontiers in Human Dynamics*, and *Geopolitics*, among others. Her co-edited books include *The Dynamics of Regional Migration Governance* (Edward Elgar, 2019 with Andrew Geddes, Leyla Hadj Abdou and Leiza Brumat), *Latin America and Refugee Protection: Regimes, Logics and Challenges* (Berghahn Books, 2021 with Liliana Jubilut and Gabriela Mezzanotti), and *Movilidades y COVID-19 en América Latina: inclusiones y exclusiones en tiempos de 'Crisis'* (UNAM, 2022 with Gisela Zapata and Luciana Gandini).

Sofia Zaragocin is an assistant professor at the geography department at the University of Illinois at Urbana Champaign. Previously she was an associate professor at the International Relations department at the Universidad San Francisco de Quito, Ecuador. She is a member of the Critical Geography Collective of Ecuador and is the international councilor for the American Association of Geographers. Her work lies at the intersections of decolonial feminist geography, anti-racist geographies, and connecting Latinx and Latin American geographical perspectives.

Perla Zusman is Professor of Geography at the Universidad de Buenos Aires (UBA). She holds a Master's degree in Latin American Integration at the University of São Paulo (Brazil) and a PhD in Human Geography at the Universidad Autónoma de Barcelona (Spain). She currently teaches Introduction to Geography at the Universidad de Buenos Aires UBA. She is vice-head of the Institute of Geography of the University of Buenos Aires. She has done research in the fields of geographical thought, cultural geography, and historical geography.

Acknowledgements

Thank you to Andrew Mould for instigating this project, and to Claire Maloney, Niamh Hitchmough and Kate Fornadel for their editorial support. Anonymous reviewers provided helpful suggestions for improving the text, and Rogério Haesbaert provided guidance in the early stages. The book is a collective effort, and I am grateful to all the authors for investing in it. I continue to learn a lot from them all; thank you. I am also grateful to colleagues from the Latin American Geographies Research Group of the Royal Geographical Society for their wider support of the project, as well as to my family for making everything possible for me. The book is dedicated to all Latin American geographers who have worked tirelessly over recent decades, including Carlos Walter Porto Gonçalves, who passed away during its creation.

CHAPTER 1

Introducing Latin American geographies

Sam Halvorsen

Introduction

Latin American Geographies is one of the first comprehensive attempts to provide students and scholars of geography with a set of pedagogical tools to study from a non-Anglophone starting point. It aims to learn from, and with, a range of empirical, theoretical, and practical innovations that have emerged in and through Latin America over recent decades. It does not bracket off "Latin America" as outside or beyond the West or the Global North, but instead appreciates its diverse historical and geographical dialogues and relations that make it such a productive area to engage with. The book thus avoids any attempt to define the region in a bounded or closed sense and aims to provide starting points rather than having the final say. It is a collective effort, consisting of 30 authors, over half of whom are based in Latin America. In so doing it seeks to highlight the richness of knowledge produced from and with the region, guiding the reader across key geographical innovations that have reshaped our understandings and relations with the world.

There is a glaring contradiction at the heart of Anglophone geography: it is a discipline that claims to know and understand the world, yet it has only been able to do so through the English language and through the lens of Western researchers and institutions (Müller, 2021). *Latin American Geographies* is an opportunity to decentre the discipline towards a more worldly perspective. From the vantage point of Anglophone disciplinary geography – i.e. the subject taught and researched through universities and related institutions – there is a growing recognition that starting points are never innocent (Jazeel, 2016). How we approach geography is a politically charged decision, that is in turn bound up with a range of structural conditions. Language and the Eurocentric basis of knowledge production are two key barriers that have posed a limit to what or who Anglophone scholars engage with (Bański and Ferenc, 2013). The book aims to provide a foundation in key epistemologies upon which Latin American geographies are founded, while also making available a diverse Latin American scholarship to an Anglophone audience.

DOI: 10.4324/9781003430926-1

This introductory chapter sets the scene for the book ahead, providing the reader with key background information with regards to how Latin American geographies have been institutionalised. Underlying the discussion is the challenge of overcoming the linguistic and epistemological barriers that have thus far kept Latin American geographies distant from students and scholars working in AngloAmerica. It is my hope that this will be the start of a fruitful journey in which the reader expands their geographical horizons and is inspired to consider alternative starting points for how they "do" geography. A broader aim is thus to take seriously the geography of geography and to move beyond the Eurocentric nature of our discipline.

As Brazilian geographer Carlos Walter Porto-Gonçalves (2009, 38) argues, decentring Anglophone/Eurocentric geography is not to undermine *any* attempt at generating universal knowledge and the quest to build theories and ideas that can travel. Rather, the book's aim is to (i) counter the uni-directional nature of geographical knowledge, which travels from a Eurocentric core outwards, and to (ii) affirm that different places in the world provide alternative rationalities through which to constitute universalizing knowledge. The book is thus a concerted effort in creating and deepening a *diálogo de saberes* – a dialogue of knowledges – that denies there can only ever be one starting point for how we understand and do geography. By placing dialogue at the heart of the book, and working with scholars based both inside and outside the Latin American region, *Latin American Geographies* seeks to avoid the pitfalls of simply extracting knowledge for the benefit of a few (Halvorsen, 2018).

Latin America is one of many possible places from which to study geography, but it has some particularities that make it an especially rich one. Not only does the region contain a plethora of political, cultural, and ecological formations that are of innate interest to any geographer, but it has also actively constructed ways of understanding and practicing the relationship between people and the world that have clear pedagogic potentials. The book is saturated with ideas such as: multiterritoriality, dependency, postneoliberalism, *cuerpo-territorio*, informality, Indigenous relational spatialities, and many others, all of which have been produced through dialogue from and with the region. Moreover, in recent debates, the struggle *over* geography – territory, land, place, and the body – has become a central, perhaps the central, battleground for social movements and marginalized communities in the region. Latin American geographies are not just a theoretical idea, they are real world practices with, at times, life and death outcomes (**see Chapter 23**).

Latin American geographies in Anglophone academia

Although Latin America is a relatively peripheral place for British academia, it is more prominent in North American geography; yet in both cases, it has been historically treated as an empirical case study rather than a site of knowledge production. In 1970, the US-based Conference of Latin American Geography was founded, which set up the Journal of Latin American Geography, providing a key set of institutions that were later joined by speciality groups of the Association of American Geographers. The UK has been much slower, and only in 2020 was the Latin American Geographies Working Group (since 2022 a research group) founded in the Royal Geographical Society. While these trends may have led to an increase in research published *on* the region, it is evident that scholarship *from* the region remains scarce. Recent bibliometric studies demonstrate a very low rate of publication from scholars based in Latin American and the Global South (see Müller, 2021). Moreover, translation of texts from prominent Latin American geographers has been slow and almost non-existent until very recently. While students and academics based in Latin America are expected to engage with Anglophone scholarship, this dialogue is not reciprocated, and knowledge of Latin American geographies remains marginal.

It is important to recognize that the term "Latin America" is closely bound up with the region's (post)colonial history. The idea emerged during the 19th century and was particularly associated with a French vision of maintaining European control over the Americas. As Mignolo (2005) has argued, the idea of Latin America is a geopolitical invention, tied to the colonial and nation-building projects of Europe, and one that has had a profound impact on the racialized identities of the area. What we today call Latin America is itself a diverse region that usually represents Mexico, Central America, and South America, in addition to the Caribbean. The latter, however, is often unevenly placed in relation to the other parts, due to a combination of historical and linguistic differences that tend to pull it apart. Moreover, there has been an explicit attempt by many scholars working in and with the region to push back against the label of Latin America and instead work with Indigenous regional concepts such as Abya Yala. Escobar (2019) notes that growing decolonial epistemic movements have generated a new lexicon for the region, summarized as: Abya Yala/Afro/Latino-América. To this could be added the role of Latinx, especially in the diaspora, and the ongoing feminization of regional knowledge. While many of the authors in the book highlight the colonial "invention" of Latin America, it remains in the title due its capacity to speak to, and interpellate, as a broad an audience as possible. Moreover, the book acknowledges the often ambiguous and excluded positionality of the Caribbean vis-à-vis Latin America, recognizing that future work needs to be done for a greater incorporation of Caribbean geographies.

Latin American geographies emerge from a set of historically and geographically specific conditions, and the ongoing attempt to make sense of and respond to deep structural challenges. The region is marked by extreme socio-spatial inequality, itself a direct legacy of colonial rule and slavery. The geographies of Indigenous peoples and Afro-descendants mark a central fault line in the post-colonial region and have provided an important inspiration for geographical scholarship. In turn, the history of extractivism, land ownership, and (lack of) distribution of wealth has generated ongoing geographical conflicts that are increasingly central to the growing ecological crisis. Against such a backdrop, it is hardly surprising that the struggle over land and territory has taken centre stage in contemporary geographical scholarship. Moreover, since the mid-20th century, the region has rapidly become the most urbanized in the world, itself posing new geographical inequalities and challenges. Although none of these challenges and histories are alien to Anglophone geography – indeed, the colonial histories of Western Europe are intimately bound up with them – there has been a failure to take seriously the *epistemological* in addition to empirical basis to Latin American geographies. In sum, Latin America is a crucial site of academic knowledge production, yet this remains largely detached from Anglophone academia.

Geography as a discipline in Latin America

Geography is today an important discipline across the region, although with significant variation between countries. Since the first geography degree was established in the 1930s, there were, according to 2011 statics, 181 degree programmes in 91 departments across 14 countries (Prieto, 2011). These figures have since increased. Brazil is the undisputed powerhouse of geography in the region, home to by far the largest number of departments and degrees, including a large undergraduate programme, with 76 (post)graduate programmes (MEC 2019; see also Novaes and Lamego, 2022). Argentina also contains a well-established graduate and postgraduate discipline across its universities (Zusman and Luz Bietti, 2017; Zusman et al., 2017). This is followed by Mexico, Colombia, and Chile, each with modest yet influential geography centres and scholars: Mexico in particular is renowned for the prestige of its geographical scholarship despite relatively small numbers of graduate programmes. In addition to taught programmes, there are dozens of geography journals edited out of Latin American universities, with geography conferences developing in the second half the 20th century (Urquijo Torres and Bocco Verdinelli, 2016), sometimes in search of critical spaces in times of military dictatorships (Ferretti, 2023). Since 1987, EGAL, now called EGALC (the Meeting of Geographers in Latin America and the Caribbean), has provided a key fixture for geographers across the region, with the 19th edition taking place in the Dominican Republic in 2023 (see Figure 1.1). Hence, Latin America is a region in which geography has firmly established itself as an academic discipline.

When studying the geographies of Latin America, the most obvious starting point is thus to consult the work of scholars based in the region. Far too often this is not the case, as Latin American scholars remain under-cited in AngloAmerican publications. Although there remain certain barriers, not least linguistic, as well as the challenges of engaging local scholarship in parts of the region where geography is less

Figure 1.1 The 19th edition of EGAL in Santo Domingo, 2023.
Source: EGALC.

well established, this is a remarkable failure that is to the detriment of all. Of the 30 authors in this book, 16 are based in the region, and a strong effort has been given to raise their prominence in the Anglophone discipline. At the same time, however, Latin American geography is inconceivable outside of the dialogues across global regions and languages, and the book has deliberately not sought to create an artificial boundary around knowledge produced in the region.

The substantial work of geographers in Latin America has historically developed through dialogue with geographers elsewhere, most notably France, but also with AngloAmerican and European scholarship. Although there was a tendency for earlier geographical research to focus on empirical content, drawing on Europe for episte-mological approaches (Urquijo Torres and Bocco Verdinelli, 2016), across the 20th century, and particularly in the second half, the region produced key geographical thinkers, such as the Brazilian Josué de Castro (1909–1973) (see Davies, 2023) or the Mexican Angel Bassols Batalla (1925–2012). None have become more renowned than Milton Santos (1926–2001, **see** Figure 1.2), who spent many years in exile, mingling among prominent Anglophone geographers. Only very recently has there been a concerted effort to translate his oeuvre into English (Melgaço and Prouse, 2017; Santos,

Figure 1.2 Brazilian geographer Milton Santos.
Source: Wikimedia Commons.

2021a, 2021b), a watershed moment in opening Anglophone geography to the theoretical insights of Latin Americans. This book is thus being written at an exciting moment for enhancing dialogues with Latin American geographies, and I hope that it will provide a tipping point by making them accessible to a readership that includes undergraduate students and geographers new to the region.

Outline and structure of book

The book aims to cover most major areas of human geography, from historical and economic, social and cultural, to environmental and health, and is organized around three sections that provide the reader with the necessary tools to engage with cutting-edge scholarship on Latin American geographies. It is designed with final year undergraduate students, or master's students, in mind, but individual chapters will be of relevance to scholars looking for a way into any of the themes presented. In brief: Section I: Core Themes provides crucial historical and empirical context; Section II: Key Perspectives outlines some key epistemological approaches and social and cultural foundations; Section III: Uneven Processes delves into contemporary challenges and conflicts. Each chapter contains a brief introduction, a discussion of key ideas, and case studies, followed by some pedagogical tools including summary points, review questions, annotated further readings, and keywords. A glossary of technical terms is included at the end of the volume.

Section I is divided thematically into four groups. Chapters 2 and 3 lay the historical foundations for how we approach Latin American geographies. Ferretti's chapter outlines the notions of decolonization and decoloniality, highlighting why Latin America matters to historical geography and outlining the contributions of Latin American historical geography to decolonial ideas. Historical context is crucial for understanding contemporary challenges, and at a time when Anglophone geographers are calling for the "decolonization" of their discipline, much can be learned from the political and epistemic struggles of Latin America. Ferretti's chapter sets the scene for many of the topics examined in subsequent chapters, including Indigenous, Afro-descendant, and other marginalized socio-spatial positionalities in the region.

Zusman's chapter examines the historical constitution of a set of Latin American nation-states, and in so doing provides insights into how the region's modern territorial forms and their borders came to exist. From the political process of decolonization in the early 19th century through to consolidation of national imaginaries in the 20th century, Zusman guides the reader into making sense of how a core foundation of political and cultural geography was produced. Such a foundation is crucial for understanding what is at stake in contemporary movements to contest and decolonize territorial formations in the region, as discussed in several later chapters.

The following chapters introduce readers to political Latin American geographies, specifically democracy and geopolitics. Halvorsen and Torres demonstrate how, across the 20th century and into the 21st, democracy has been a geographically unevenly produced idea and practice that has suffered a number of setbacks. Most centrally, they provide an introduction into key regional concepts and debates, namely democratization, populism, participatory decentralization, and pluri-nationalism. They outline what is at stake intellectually and politically in the struggles to pursue these forms of democracy, all of which have a broader significance to human geography around the world.

Benwell and Núñez then delve into the geopolitics of the region. They start by outlining the role of US influence in shaping geopolitical relations during the 20th century before moving on to discuss key advances from recent scholarship on Latin American geopolitics, centred around a discussion of feminist and everyday approaches. In so doing they shift attention from the highly visible political actors and institutions and towards embodied and everyday sites and spaces of geopolitics. These ideas are further developed through a discussion of two case studies. First, Benwell and Núñez examine young people's everyday geopolitics of the home in relation to the dispute over the Falklands/Malvinas dispute. Second, they explore the role of everyday geopolitical objects – flags – in the context of Chilean protest in 2019–20.

Chapters 6 and 7 move into the realm of economic and urban geographies. Sariego-Kluge shifts vantage point to Central America in order to introduce readers to economic geographies. Exploring key concepts such as production, trade, consumption, and finance, the chapter uses Central America to unravel key debates surrounding economic globalization, industrialization, protectionism, and strategies for growth from the periphery. Recent experiences of global integration and neoliberal development are given close attention to understand how the experiences of the region are enlightening for how we understand economic processes more generally.

Kanai then introduces large-scale dynamics of urbanization since the early 20th century. The chapter reviews the main trends of three distinct periods of metropolitan expansion, neoliberal unravelling, and contemporary territorial disputes in the context of a thoroughly yet unevenly urbanized continent. For each period, a general political economy context is presented before delving into its implications for urbanism and urban life. Attention is paid to the influential ideas that shaped city-building in different countries and the critiques of oft-imported models by social scientists and discontented actors.

The final three chapters of Section I examine contemporary themes of (post)neoliberalism, migration/displacement, and sustainable development, three processes that have come to define key transformations in recent years. Grugel and Riggirozzi provide an overview of how the idea and practice of neoliberalism, which swept across much of the region in 1980s, was eventually contested and updated into a strategy of postneoliberal governance. Led by left of centre governments, the latter sought to bring back the welfare regimes, renegotiate state-market relations, and renationalize key economies. The authors provide a detailed account of key characteristics of neoliberal and postneoliberal development models, drawing our attention to key challenges and dilemmas. In so doing, they outline the contours for how Latin American geographies have been impacted by development strategies, e.g. through a discussion on the scales of citizenship, inclusion and ethnicity, and (neo)extractivism.

Vera Espinoza and Reyes Muñoz then explore how migration, including different forms of mobility, is currently understood in the regional context, by paying attention to the spaces, places, trajectories, and actors that are shaping these debates. They begin with a brief overview of key migration dynamics in the region, with an emphasis on the migration patterns and drivers that have taken shape over the last decade, including the current displacement of Venezuelans and the mobility of Haitians and people from

Central America, as well as the impacts of climate change. The chapter then discusses the extent to which dynamics of (im)mobility are being shaped by bordering practices, both within and outside the nation-states, by exploring key aspects of regional migration governance, such as an increased securitisation of migration and the context of multiple "crises". The chapter identifies key characteristics of migration in Latin America, such as the feminization of migration, the transnational practices of care, and the influence of the political lives of migrants ("luchas migrantes").

Finally, Hope discusses the agenda of sustainable development, tracing it as both an idea and a core practice promoted through the UN's Sustainable Development Agenda. By critically examining the rise of sustainable development discourse from the perspective of Latin America, Hope delves into key geographical debates, including those of post-development and political ontology. She weaves Indigenous concepts such as *Buen Vivir* and the pluriverse with a discussion of how sustainable development felt on the ground. This discussion is brought to life through the case study of the TIPNIS conflict in Bolivia, where the pluri-national and decolonial agenda of the Evo Morales government came across its own limitations as it moved ahead with plans to construct a highway through an Indigenous national park in order to promote its extractivist development model.

Section II shifts gear and provides readers with an overview of key historical, economic, political, and social perspectives to Latin American geographies. These can be read as core concepts, ideas, and theories that have informed how scholars, practitioners, and activists have engaged with geographical debates in the region. They represent key debates and concerns of those working in and with the region.

Hesketh begins with an introduction to one of Latin America's most significant exports to global political economy: dependency theory. He begins the chapter by outlining how scholars grappled with understanding macro development in the wake of the Second World War, and introduces the emergence of dependency theory as a diverse and heterogenous set of critical ideas for understanding the geography of capitalist development. Centred on an understanding of core-periphery relations in the world, dependency theory not only shaped intellectual agendas in the 1960s and 70s, but remains a key body of thought today. Hesketh helps the reader grapple with this theoretical discussion through a case study of import-substitution industrialization, a core development strategy in the post-war era.

Haesbaert and Halvorsen then shift gears and provide an introduction to one of the central themes of Latin American geographies in recent decades: the movement of decolonizing territory in thought and practice. In so doing they demonstrate how marginalized and grassroots actors and communities in the region have been at the forefront of challenging dominant notions of territory, inherited through the post-colonial state. The chapter is organized around four sections. First, Haesbaert and Halvorsen explore the struggles of movements to gain autonomy in and through territory. Second, they consider the re-scaling of political struggles to the site of the body, or body-territory. Third, they explore the political ontological dimension to territory and its socio-ecological relations. Finally, they elaborate the related ideas of multi/transterritoriality to further decentre Western narratives around territory.

Ulloa (re)introduces Latin American geographies through the lens of Indigenous people, providing an approach that she defines as Indigenous relational spatialities.

She focuses on how Indigenous ideas are reshaping familiar geographical categories by breaking down distinctions between human and non-human beings and promoting relational understandings of time and space. New and contrasting worldviews are emerging that demand a rethinking of political and spatial practices, responding to calls for self-determination for Indigenous territories. Ulloa draws on a range of examples of Indigenous thought and practice to introduce readers to their geographical worldviews.

Zaragocin then provides an overview of feminist approaches to Latin American geographies. Her chapter charts the history of feminist thought in the region's feminist disciplinary spaces and examines how these have related to feminist struggles around territory, their bodies, and body-territory: *cuerpo-territorial*. The chapter specifically draws on three main debates happening in the region: geographies of gender-based violence, feminist geography activisms, and, finally, contemporary feminist debates on territory. These topics are framed in relation to their discussions within academic circles, activist and collective groups, and public policy practitioners.

In the final chapter of this section, Zavala Guillen and Mosquera Muriel examine the geographies of Afro-descendants and their social movements, decentring the implicit whiteness in Latin American geographies. Their chapter introduces key geographical ideas from the vantage point of Afro-descendant movements in Colombia and Brazil. *Vivir sabroso*, or living richly, is discussed as a concept that weaves together ideas of place-making, spiritual notions of territory, and alternative articulations around territory. The *vivir sabroso* worldview provides significant insights into territorial relations and knowledges and overlaps with related notions such *as vivir enmontado* (living under siege). The remainder of the chapter provides an introduction to the key idea and practice of Marooning: a resistance strategy of African slaves seeking liberation in and through space. This is further elaborated through a discussion of the related idea of *quilombismo*.

Section II: Uneven Processes provides several situated analyses of key processes of inequality and resistance in the region, covering a broad area of ecology, urban, agrarian and health geographies. Souza opens the first subsection, Ecologies, with a chapter on the political ecology of cities, drawing attention to the ecological processes that underpin Latin America's cities and the many inequalities and crises that are produced. The chapter starts with a discussion of environmental suffering and the rapid growth of "sacrifice zones" where multiple injustices intersect. Peripheral and semi-peripheral urban neighbourhoods increasingly experience economic poverty alongside environmental degradation. This is elaborated through a discussion of a steel plant in Rio de Janeiro, and the conflict between residents and the corporate owners. This leads to a discussion of ecofascism and the growing fear of cities and their perceived violence (phobopolis). Once again, Souza elaborates through a discussion of Rio de Janeiro and the governmentalization of a national park, where ecofascism and fear implicate what is and is not deemed possible in urban space.

Svampa's chapter turns our attention directly to the global socio-ecological crisis, recently heightened by the COVID-19 pandemic, and explores what a Latin American and Southern perspective provides our understanding of this geographical process. Having outlined the shift to the Anthropocene and what implications this has for global ecological geographies, Svampa's chapter is dedicated to the search for alternatives: lexicons and approaches to a transition out of socio-ecological crisis. By shifting our starting point from the Global North to Latin America and the Global South, Svampa starts from a critique of so-called bad development models and looks towards alternative political economic and ecological imaginaries. The chapter shines light on the role of socio-environmental movements in pushing the transition agenda and innovating with ideas and practices for implementing it. Svampa also discusses the work of global networks of scholars and activists developing new geographical ideas for the design and implementation of ecosocial transitions.

The is followed by Carter's chapter on health, environment, and disease, which introduces readers to key geographical concerns relating to public health issues in the region. The chapter starts by critically exploring how the geographical imaginary of 19th-century Latin America represented it as a site of dangerous tropical diseases. It then moves on to examine how urbanization in the 20th century led to acute environmental crisis, including air and water. Carter examines the case studies of Mexico City and Buenos Aires, two large cities where environmental degradation has disproportionately affected the most marginalized communities. The chapter ends with a discussion of intercultural health in Andean countries, returning to themes around relational ontologies and Indigenous communities.

The following sub-section examines urbanization, recognizing that Latin America is the most urbanized region on the planet, a process that has strongly informed its geographies. Richmond's chapter provides the reader with an overview of the concept of urban peripheries, a key geographical idea for the spatial organization and development of cities. The chapter opens with an historical overview of debates around the theorization of urban peripheries in the 1950s–80s, a time when Latin American cities were growing rapidly, with informal and squatter settlements as notable features. Richmond then explores more recent transformation in peripheries, including the partial provision of services and the mass production of state-subsidized housing. The chapter ends by exploring these transformations through the lens of Fazenda de Juta, a peripheral urban neighbourhood in São Paulo.

Crossa's chapter switches the focus to debates on urban informality and public space. The chapter introduces readers to the formal/informal divide, as it has impacted understandings of the geographies of urban economies, highlighting the significance of Latin America for these debates. Crossa then delves into the significance of public space, a key geographical site through which cities are understood to come alive. The remainder of the chapter uses the case study of street vendors in Mexico City to understand what is at stake in terms of how we understand the informal and public space, and what this tells us about a key set of urban disputes regarding Latin American geographies.

The final sub-section of three chapters centres around the broad notion of resistance and provides insights into key grassroots actors and conflicts in the region. Santos opens by providing the reader with an introduction and overview of social movements in the region, and key actors who have sought to (re)create the region's geographies in recent decades. The chapter highlights not only the significance of social movements for understanding key socio-political trends in the region, but emphasises how they can only be understood in and through their geographies. Santos explores a range of paradigmatic social movements from the region, drawing out their spatialities, and includes a brief case study into the Black Movement in Brazil.

Fernandes then examines how agrarian inequalities have defined a central geographical fault-line in the region, creating conflict and grassroots organizing. The chapter strats by exploring the historical production of agrarian inequalities. It explores key concepts in order to understand this process, beginning with colonization in the 16th century and concluding with the exploitation of natural resources in the Amazon in the 21st century. The second part moves on to analyze the prospects for overcoming agrarian inequalities. It does so by critically examining types of predatory development and sustainable development, relating them to hunger and food sovereignty.

In the closing chapter, Arzeno and Farías introduce the idea of "the politicization of life" in recent to new cycles of protest in the region. Tracing back grievances to the intense wave of neoliberalization in the 1980s and 1990s, the authors argue that life itself has become a central space for struggle, drawing on geographical concepts such as "r-existence" and "body-territory". The chapter is organized around two case studies. First, they examine the Lenca people's struggle in Honduras, which grabbed global attention following the assassination of activist Berta Cáceres. They then examine the case of the *Ni Una Menos*/Not One Women Less movement in Argentina. Both examples demonsrate the centrality of "the defence of life" to resistance.

Summary

- Geography has a significant, if uneven, institutional presence across Latin America.
- Nevertheless, Latin American geographies remain marginal to Anglophone geographers due to linguistic and epistemic barriers to dialogue.
- The book seeks to foster and deepen dialogues of knowledge: over half the 30 authors are based in the region.

Review questions

1. What reasons can you think of for the exclusion of Latin American geographies in the English-speaking world?
2. What opportunities and challenges are presented by opening dialogue between Anglophone and Latin American geographies?
3. What tactics could be used to deepen dialogue across linguistic and epistemic boundaries?

Further reading

https://muse.jhu.edu/journal/240
Journal of Latin American Geography website: the flagship journal with papers in English, Spanish, Portuguese and French. An excellent resource.

https://lagukinfo.wixsite.com/lag-uk

Latin American Geographies Research Group of the Royal Geographical Society

Useful website that contains blog posts, resources and contacts of scholars working in and on the region.

Müller, M. (2021). Worlding geography: From linguistic privilege to decolonial anywheres. *Progress in Human Geography* 45: 1440–1466.

A useful overview of the challenges of working with regions on the edge of Anglophone disciplinary spaces.

Novaes, A.R., Lamego, M.A. (2022). Brazilian universities and graduate programmes in Geography: Institutional developments and political challenges. *Transactions of the Institute of British Geographers* 47: 16–22.

A helpful analysis of the growth of graduate programmes of geography in Brazil.

Zusman, P., Castro, H., Soto, M. (2007). Cultural and social geography in Argentina: Precedents and recent trends. *Social & Cultural Geography* 8(5): 775–798

A rare paper in English that maps progress in geography as it has been produced from within Latin American

Keywords

Anglophone hegemony: the dominance of the English language in disciplinary spaces (especially journal and book publications).

Dialogues of knowledge: the process of exchanging ideas and world views without a dominant theory or language.

References

Bański, J., Ferenc, M. (2013). "International" or "Anglo-American" journals of geography? *Geoforum* 45: 285–295.

Davies, A. (2023). *A World without Hunger: Josué de Castro and the History of Geography.* Liverpool University Press.

Escobar, A. (2019). Desde abajo, por la izquierda, y con la Tierra: SUReando desde Abya Yala/Afro/Latino/América. *Revista Interdisciplinar Sulear.* Available at: https://revista.uemg.br/index.php/sulear/article/view/4141 (accessed 13/11/2024)

Ferretti, F. (2023). South of the south: Political dissidence, exile, and Latin American transnationalism around the "new geography" meetings in the southern cone (1960s–1970s). *Annals of the American Association of Geographers* 113(8): 1745–1761.

Halvorsen, S. (2018). Cartographies of epistemic expropriation: Critical reflections on learning from the south. *Geoforum* 95: 11–20.

Jazeel, T. (2016). Between area and discipline: Progress, knowledge production and the geographies of geography. *Progress in Human Geography* 40(5): 649–667.

Melgaço, L., Prouse, C. (Eds.) (2017). *Milton Santos: Pioneer in Geography.* London: Springer.

Mignolo, W. (2005). *The Idea of Latin America.* John Wiley & Sons.

Müller, M. (2021). Worlding geography: From linguistic privilege to decolonial anywheres. *Progress in Human Geography* 45: 1440–1466.

Novaes, A.R., Lamego, M.A. (2022). Brazilian universities and graduate programmes in geography: Institutional developments and political challenges. *Transactions of the Institute of British Geographers* 47: 16–22.

Porto-Gonçalves, C.W. (2009). De saberes y de territorios: Diversidad y emancipación a partir de la experiencia latino-americana. *Polis* 8(22): 121–136.

Prieto, J.P. (2011). Los estudios de Geografía en las universidades de América Latina; desarrollo, situación actual y perspectivas. *Investigaciones Geográficas* (74): 107–124.

Santos, M., (trans Baletti, B.) (2021a). *The Nature of Space.* Duke University Press.

Santos, M., (trans Davies, A.) (2021b). *For a New Geography.* University of Minnesota Press.

Urquijo Torres, P., Bocco Verdinelli, G. (2016). Pensamiento geográfico en América Latina: retrospectiva y balances generales. *Investigaciones Geográficas* 90: 155–175.

Zusman, P., Castro, H., Soto, M. (2007). Cultural and social geography in Argentina: Precedents and recent trends. *Social & Cultural Geography* 8(5): 775–798.

Zusman, P., Luz Bietti, G. (2017). La trayectoria de la(s) geografía(s) en Argentina. En: Rucinque, H.F., Muñiz-Solari, O., Zapata Salcedo, J.L. (Eds.), *Historia de la Geografía en América Latina y el Caribe.* Asociación de Geógrafos Colombianos.

PART I

Core themes

Historical and colonial geographies

CHAPTER 2

Decolonization and decoloniality

Federico Ferretti

Introduction: other histories for other geographies

Until recent times, historical geography was considered as a research field not particularly rich in innovative intellectual stimulations. Although classical research on the history of territories often included politically progressive or even radical works (Baker 2003; Cosgrove 1984), its main focus has traditionally been historical landscapes that were overwhelmingly located in some regions of Western Europe. Thus, for a period, historical geography seemed to offer little attraction to critical geographers. In the UK, evidence shows that research tends to focus on the history of British rather than international or global geographies (Newman et al. 2022). Nevertheless, there is a growing interest in and of the Global South, with a great potential for historical geography to contribute to wider calls to decolonize geography, theory and practice (Esson et al. 2017; Radcliffe 2017).

This trend can also be seen across the discipline of geography, where special attention is being paid to so-called 'other geographical traditions' (Ferretti 2019). That is leading to the rediscovery of authors of the past who had been marginalized by mainstream academics because they were political dissidents, or because their gender, ethnicity, language, social or national origins did not match the required standards. This has implied the rediscovery of several Latin American geographers and their neglected worldwide impact, the most famous example being that of Brazilian radical geographer Milton Santos, whose works have been recently translated into English (see Further reading), as well as new attention to women in South American geographies (see Further reading).

Latin American geographies, including specialized networks such as *Rede Brasilis*, the Brazilian Network of Historical Geography and the History of Geography (see Further reading), are growing. They increasingly draw the attention of geographers from the Global North to new ideas about considering Indigenous knowledges and environmental ethics as alternatives to Eurocentric models as detailed below. Historically, the shift from Indigenous territoriality to territorial states (**see Chapter 12**) passed through the colonization operated by European empires, but also through the successive decolonization, a phenomenon that likewise increasingly interests geographers.

DOI: 10.4324/9781003430926-4

The differences between this kind of political decolonization and a fuller decolonial turn are explained in the following sections.

Political independence versus real decolonization

Between the fifteenth and the eighteenth century, the Iberian empires (Spain and Portugal) were responsible for the well-known and disgraceful process of the *Conquista*. Across this period, the Iberian domination of Southern, Central and (partially) Northern America led to the Indigenous genocide through war, enslavement, exploitation and diseases. These early European empires also inaugurated another shameful phenomenon: the deportation of African slaves from the other side of the Atlantic Ocean to serve in the plantations, where exploitation was ruthless and mortality high. Yet, rather than insisting on victimization and mere deprecation of colonial crimes, current geographical scholarship increasingly focuses on Black and Indigenous resistance and insurgencies. As Brazilian historical geographer Manuel Correia de Andrade (1980) argued, slaves and Indigenous communities were never docile and always revolted in several ways, sometimes taking up arms against their masters, sometimes fleeing from the plantations, sometimes operating various forms of daily insubordination.

An important phenomenon was the formation of *quilombos*, or *palenques*, communities of rebel slaves that found sanctuary in the hinterland (**see Chapter 15**). Later, these settlements moved closer to the main cities as slavery became less frequent until its definitive abolition in the nineteenth century. In Brazil, the most famous historical experience of this kind, the *quilombo* of Palmares in the North-eastern state of Alagoas, acquired political prominence between the seventeenth and the eighteenth century. In 1894, Palmares was mentioned by anarchist geographer Elisée Reclus as an outstanding example of people's self-determination: 'Nowhere else, in Brazil, fields were better cultivated' (Reclus 1894, 228).

Before Reclus, another European geographer who advocated for the rights of Indigenous peoples and for the abolition of slavery was Alexander von Humboldt, still considered a 'founding father' for the newly-decolonized republics in Southern and Central America (Mendoza Vargas 2007). Humboldt brought the ideas of the French Revolution to Latin America, arguing for the equality of all humans' rights. Yet, these values had already been contradicted by the behavior of the 1789 Paris revolutionaries, who did not extend the abolition of slavery to the colonies, including the Caribbean island of Haiti. This means that the rights that white people enjoyed in Europe were not applied to non-white people outside the region. In this vein, the Haitian revolution of 1791–1804 is still considered an exemplarily experience in decolonial agency (Mignolo 2010), as Haitian slaves did not wait to be redeemed by external 'liberators'. Indeed, they literally applied the idea of equal rights and took their emancipation by themselves before their movement was repressed by the Napoleonic armies.

Nevertheless, the political decolonization of Latin America did not take place through these kinds of subaltern revolts. Like in the case of the United States, which gained independence from the British Crown in 1776 after a revolt of European (mostly Anglo-Saxon) settlers, the independences of Latin American republics from Spain, and of the Brazilian Empire from Portugal, obtained in the 1810s and 1820s, resulted from a movement that was dominated by elites of European origin. Although inspired by progressive ideals such as republicanism, federalism and equality of all citizens' rights, which clashed with the feudal-aristocratic Restoration that characterized Europe since 1815, the decolonizers, led by legendary figures such as Simón Bolivar (1783–1830, see Figure 2.1), were caught in contradictions similar to those of the aforementioned Paris revolutionaries. Decolonial scholars currently identify these through the notion of 'double consciousness'.

For Walter Mignolo (1993), the 'double consciousness' of Latin American creole elites of European origin marked political decolonization in two ways. On the one hand, these elites opposed European colonial domination. On the other, they maintained their European cultures, together with forms of economic and political domination of European origin such as capitalism and the territorial state. Thus, political decolonization did not equate with emancipation of 'non-European' Americans, such as Indigenous and Afro-descendants, who remained in most cases subject to exploitation, socio-spatial marginalization and racial stigma. Conversely, it reproduced coloniality by marginalizing Indigenous and Afro-descendant communities within the new setup of national states that were based on European territorialities, which implied bounding national and administrative frontiers and protecting private land property.

Figure 2.1 Portrait of Simón Bolivar by Francis Martin Drexel (1825).

Source: https://commons.wikimedia.org/wiki/File:Sim%C3%B3n_Bolivar,_Francis_Martin_Drexel_(1792%E2%80%931863).jpg

In these processes, geography was instrumental in consolidating the power of the state by drawing its boundaries and national imageries.

Among those who materially drew these new frontiers, inaugurating the new professional figure of the 'geographer-engineer' in several states including Mexico (Mendoza Vargas 2007, vii), there were former European officers of the Napoleonic armies, such as the Italian Agostino Codazzi (1793–1859). His case was exemplary because Codazzi was a representative of the politically progressive milieus that were inspired by the French revolution and opposed aristocracy in Europe. Yet, their progressivism did not always apply to non-European people, as Codazzi worked for the republics of Venezuela and later Nueva Grenada (Colombia) to demarcate borders, keeping a paternalistic attitude towards Black and Indigenous communities (Morelli and Venturoli 2021).

In the decades following decolonization, the demarcation of borders between South American states often took the forms of conflicts and border controversies (**see Chapter 3**), especially around Brazil. This country differed from its neighbors not only by speaking Portuguese instead of Spanish, but also because early independent Brazil took the institutional form of an Empire, which only abolished slavery in 1888, before becoming a Republic in the following year. Especially in the Amazonian area, geography was a politically performative exercise, as the agreements between states were first represented by diplomats' traits on maps. Only later did these boundaries become real on the ground through violent occupation of Indigenous land, as establishing frontiers meant settling posts and new inhabitants to push Indigenous communities more and more into the hinterland (Burnett 2000). Thus, the 'double consciousness' of creole elites produced formal political decolonization of Latin American states, but did not question European territorial models or colonial relations of exploitation and subalternity for non-European communities. In the following section, I explain what we mean by fuller decolonization, including epistemic decolonization and decoloniality.

On decoloniality: epistemic decolonization is not a joke

Decolonial scholars consider the very notion of 'Latin America' a geopolitical concept, and more specifically an exclusionary one. Indeed, that definition was adopted in the second half of the nineteenth century in the context of an idea of 'Latinity' that was then serving French diplomacy in its imperial rivalry with the Anglo-Saxons. Thus, European and American 'Latin' identities were in some way equated, by operating once more an exclusion: that of the communities of non-European origins, namely the Indigenous and Afro-descendants. These people could not be represented by a 'Latin' identity that only came with the *Conquista* and its Portuguese and Spanish settlers. Still today, colonialism finds a continuity under the form of ongoing internal colonialism, social oppression and extractive destruction of environments. For this reason, we normally add to Latin America the definition of *Abya Yala*, one of the possible Indigenous names of the continent.

Decolonial scholarship differs from so-called 'postcolonial studies'. Widely developed since the 1980s and 1990s after the works of American-Palestinian author Edward Said (1978), postcolonialism contributed to a critique of colonization by revealing the mechanisms through which the Orient was 'invented' by diffusing commonplaces on different peoples in literature and media. Conversely, decolonial scholars put special emphasis on practices and concrete links with grassroots movements and emancipation struggles in the Global South. Some bases of decolonial thought were established by authors who worked on the notion of 'post-development', such as Wolfgang Sachs, Gustavo Esteva and Arturo Escobar (see Further reading).

These scholars defined the very ideas of development and 'underdevelopment' as discourses led by Northern authors and politicians to substitute political colonialism with neo-colonialism since the aftermath of the Second World War, imposing recipes that were based on Western ideas of merely quantitative economic growth and on indicators such as GNP. Far from improving people's well-being, these models of development produced increasing hunger and misery in Southern countries, as denounced by authors such as critical geographer Josué de Castro (Davies 2023). Today, decolonial scholars and activists propose notions such as Pachamama and *Buen Vivir/Sumak Kawsay*, an Indigenous ethics based on slowness and mutual respect between humans and environments, as alternatives to Northern-led forms of development.

Thus, the 'decolonial turn' argues for rediscovering Indigenous and in general non-European and non-Eurocentric cultures that were destroyed or are still threatened by colonialism and Euro-centric cultural domination, which were responsible for the destruction of different ways of thinking all over the world (Sousa Santos 2016). Furthermore, decolonial authors contend that Europe does not have a monopoly on critical theory. For this reason, Latin American geographers today lament that most of the inspirers of the different tendencies of so-called 'critical' scholarship were white, male, European and middle-class authors. They propose to take theoretical inspiration from alternative sources, including marginalized and racialized authors from the Global South, as well as grassroots movements and activist experiences.

For Mignolo, a tentative list of these alternative references may include: 'Individual thinkers and activists like Waman Puma de Ayala in colonial Peru, Ottobah Cugoano, in British Caribbean and then in London, in the eighteen century; Mahatma Gandhi in nineteenth-twentieth century India; Amilcar Cabral in the Portuguese colonies of Africa; Aimé Césaire and Frantz Fanon in the French Caribbean; W.E.B. Dubois and Gloria Anzaldúa in the US . . . countless uprising and social movements that, today, have in the Zapatistas and the Indigenous movements in Ecuador, Bolivia as well as Indigenous activists in New Zealand, Australia, Canada and the US' (Mignolo 2010, 16), in addition to Haitian Black revolutionaries such as Jean-Jacques Dessalines and Toussaint Louverture.

From a decolonial perspective, colonialism is far from being a phenomenon of the past, as colonial relations are still working through the mechanisms that were highlighted above. In this vein, the notion of 'epistemic decolonization' means that one should listen to different voices. That is, it should not be an external intellectual who speaks on behalf of a marginalized community from a position of privilege: it is subaltern people who have to speak with their own voices. In this vein, a key principle of decoloniality is the need for mutual acknowledgement between different cultures and different activist circuits. This implies the need to acknowledge the position of the Other, so that dialogue can undermine the establishment of any unique or dogmatic thought. This notion is expressed by the definition of 'pluriverse'.

According to the Neo-Zapatistas who famously revolted in Chiapas in the 1990s against neoliberalism and the depredation of Indigenous land, 'pluriverse' is a world where more worlds can fit, or 'the world of many worlds' (De la Cadena and Blaser

2018). The lesson of the Zapatistas is very significant, as they declared since the beginning that the aim of their revolt was not to seize power, but just to take control of their lands and their lives. That is, decolonizing social practices also means giving up the European idea of the vanguard party leading revolutionaries to 'storm the palace' and take 'the place of the King' without deeply questioning social privilege.

Thus, the pluriverse respects all voices and all cultures, against the notion of ethnocentrism, which characterizes communities who believe themselves to be the center of the world as Europe has done for centuries. This does not mean that cultures should refuse mutual dialogue or give up common points. Instead, dialogue is fundamental to reach what French-Caribbean anti-colonial thinker Aimé Césaire called a 'universality . . . rich with all that is particular'. Decolonial scholars define pluriverse as the idea that there is not a sole way to fight capitalism, racism, patriarchy, internal colonialism and environmental destruction (Kothari et al. 2019). That is, activists and scholars with different backgrounds can share some common points giving their diverse, but equally useful, contributions to the common cause of justice.

Geography is interested in decolonial approaches in response to how space and territory have become key strategies of social movements in Latin America (see **Chapter 21**). In the next section, I discuss a couple of Brazilian examples from historical geographical research that can potentially nourish a historical-geographical decolonial scholarship.

Towards decolonial historical geographies

There are historical experiences from the ideas and practices of Latin American geographies that can help us, today, make sense of ongoing struggles around decolonization. This section presents two examples in the work of Latin American geographers, the first the 'geography of *quilombo*' by Manuel Correia de Andrade, and the second the 'geography of *favela*' by Mauricio Almeida Abreu and Andrelino Campos, which both pioneered recent studies on these matters.

A geographer from North-Eastern Brazil, traditionally considered the less 'developed' region of the country, Manuel Correia de Andrade (1922–2007) was a key figure for both Brazilian historical geographies and for the rise of critical geographies in Latin America. For Andrade, dealing with historical revolts of marginalized communities from his region was a way of fostering new social struggles. Andrade was one of the earliest supporters of the movement of landless rural workers (*Movimento dos Trabalhadores Rurais sem Terra* – MST) that he saw as the continuation of historical efforts for obtaining an agrarian reform (see **Chapter 22**). Politically persecuted by the Brazilian dictatorship from 1964 to 1985, Andrade participated in the rediscovery of subaltern territorialities that followed the promulgation of the Brazilian constitution in 1988. This text granted the right for the descendants of Indigenous and *quilombola* communities to the devolution of their ancestors' lands, anticipating current worldwide decolonial debates on the need for returning Indigenous lands that were expropriated by settlers, as 'the gravity of history should not be underestimated' (Clayton 2021, 1696).

While real devolution of lands only happened in a few cases, it is worth noting the importance of knowing the geographies of past Indigenous and *quilombola* settlements to enforce the rights of their descendants. Crucially, a characteristic of *marronage* was that of being constantly present all over the Americas since the *Conquista*. The 'geography of *quilombo*' that Andrade inaugurated remains an important part of Latin American historical geographies, inspiring similar studies all over the region (see Chapter 15). As a staunch critic of internal colonialism, Andrade understood *quilombos* as a phenomenon that lasted well beyond the abolition of slavery, constructing spaces of resistance to capitalist and colonial oppression (see Figure 2.2). He connected these stories to the political needs of post-dictatorship Brazil.

For Andrade, the phenomenon of *quilombos* was incredibly varied in terms of their size, duration, location and socio-spatial organization, which calls for detailed empirical studies rather than for generalizations. Yet, there were some common points. First, *quilombos* were not only composed by Black people, as they could also be participated in by Indigenous and other marginalized people, being places for an early 'convergence of Blacks and Indians against the colonizers' (Andrade 2001, 80). Second, these insurgent phenomena were the real drive for abolition. Andrade countered all ideas that the abolition of slavery resulted from the spreading of humanitarian values among white elites, being instead the outcome of the fear that ruling classes had of subaltern insurgence, after the example of Haiti.

Finally, Andrade considered abolition to be work in progress, given that the formal abolition of slavery in 1888 did not grant the land to former slaves and exploited

Figure 2.2 Brasilia, one of the numerous monuments that are dedicated to the mythic figure of Zumbi, leader of the ancient *quilombo* of Palmares, in Brazil.

Source: https://commons.wikimedia.org/wiki/File:Zumbidospalmares.jpg

people, and did not eliminate social and racial discrimination. Thus, *quilombos* still exist due to the lasting nature of colonial relations after formal independence and abolition, and even after the establishment of a federal republic in Brazil. Against that, for Andrade, scholars should value the direct action of subalterns who speak by their acts, and directly support social movements.

Following Andrade, Rio de Janeiro-based historical geographer Mauricio de Almeida Abreu (1948–2011) launched the idea of investigating the formation of space that is subject to social, racial and security stigma, eventually in *favelas*, Brazilian versions of the worldwide phenomenon of slums, shantytowns and bidonvilles (see **Chapter 19**). Abreu argued for giving attention to 'forgotten histories', focusing on the Rio case, to clarify the formation of these spaces, which are still the object of commonplaces about danger, drug trafficking and lack of basic facilities (see Further reading).

Investigating new historical sources since the end of the nineteenth century, Abreu inserted *favelas* into the context of struggles for the 'right to the city' that accompanied the social tensions of republican Brazil since the end of the nineteenth century (Abreu and Le Clerre 1994). *Favelas* started to be built as informal settlements in the outskirts of the city, namely its surrounding hills, inaugurating the identification between the *favela* and the Portuguese term *morro* (hill) as a specific geographical connotation of *favelas* in Rio and in other major Brazilian cities. For Abreu, the growth of these settlements paralleled the social and racial stigma that was thrown on their inhabitants by both ruling classes and white popular milieus, as it was mostly people from racialized groups who first occupied these hills.

While Abreu noted that *favelas* were not an issue for growing capitalism, as they provided a reservoir of cheap workforce close to the city, an author who was strongly inspired by his work, Andrelino de Oliveira Campos (1949–2018) addressed the problem of the formation of *favelas* as racialized urban spaces. An Afro-Brazilian geographer and an activist for Black rights, Campos contributed to what Marcelo Lopes de Souza calls the 'blackening' of Brazilian geography through an 'Afro-descendant outlook' (Souza 2012, 13) to Brazilian cities. All his life, he argued that 'knowledge should be at the service of the training of critical citizens' including 'politicizing the agency of a Black field' (Oliveira 2016, 12) in Brazil.

Campos found the origins of *favelas* in the *quilombo*, matching Andrade's idea that they were spaces of resistance for marginalized groups (see **Chapter 23**). Campos importantly stressed how these structures of resistance participated in the creation of urban space as they got closer to urban centers. This implies the possibility of linking *quilombo* and *favela* 'in one process of formation of Rio's urban space [as] a product of Afro-descendant diasporas' (Campos 2012, 29). For Campos, '*Quilombo* people were in the past a threat to Empire; *favela* people became unwanted subjects after the proclamation of the Republic' (Campos 2012, 64). If we consider that the Brazilian Empire followed the end of Portuguese colonial rule in 1821 and the Republic followed the end of the Empire in 1889 (and of abolition in 1888), the decolonial lessons remain the same: colonial relations of oppression lasted after all changes in political regimes, and the sole effective response to this condition was subaltern agency. Marginalized and stigmatized communities were the protagonists of socio-spatial change more often than is commonly believed.

Conclusion

Scholars are increasingly advocating for the need to decolonize geography: starting from its histories, and from the history of places and territories. This task can be performed by analyzing new archives and listening to voices that had been silenced by the mainstream academy. Thus, geography as a whole is increasingly inspired by and learning from Latin America and its historical geographies, while decoloniality is increasingly informed by matters of space, places and territories. A key point to understand are the differences between political decolonization, which led (in Latin America and elsewhere) to the reproduction of socio-spatial domination and colonial territorialities such as the nation-state, and decoloniality, advocating for decolonizing epistemes starting from concrete social praxes. Decolonization cannot be only theoretical; it should be built in everyday practices, otherwise coloniality will continue to reproduce itself.

Summary

- Historical geography is no longer based on descriptive or conservative methods, being now increasingly informed by critical thought and decoloniality, including contributions from the 'Souths'.
- Political decolonization does not equate with full decolonization. In most cases, it reproduced colonial social relations.
- Decoloniality advocates the end of ongoing colonial relations as well as epistemic decolonization, which means starting to think in completely different ways from the common Northern/Western models.
- Studying spatial histories of socially, spatially and racially marginalized communities is a key contribution to historical geography from the South, and from historical geography to the broader field of decoloniality.

Review questions

1. What are the main limitations of political decolonization from a decolonial standpoint?
2. Why is historical geography an especially useful field to deal with current decolonial matters?
3. Why is Latin America considered an exclusionary concept by decolonial scholars?

Further reading

Geographers biobibliographical studies, vol. 40, 2022. https://bloomsburycp3.codemantra. com/viewer/6296a4c85f150300016f11c1
Collection of geographers' biographies from all over the world, a key instrument for the history of geography

Kothari, A., Salleh, A., Escobar, A., Demaria, F., and Acosta A. (eds.). (2019). *Pluriverse, a post-development dictionary*. Tulika Books.
Short and simple readings on the pluriverse
Perlman, J. (2010). *Favela, four decades of living on the edge in Rio de Janeiro*. Oxford University Press. http://abahlali.org/files/favela.pdf
Key work on favelas' recent history
Santos, M. (2021). *For a new geography*. Minnesota University Press.
Pioneering text on international histories of geography
Santos, M. (2021). *The nature of space*. Duke University Press.
Foundational text in Brazilian geography
Terra Brasilis
Multilingual journal of *Rede Brasilis*, the Brazilian network of historical geography and the history of geography. https://journals.openedition.org/terrabrasilis/

Keywords

Decoloniality: the idea of fighting today's colonialism by rediscovering Indigenous thoughts, epistemes and praxes and supporting social movements in the Souths.

Decolonization: a world that can be associated with various adjectives, generally indicating the political independence of former colonized countries in the Global South.

Double consciousness: for leading Black scholar W.E.B. DuBois, the dilemma of the colonized, caught between two or more cultures. For decolonial authors, it was the attitude of (North and South) American white creole elites rejecting European colonial domination but reproducing coloniality under the fork of domination and marginalization of non-White communities.

Epistemic decolonization: thinking in different ways to remediate epistemicide, that is, the destruction of Indigenous thoughts and cultures by colonial oppression.

Historical geography: the study of the past of territories, and more broadly of the roles possibly played by spaces and environments in historical processes. It sometimes overlaps with the history of geography.

Pluriverse: the world in which several worlds can fit; that is, there is not just one way to reach ideas of social justice and freedom.

Subalternity/subaltern: famously inaugurated by Antonio Gramsci, this definition indicates the pre-industrial revolutionary subject, with special reference to peasants and to oppressed classes which are denied their voices because their history is written by the ruling classes.

Bibliography

Abreu, M.A., and Le Clerre, G. (1994). 'Reconstruire une histoire oubliée: origine et expansion initiale des "favelas" de Rio de Janeiro'. *Genèses*, 16, 45–68.

Andrade, M.C. (1980). *The land and people of Northeast Brazil*. University of New Mexico Press.

Andrade, M.C. (2001). 'Geografia do quilombo'. In C. Moura (ed.), *Os quilombos na dinâmica social do Brasil*. Maceió: EDUFAL, pp. 75–86.

Baker, A.R.H. (2003). *Geography and history: Bridging the divide*. Cambridge University Press.

Burnett, D.G. (2000). *Masters of all they surveyed: Exploration, geography, and a British El Dorado*. University of Chicago Press.

Campos, A. (2012). *Do quilombo à favela. A produção do "espaço criminalizado" no Rio de Janeiro*. Bertrand Brasil.

Clayton, D. (2021). 'Historical geography I: Doom, danger, disregard – towards political historical geographies'. *Progress in Human Geography*, 45(6), 1692–1708.

Cosgrove, D. (1984). *Social formation and symbolic landscape*. Croom Helm.

Davies, A. (2023). *A world without hunger, Josué de Castro and the history of geography*. Liverpool University Press.

De la Cadena, M., and Blaser, M. (eds.). (2018). *A world of many worlds*. Duke University Press.

Esson, J., Noxolo, P., Baxter, R., Daley, P., and Byron, M. (2017). 'The 2017 RGS-IBG chair's theme: Decolonising geographical knowledges, or reproducing coloniality?'. *Area*, 49(3), 384–388.

Ferretti, F. (2019). 'Rediscovering other geographical traditions'. *Geography Compass*, 13(3), e12421.

Mendoza Vargas, H. (2007). 'La geografía en el siglo XIX mexicano'. In H. Mendoza Vargas (ed.), *Lecturas geográficas mexicanas, siglo XIX*. UNAM, pp. vii–xxviii.

Mignolo, W.D. (1993). 'La colonialidad a lo largo y a lo ancho: el hemisferio occidental en el horizonte colonial de la modernidad'. In E. Lander (ed.), *La colonialidad del saber: eurocentrismo y ciencias sociales. Perspectivas latinoamericanas*. CLACSO, pp. 52–82.

Mignolo, W.D. (2010). 'Introduction. Coloniality of power and decolonial thinking'. In W. Mignolo, and A. Escobar (eds.), *Globalization and the decolonial option*. Routledge, pp. 1–21.

Morelli, S., and Venturoli, F. (2021). *Geografia razza e territorio. Agostino Codazzi e la Commissione corografica in Colombia*. Il Mulino.

Newman, B., Martin, P., and Crawford, L. (eds.). (2022). *Environments, spaces, knowledges. New and emerging research in historical geography*. RGS-IBG Historical Geography Research Group.

Oliveira, D.A. (2016). 'Legado de um professor: uma homenagem a Andrelino de Oliveira Campos'. *Synthesis*, 9(2), 9–15.

Radcliffe, S. (2017). 'Decolonising geographical knowledges'. *Transactions of the Institute of British Geographers*, 42(3), 329–333.

Reclus, E. (1894). *Nouvelle Géographie universelle*, vol. XIX. Hachette.

Said, E. (1978). *Orientalism*. Random House.

Sousa Santos, B. (2016). *Epistemologies of the south: Justice against epistemicide*. Routledge.

Souza, M.L. (2012). 'Prefácio'. In A. Campos (ed.), *Do quilombo à favela. A produção do "espaço criminalizado" no Rio de Janeiro*. Bertrand Brasil, pp. 13–17.

CHAPTER 3

Borders and territory

Perla Zusman

Introduction[1]

Latin American countries began to form in the wake of their independence from European powers. What role did territories play in the constitution of states? State formation was based on colonial administrative units, but ruling elites proposed to occupy the spaces controlled by Indigenous populations. The elites presented these spaces as unoccupied and available for appropriation. At the same time, the delineation of boundaries between state territories reshaped the mobility of societies now living in border areas. These territorial practices aimed to establish closed states, with a homogeneous identity. However, the multicultural nature of society could not guarantee identity homogeneity. In fact, society comprised individuals who had migrated from Europe or Asia, criollos born in the region who had European ancestors, multiple Indigenous nations, mestizos, and forcibly enslaved Africans brought to the region. In this context, the idea of territory emerged as providing symbolic unity and cohesion to country inhabitants, mitigating gender, class, and ethnicity differences.

The chapter is divided into four sections. Section one will describe the characteristics of the independence process and its implications for the construction of new territorial designs. In addition, it will highlight the limited divorce of these proposals from European ideas and values regarding civilization. Section two will identify state practices to occupy territories under Indigenous control, a process that concurred with the definition of international boundaries. Then, section three will present *uti possidetis* and natural boundaries, the criteria used to define international boundaries, along with the strategies adopted to face implementation challenges. Section four will focus on bordering practices during the 1930s, which tightened border controls and regulation, and established protected areas. Finally, the concluding remarks aim to shed light on a re-signification of the dimensions ascribed to borders in the process of state formation, in the current global scenario. The analysis will focus on territory and border formation processes in South America during the 1870–1940 period.

DOI: 10.4324/9781003430926-5

Emancipation processes and territorial design

The Latin American states as we know them today are the result of revolutionary processes that began in the late 18th and early 19th centuries. The local elites, inspired by the Enlightenment ideas and values from France and the United States, sought to become independent from the domination of the European powers (England, France, the Netherlands, Spain, and Portugal). For European cities, the region's territories acted as suppliers of precious metals (gold and silver) and agricultural products from plantation systems (sugar, cotton, and tobacco). In turn, they were also used as markets to place manufactured goods.

The first revolution recorded in the region was sparked by the enslaved population of Haiti who worked in sugar cane production (1791–1804). Later, South American countries gained independence, followed by those in Central America and finally some in the Caribbean (see **Chapter 2**). While the emancipation of South American countries was advanced by alliances between England and local elites, the United States played a key role in Central America and the Caribbean (Etcheverri and Soriano, 2023).

Toward the 18th century, the territories of South America were ruled by the Spanish and Portuguese crowns. Spain's dominions included the administrative units known as the Viceroyalty of New Spain (currently Mexico, the central-western part of the United States, and the Caribbean countries), the Viceroyalty of New Granada (currently Colombia and Venezuela), the Viceroyalty of Peru (Peru and Ecuador), and the Viceroyalty of Rio de la Plata (currently Bolivia, Paraguay, and Argentina). Portugal's main dominion was the Viceroyalty of Brazil.

The elites of the American colonies interpreted the invasion of the Iberian Peninsula by Napoleon in 1808 and the imprisonment of King Ferdinand VII of Spain as an opportunity for the sovereignty yielded to Spain's monarchy to return to the people (Goldman, 2008). Struggles over political and territorial control were triggered within the regional elites living in the former Spanish administrative units. The Viceroyalty of Brazil followed a different path, as the advance of Napoleon's troops (1807) determined the entire relocation of the Portuguese crown to its Latin American colony. After the return of King Joao IV to Portugal, his son Pedro II declared Brazil's independence, and the country became an Empire (1822). Former Spanish administrative units were fragmented as a result of colonial emancipation. In contrast, Brazil's independence was accompanied by a political and territorial process unifying heterogeneous geographical environments that had been the object of appropriation and valorization by settlers. These dynamics did not challenge the slavery system, which was not abolished in Brazil until 1888 (Pimenta, 2012) (see Figure 3.1).

These emancipation processes did not question Eurocentric ideas and values. The elites sought to bring the newly forming countries into the world's capitalism in line with the civilization and progress ideas and values from Europe and the United States (Moraes, 2009a). In other words, Indigenous or Afro-descendant populations were not considered inhabitants with social and political rights. Their ancestral occupation of these lands was not taken into account (see **Chapters 13 and 15**). On the contrary, these societies were seen as lagging behind, and the only way for them to overcome this situation was to be incorporated into Western lifestyles. In this framework, the actions implemented by Latin American ruling classes in areas under Indigenous control mirrored the colonial domination practices imposed by European states in Asia and Africa.

Figure 3.1 Map of South America according to latest and best authorities (1826). Anthony Finley.
Source: Digital Library of Congress, Washington. United States of America.

Occupation of territories under Indigenous control

Multiple territorial designs rooted in the preexisting colonial legal-administrative units were envisioned against the background of the independence of South American countries. The multiplicity of designs formed part of the tensions for power within regional elites. However, it is worth noting that while colonial borders were permeable and blurred, encompassing populations of different origins and ethnicities, forming state territories were conceived as closed and socially homogeneous entities that were economically integrated through telegraph and railroad networks.

These designs included the incorporation of Indigenous territories into the states, shifting away from the strategy of considering Indigenous populations as political subjects with whom specific alliances could be forged. In contrast, military action was used to exterminate Indigenous populations and displace them from their lands, forcing them to abandon their ways of life.

Before conducting military occupation, states took nominal possession of these spaces. The ideas of "desert" (Patagonia), "*sertão*" (Brazil), or "orient" (Ecuador) designated areas that were inhabited by populations with distinct political, cultural, and economic dynamics compared to Western societies (Moraes, 2009b). For example, the Pampas and northern Patagonia were inhabited by different nomadic Indigenous communities (*ranqueles, pehuenches, pampas, tehuelches, mapuches*) that relied on the trade of salt and livestock from Buenos Aires to Chile as a means of livelihood. This activity helped strengthen social interethnic bonding and develop the leaders who would negotiate with the white population.

In 1879, the Argentine State launched the so-called "Conquest of the Desert," referred to as *Wingka Malon* in Mapudungun. As a result of this expedition, the *pampas, ranquel, mapuche,* and *tehuelche* Indigenous peoples, living in the southwestern Pampas and northern Patagonia, were dispossessed of their land. Expedition members included not only military personnel but also naturalists who explored and mapped the area. The information thus gathered found its way into scholarly discourse and dissemination within scientific societies and institutions responsible for map production. It is important to highlight that the lands taken from Indigenous populations were incorporated into productive activities and, over time, contributed to positioning Argentina as one of the world's largest cereal and grain exporters until the 1930s.

In the case of Brazil, the occupation of territories within the "*sertão*" was promoted by the Strategic telegraph Lines Commission from Mato Grosso state to Amazonas state. This Commission was also known as the Rondon Commission, as it was led by military engineer Cândido Mariano da Silva Rondon from 1900 to 1930 (**see** Figure 3.2). It forced some Indigenous communities, such as the *parecís*, to work on telegraph lines. Similar to the "Conquest of the Desert" expedition, the Rondon Commission included scientists from the Paulista Museum and the National Museum to improve the knowledge of the country (Antunes Maciel, 2012).

The penetration of states into Indigenous territories triggered a process of deterritorialization among these communities (**see Chapter 12**). While some managed to relocate and maintain their autonomy, others were uprooted and forced to join military forces or work as rural laborers or domestic servants in urban households. It is also worth mentioning that some groups were confined to concentration camps and reserves, while others were displayed in museums, exhibitions, fairs, zoos, and circus shows (Escolar et al., 2015).

Part of the land appropriated from Indigenous communities was given to the military or private individuals. In some cases, this led to the concentration of property and the formation of large estates (*latifundios*) (**see Chapter 22**). In other cases, land was

Indios Parecis aldeiados pela Commissão.
Nucleo de Utiarity : antiga casa da estação telegraphica.

Figure 3.2 Parecis nation reduced to villages.

Source: Fundación Oswaldo Vio Cruz. Comisión Rondon. Obras raras.

delivered to individuals with the aim to foster colonization by Europeans. Finally, the state took possession of a third portion of the land.

The creation of national territories

States used different administrative strategies to incorporate these lands and their inhabitants (often migrants, Indigenous people, and military personnel) into the countries being formed. Some states established legal units that were devoid of autonomy to appoint their authorities and were under direct control by the Executive Power. This perspective held that only central governments could guarantee the social, economic, and political integration of these lands and individuals into the larger country. Thus, based on the administrative organizational model developed by the United States for the lands occupied during its westward expansion, the process of territorial appropriation in Patagonia, Chaco, or the Amazon led to the creation of what has been referred to as National Territories or Federal Territories (Serje, 2011; Zusman, 2011). The proximity of these lands to the boundaries of states was an additional justification for the formation of these National Territories. Indeed, it was considered that geographic areas adjacent to neighboring countries required special surveillance. Some of these political-administrative units were formed as a result of specific circumstances. For example, the

Federal Territory of Acre (1904) was organized in Brazil after the war with Bolivia over the appropriation of gold and rubber resources from the region (López Beltrán, 2001). Margarita Serje (2011), a Colombian anthropologist, pointed out that different textual and visual records, from geography textbooks to geopolitical discourse, portrayed the National Territories as expanses of land, empty, wild, lagging, difficult-to-access spaces, yet rich in lush nature.

The definition of international boundaries

While occupying territories under Indigenous control, states simultaneously delineated boundaries with their neighboring countries. Two prominent diplomatic criteria were used for this purpose: *uti possidettis juris* and natural boundaries. The *uti possidettis juris* principle had a historical basis and established that the boundaries of new states would preserve the configurations of inherited colonial units, such as Viceroyalties (administrative jurisdictions), Provinces (subdivisions within larger administrative units), Captaincies General (units formed in areas of strategic importance vis-à-vis other powers or against piracy and Indigenous populations), and Royal Audiences (Spanish judicial bodies). However, this principle alone could not always define the borders between states: the boundaries of many colonial units were not distinct. In addition, in some cases, different states emerged from former colonial units. For instance, in the framework of the Viceroyalty of New Granada (1717–1723, 1739–1810, and 1815–1822), the states of Colombia, Venezuela, Ecuador, and Panama were formed. In this context, several constitutions of Ecuador have asserted the country's territorial inheritance from the Presidency of Quito or the Audience of Quito (1563–1822), while Venezuela traces its current country format back to the Captaincy General of Venezuela (1777–1823) (Duque Muñoz, 2020).

The *uti possidettis juris* principle was complemented by the criterion of natural boundaries. Rooted in scientific rationale, this criterion aimed to establish unambiguous and precise borders between countries that would not be affected by changes over time: the line defined by a mountain range dividing watersheds or the line connecting the lowest points of a riverbed (talweg) could be used to establish the territorial limit between states.

Both criteria were taken into account as the rationale of boundary treaties and when Joint Boundary Commissions implemented diplomatic decisions on the ground. It is important to mention that the members of Boundary Commissions could also produce information through their interaction with local populations, being guided and assisted by Indigenous people and peasants. Additionally, astronomical or topographic measurements were carried out using specific instruments such as telescopes, barometers, compasses, chronometers, theodolites, and goniometers, among others (Tamayo, 2015).

The result of these procedures by the Boundary Commissions entailed the placement of boundary markers (*mojones*) in the field, along with the desk production of cartographic sheets reflecting the actual boundary. In particular, cartography provided the imaginary line with a contiguity that it did not have on the terrain. On the ground, the boundary was permeable, allowing for the movement and exchange of licit and illicit goods from one side to the other. In some cases, such as the Chile-Bolivia border during the 1910s, Indigenous populations could alter the location of boundary markers to recreate dynamics from colonial times and gain access to pastures, cattle ranches, or mining sites now located on the other side of the border (Castro, 2014). These types of connections and mobilities played a role in consolidating the so-called twin cities located on both sides of the boundary, including Leticia (Colombia) and Tabatinga (Brazil), Arica (Chile) and Tacna (Peru), and Colon (Argentina) and Paysandu (Uruguay) (Osorio Machado, 2005).

During the state formation process, it was frequently challenging for countries to agree on boundary demarcation. This difficulty was particularly evident when criteria such as *uti possidettis* or natural boundaries proved insufficient in areas characterized by intricate physiography, or when there was ambiguity in the naming of geographic features that could support boundary demarcation. In these cases, Latin American countries perceived the benefits of resorting to arbitration by other countries to reach an amicable settlement. In many cases, the United States or England were selected to act as arbitrators, in pursuit of furthering the development of fluid financial and trade relations with them.

For example, defining the boundary between Argentina and Chile in the Puna de Atacama area was an intricate case. Puna de Atacama is a high-altitude Andean plateau nearly 4,000 m high, extending across an area of almost 80,000 km² that includes low-elevation ridges running from south to north and from east to west. Here, high summits and watershed divides are hard to identify, making it difficult to implement the criterion of high watershed divides used in other areas of the Andes. Moreover, the strategic significance that the area gained during the development of the saltpeter economy increased the complexity of this scenario (see Figure 3.3).

The proposal to resolve the conflict by William Buchanan, the United States Minister in Argentina, in 1899, involved dividing the disputed area into seven parts. It ignored the criterion of natural boundaries, using instead a straight demarcation line to divide the two countries. This approach disregarded the mobility of preexisting populations within these geographic areas, such as caravan merchants who carried products across different ecosystems in the Andean region, including fabric, *charqui* (dried meat), feathers, potatoes, corn, salt, or livestock (Vila, 2019). Buchanan sought to impose a "rational" and "abstract" criterion that would overcome the positions of each of the warring countries (Zusman and Hevilla, 2014).

Despite efforts to demarcate and separate the territorial domains of states, it was not always possible to define clear and distinct international boundaries. Some special circumstances make this matter even more visible. For example, the border dispute between Peru and Ecuador over Cordillera del Condor led to an armed conflict in 1995. Political interpretations suggest that the governments of both countries brought the territorial conflict to public attention as a means to win social support amidst the economic or social problems of the time.

The reinforcement of borders and the creation of protected areas

During the late 1920s and 1930s, various forms of nationalist movements emerged in South American countries. These were driven by ruling elites as they noticed the outcomes of the policies implemented since the late 1870s to encourage migration from Europe or Asia. The heterogeneous ethnic and political character of the societies in these countries called into question the ideas and values of the dominant elite regarding the formation of homogeneous countries from a cultural perspective. In this context, heightening the

Figure 3.3 Caricature of then-president of Argentina, Julio A. Roca, resembling Louis XV of France within the framework of the Puna Conflict.

Source: Caras y Caretas October 29, 1898. Biblioteca Nacional de España, Madrid.

link between territory and nation could help dilute social disparities. Multiple territorial policies materialized this sentiment, particularly those aimed at ensuring the state's presence in border areas. While the government of Getulio Vargas in Brazil proposed a "March to the West" to integrate the coastal regions with the West, Center, and North of the country (De Almeida, 2019), other countries pursued policies to consolidate the state's presence along borders and exert tight controls over mobility. This explains the creation of border military forces in Chile (the "Carabineros" in 1927) and Argentina (the "Gendarmería" in 1938), the interest in establishing public schools, or the extension of road and railway infrastructure into these areas (Grimson, 2003).

These policies were accompanied by a process that highlighted the touristic value of border landscapes as beautiful and exceptional national icons, extensively promoted at the time on postcards and in illustrated magazines targeting the urban middle class (Silvestri, 2011).

Many of these landscapes were incorporated into protected areas that were created during this period, mirroring the establishment of Yellowstone Park in the United States (1872). The organization of protected areas as national parks ensured management by the central government and pursued geopolitical (effectively establishing the state's presence along borders), educational (raising awareness about the existence of these areas among students and tourists), and scientific (carrying out activities that would help recognize the economic and pharmaceutical potential of species) goals. Under some circumstances, the idea of fostering development and colonization outweighed the preservation policy itself. For example, exotic animal and plant species were introduced in the Nahuel Huapi National Park (Province of Rio Negro, Argentina), while deforestation paved the way for infrastructure development in this area to facilitate access and contemplation (including hotels and walkways). These actions do not seem to be consistent with an idea of protection. In all cases, proposals for the demarcation of these protected areas included the coercive resettlement or the invisibilization of the original population that inhabited these lands on the grounds that this population was "part of nature." Some of the national parks created in late 1920s and 1930s along border areas in the Southern Cone include Perez Rosales (Chile, 1926), Nahuel Huapi and Iguazu (Argentina, 1934), Iguaçu (Brazil, 1939), and Samaja (Bolivia, 1939) (De Marchi Moyano, 2020).

Conclusion

For international recognition, a state needs to show exclusive and absolute authority over a specific territory. This is why territory has been a significant component in the process of state formation across Latin America. Additionally, displacing Indigenous populations was driven by a desire not only to expand the territory under control, but also to thwart any potential forms of territorial organization that could challenge the state's dominance. Former Indigenous territories were regarded as areas requiring direct control from the central government. Therefore, different administrative structures were organized over time, such as National or Federal Territories, or protected areas. The state further sought to assert its presence through military or educational establishments. Given that territories were conceived as closed spaces, the definition of international boundaries also played a significant role in their constitution. Once defined, the state would encourage the mobility of certain people and goods to the detriment of others, deciding which ones would be legal and which ones not.

The construction of territories and borders in Latin America did not end in the 1940s. Rather, it has remained an ongoing process by which states redefine their territories based on the meanings they attribute to them. In today's globalized world, Latin American borders have taken on new dimensions. They are no longer regarded as mere national margins but have become critical areas for capital. On the one hand, these areas attract transnational capital for open-pit mining or large-scale infrastructure projects such as dams. At the same time, the rise of ethnic awareness has led Indigenous populations to advocate their right to live in these areas of interest to capital.

Many protected areas have also been subject to globalization when UNESCO recognized them as World Heritage Sites. This has increased tourism from around the world. Indigenous communities are claiming their rights over these areas, originating some collaborative management efforts between the state and these populations. These practices reflect one more way for states to relate to their territory in a scenario where multiple actors exert power over a domain that, before the conclusion of the Cold War, was considered to be solely a state domain.

Summary

- Latin American independence processes led to the formation of states, the territorial configuration of which became a subject of dispute among elites.
- The processes resulting in the territorial formation of Latin American states involved the occupation of lands under Indigenous control and the definition of international boundaries. To achieve territorial occupation, the region's ruling elites sought to implement actions mirroring the practices imposed by European powers in colonies across Asia and Africa.
- Lands under Indigenous control were occupied through military action, displacing Indigenous communities from the areas where they carried out their daily activities.
- International boundaries were defined considering the criteria of *uti possidetis* and natural boundaries. If the establishment of unambiguous and precise borders was unsuccessful using these criteria alone, field reconnaissance was performed. When boundary definition could not be achieved by means of diplomatic efforts or fieldwork, arbitration by a third country was sought.
- States developed different strategies to assert control and authority over their borders, such as the creation of national territories or the establishment of protected areas.

Review questions

1. What were the territorial implications of independence processes in Latin America?
2. What practices did the newly formed states carry out to appropriate territories under Indigenous control?

3. What criteria were considered to define the international boundaries of Latin American countries?
4. What characteristics did the protected areas organized along the borders acquire in the 1930s?
5. What kind of territorial configurations could have been designed in Latin America if the elites had respected the spaces under Indigenous control?

Further reading

Freitas, F. (2021). *Nationalizing Nature: Iguazu Falls and National Parks at the Brazil-Argentina Border*. Cambridge University Press.
The book establishes links between conservation practices and the definition of National Parks on the border between Argentina and Brazil.
Hevilla, C. and Zusman, P. (2009). 'Borders which unite and disunite: Mobilities and development of new territorialities on the Chile - Argentina frontier'. *Journal of Borderlands Studies*, 24(3), 83–96. https://doi.org/10.1080/08865655.2009.9695741.
The article is interested in articulating the processes of defining the boundaries between Chile and Argentina and their implications on transhumance practices in the past and present, considering the globalization of the border.
Mignolo, W. (2001). 'Coloniality at large: The Western hemisphere in the colonial horizon of modernity'. *The New Centennial Review*, 1(2), 19–54.
By discussing the idea of coloniality of power, the text addresses how the processes of Latin American independence did not imply a rupture with Eurocentric ideas and values.
Oszlak, O. (1981). 'The historical formation of the state in Latin America: Some theoretical and methodological guidelines for its study'. *Latin American Research Review*, 16(2), 3–32.
The text offers conceptual elements from which to interpret the process of constructing the Latin American states.
Radcliffe, S. A. (1998). 'Frontiers and popular nationhood: Geographies of identity in the 1995 Ecuador–Peru border dispute'. *Political Geography*, 17, 273–293.
Through the analysis of the territorial conflict that took place between Ecuador and Peru, it examines the dynamics and complexities of border disputes in the region.

Keywords

Border: an environment where societies with heterogeneous socio-economic dynamics, spatialities, and temporalities both converge and diverge. A border can emerge from international boundaries or when one form of territorial utilization outweighs another.
Bordering practices: different ways in which the state or other social groups seek a distinction between an inside and an outside and, at the same time, between "we" and "others."

> **International boundary:** a linear device between neighboring states to establish the territorial domain of each.
>
> **State:** a political-legal unit considered to be the preeminent actor in the international system. The Treaty of Westphalia, signed in 1648, established that, to achieve recognition, a state needed to exercise complete and exclusive authority over a territory. To fulfill some of its functions, such as tax collection, the state requires a bureaucratic apparatus. At the same time, it also needs a coercive apparatus to effectively establish its institutional power. Furthermore, the construction of an imagined common community ensures that everyone living within a territory under a state's domain can feel part of the same nation.

Note

1 The work has been carried out as part of a research project supported by the Spanish State Research Agency under Grant Number PID2020–114088GB-I00/ AEI/10.13039/501100011033.

References

Antunes Maciel, L. (2012). 'A Comissão Rondon e a conquista ordenada dos sertões: espaço, telégrafo e civilização'. *Projeto História: Revista Do Programa De Estudos Pós-Graduados De História*, 18. https://revistas.pucsp.br/index.php/revph/article/view/10994

Castro, C. (2014). 'La conformación de la frontera chileno-boliviana y los campesinos *aymaras* durante la *chilenización* (Tarapacá, 1895–1929)'. *Historia Crítica*, 52. http://journals.openedition.org/histcrit/999

De Almeida, T. F. (2019). 'Quando no Oeste construía-se uma Nação: os Povos Indígenas e a formulação de novos projetos nacionais (1937–1948)'. *Temporalidades. Revista de História*, 11(3), 453–472.

De Marchi Moyano, B. (2020). 'Áreas protegidas y fronteras internacionales de Bolivia en perspectiva histórica'. *Revista Ciencia y Cultura*, 24(44), 217–246.

Duque Muñoz, L. (2020). *De la Geografía a la Geografía Política. Discurso geográfico y cartografía a mediados del siglo XIX en Colombia*. Pontificia Universidad Javierana.

Escolar, D., Salomon Tarquini, C. C. and Vezub, J. E. (2015). 'La "Campaña del Desierto" (1870–1890): notas para una crítica historiográfica'. In Lorenz, F. (ed.), *Guerras en la Historia Argentina*. Ariel, 223–247.

Etcheverri, M. and Soriano, C. (2023). *The Companion to Latin American Independence*. Cambridge University Press.

Goldman, N. (2008). 'Introduccion. El concepto de soberanía'. In Goldman, N. (ed.), *Lenguaje y revolución. Conceptos políticos clave en el Río de la Plata, 1780–1850*. Prometeo, 9–18.

Grimson, A. (2003). *La nación en sus límites: contrabandistas y exiliados en la frontera Argentina-Brasil*. Gedisa.

López Beltrán, C. (2001). 'La exploración y ocupación del Acre (1850–1900)'. *Revista De Indias*, 61(223), 573–590. https://doi.org/10.3989/revindias.2001.i223.573.

Moraes, A. C. R. (2009a). 'Ocidentalismo e História da Geografia Brasileira'. In *Geografia Histórica do Brasil*. Annablume, 11–35.

Moraes, A. C. R. (2009b). 'O sertão um outro geográfico'. In *Geografia Histórica do Brasil*. São Paulo: Annablume, 87–101.

Osorio Machado, L. (2005). 'Estado, territorialidade, redes. Cidades-gêmeas na zona de fronteira sul-americana'. In Silveira, M. L. (org.), *Continentes em Chamas. Globalização e Território na América Latina*. Civilização Brasileira, 246–284.

Pimenta, J. P. (2012). *Estado-Nación al final de los Imperios Ibéricos. Río de la Plata-Brasil (1808–1828)*. Ed. Sudamericana.

Serje, M. (2011). *El revés de la Nación. Territorios salvajes y tierras de Nadie*. Universidad de los Andes.

Silvestri, G. (2011). *El lugar común. Una historia de las figuras del paisaje en el Río de la Plata*. Ed. Edhasa.

Tamayo, L. M. (2015). 'La Comisión Mexicana de Límites y la definición de la frontera sur del país'. *Revista de Geografía Norte Grande*, 60, 115–134.

Vila, B. (2019). *Caravanas en las alturas*. Luján: VICAM.

Zusman, P. (2011). 'La Alteridad de la Nación: La formación del Territorio del Noroeste del Río Ohio de los Estados Unidos (1787) y de los Territorios Nacionales en Argentina (1884)'. *Documents d'Anàlisi Geogràfica*, 56(3), 503–524.

Zusman, P. and Hevilla, M. C. (2014). 'Panamericanismo y arbitraje en conflictos de límites: la participación de Estados Unidos en la definición de la frontera argentino-chilena en la Puna de Atacama (1899)'. *Cuadernos de Geografía: Revista Colombiana de Geografía*, 23(2), 95–106. https://doi.org/10.15446/rcdg.v23n2.

Political geographies

CHAPTER 4

Democracy

Sam Halvorsen and Fernanda Valeria Torres

Introduction

Democracy, from the Greek "rule" of the "people", was adopted, adapted and rejected across Latin America's post-colonial history before becoming widely accepted in the late 20th century. To this day, many use democracy to refer exclusively to the institutional set of rules that corresponds to Schumpeter's (1942) famous minimalist definition of governments chosen through competitive elections. However, the experiences of Latin America have challenged this liberal understanding in both thought and practice. Free and fair elections are no guarantee that democracy will have any substance or promote a citizenship whose needs are represented. Argentine political scientist Guillermo O'Donnell (1994) famously lamented the delegative nature of democracy in what turned out to be, following democratization, hollow nation-states based on strong Presidential powers. Yet others have highlighted the need to actively mobilize democracy (Avritzer, 2002), whether it be through institutional innovations such as participatory budgeting or through populist linkages between the people and their leaders (Roberts, 2006).

This chapter provides an overview of the post-19th-century transformations of democracy in Latin America by paying attention to its geographies. Democratic traditions have a longer history in the region, and the post-colonial era saw the uneven establishment of civil and political society with a consolidating public sphere (Forment, 2003). It is widely assumed, yet little commented, that democracy inherently relies on geographical notions, such as territory, boundaries and scale (see Barnett and Low, 2004). Indeed, democracy's modern history is inextricably bound up with the rise of the nation-state and its territory (see **Chapter 3**) including national economic formations (see **Chapter 11**). Geography is not just a backdrop, a container in which the people struggle for how they are ruled, it is an active component in shaping the possibilities and limits of democracy.

The chapter begins by providing a brief discussion of the geographically uneven and nonlinear uptake of electoral democracy, highlighting some key examples, ending with

DOI: 10.4324/9781003430926-7

the so-called third wave of democratization in the 1980s. The following three sections each identify a core strand of democracy in the region and outline their key geographical contours. First, populism established itself as a political strategy from the early 20th century, both deepening and stretching the limits of (liberal) democracy, and has tended to take on a highly nation-centric approach. Second, post-authoritarian democratization, combined with neoliberal restructuring, led to a process of decentralization that shifted administrative, fiscal and/or political power away from central governments and towards local levels, in turn opening up new possibilities for citizen participation. Finally, ongoing struggles to decolonize the state and territory (see Chapter 12) created opportunities for plurinational democracies that challenge colonial legacies of citizenship and informed a new geography of democracy.

Democratization

Authoritarian regimes, usually initiated via a military coup and often with some support from the USA, dominated politics for much of the region throughout the 20th century. This dramatically changed with the simultaneous downfall of multiple dictatorships in the late 1970s through the early 1990s. This period saw the region go from approximately 20% living under an electoral democracy to over 90% (Hagopian and Mainwaring, 2005). Latin American democratization was part of a global transition to democratic regimes, labelled by Huntington (1991) as the "third wave", which included the collapse of the Soviet Union. Since then, democracy has remained relatively stable, although ongoing concerns over backsliding and institutional erosion remain prevalent. Latin America's path to the seeming universalization of electoral democracy was neither historically linear nor geographically congruent and contains much unevenness.

The early decades of the 20th century saw successive waves of democratization as political parties institutionalized themselves, both establishing roots in a nascent civil society and competing with opposition in congress and legislatures. This brought about key developments such as women's suffrage, first enacted in 1929 in Ecuador, as well as the recognition of unions and other political organizations of the popular, working-class sectors (Collier and Collier, 1991). Yet this did not lead to a linear process of democratization. The Cuban Revolution (1959) used arms to set up a one-party communist state that, despite notable socio-economic advances, prevented free and open elections. The earlier Mexican revolution of the 1910s also led to the consolidation of a one-party state under the PRI (Institutional Revolutionary Party), generating a "hegemonic party system", in Sartori's (1976) words. In this case, elections were never banned, yet no opposition won the Presidency until 2000, with the party instead choosing its successors through informal mechanisms.

Authoritarian regimes led by military juntas came and went throughout the 1950s–1980s. In some cases, such as Chile's dictatorship led by Augusto Pinochet, these regimes were long-lived and came to an end in a gradual process of opening up to democratic elections. In other cases, they came under crisis and collapsed. This was the fate of the Argentine junta following the disastrous war of the Malvinas/Falklands and growing mobilization by human rights organizations such as the Mothers of the Plaza de Mayo. Elsewhere, democratic elections ushered in authoritarian rulers that would close down democratic institutions, such as Peru's Alberto Fujimori who, following his 1990 election, led a so-called self-coup and shut down congress in 1992. Moreover, democratically elected governments in Venezuela, under Nicolás Maduro (2013–), and Nicaragua under ex-revolutionary Daniel Ortega (2007–), have both come under harsh criticism of democratic backsliding including the persecution of political oppositions. In Brazil, the democratically elected President Dilma Rousseff was impeached in 2016 and ex-President Lula imprisoned in 2018, both acts seen by many observers as politically charged events that have tested the stability of institutional democracy in the region's largest country.

The transition to democracy is thus non-linear and geographically uneven. Although scholars have sought to account for this based on historical readings of modernization and bureaucratization in state regimes (O'Donnell et al., 1986) as well as the building and survival of political party organizations (Levitsky et al., 2016), the geography of democratization is often sidelined or left implicit, although recent structural transformations have generated some interest. The widespread trend of decentralizing democratic institutions across much of the region following the consolidation of electoral competitions led to a greater recognition of sub-national variation. Moreover, a new wave of populist governments in the early 20th century re-ignited national imaginaries of popular-democratic movements, often in tension with new forms of localism. The rise of Indigenous politics has also provoked attention to their democratic geographies.

Populism

In the mid-20th century, several Latin American countries witnessed the emergence of governments that would be labelled as populist based on distinctive characteristics that are usually centered on a clash between "the people" and an elite or oligarchy. Populism can be interpreted as a particular type of political participation and representation that arose in the context of a crisis of liberal democracy after the First World War, under the expansion of fascism and the Russian revolution. The paradigmatic cases of classical populism in Latin America refer to the governments of Lázaro Cárdenas in Mexico (1934–1940), Rómulo Betancourt in Venezuela (1945–1948), Getúlio Vargas in Brazil (1930–1945) and Juan Domingo Perón in Argentina (1946–1955) (Drake, 1982). These experiences had their own particularities, and it is important to acknowledge their geographical specificity in relation to more global processes.

The archetypal case of populism is Argentine Peronism, which highlights key features of this epochal phenomenon and has had significant global reverberations in terms of both style and content (de la Torre, 2015). First, populism constituted a coherent response to the processes of acceleration of industrialization, social differentiation and urbanization in Latin America (see Chapter 7). An active process of upward social mobility was, unlike the previous era, directly promoted by the State, together with an expansion of the national economy through import-substitution industrialization (ISI) (see Chapter 11). This led to the rapid growth of the Argentine working class: 500,000 jobs were created between 1947 and 1960 with almost 50% engaged in the manufacturing industry. From one census to the next, the number of industrial workers went from 1,000,000 to 1,200,000 (Torre and Pastoriza, 1989). If the period prior to Peronism is taken as a reference, the data are even starker: the number went from 435,816 industrial workers in 1930 to 1,056,673 in 1946 (James, 1988). This paved the way for the incorporation of the "popular sectors", by which we mean the large mass of working-class people, into systems of economic (re)distribution and, above all, political participation (Collier and Collier, 1991). Perón's government instigated women's suffrage in 1947, in addition to legally recognizing trade unions, and the share of wage earners in the distribution of national income reached 50%. At the same time, a new form of democratic participation was inaugurated through mobilization that, while fulfilling a central place in the plebiscitary legitimation of the leader, defined the identity of the movement (Romero, 2001).

The Peronist government's redistributive action led to clear material transformations. This included an increase in real wages and the "relative price policy" that stabilized the family consumption basket. It also sanctioned regulations to guarantee labor rights, including paid vacations, Christmas bonuses and collective labor agreements. This went hand in hand with a massive unionization process, increasing the power of the General Confederation of Labor (CGT) and the political weight and functions of the mass industrial unions. The social security system, until then extremely deficient, was also strengthened, guaranteeing retirement and pension funds, under a solidarity system. In the field of education, government action had an unprecedented scale in the country lowering illiteracy rates, together with an expansive trend in enrolment in primary, secondary and university education.

The geographies of Perón's populism have emphasized, above all else, the centrality of national political identification and mobilization. Peronism's socio-economic transformations were led by a strong centralized state and, moreover, depended on the practice and an affective imaginary of a national political body (Laclau, 2005). Through the leadership of Juan Domingo Perón and his wife Eva (see Figure 4.1), who built strong affective ties to the popular sectors, which she referred to as the "descamisados" (shirtless ones), Peronism sought to generate "political community" that would have dignity through labor and be loyal to the Argentinian nation.

The Argentine sociologist Gino Germani (1962) analyzed the "national-popular movement" of Peronism, examining how populism became fused into a national political project. This relied, he argued, on internal migration from the countryside to the city in order to meet the demand for labor in the nascent industries. A "new working class" was created that provided an "available mass" because they did not have pre-existing identities or organizations that represented them. These masses would become the object of domination (and manipulation) of the national-popular leaders (Germani, 1962).

Another Argentine scholar, Ernesto Laclau (1977, 2005), provided an alternative reading of Peronism in which he argued that populism involves the generation of demands from below, by the working class, that are presented in antagonism to the dominant ideology. Populism thus takes on the form of a social division between two poles: the elite and the people, confronted without the possibility of consensus because they articulate antagonistic projects and identities. The "national" provides a key device, or signifier, around which a range of different demands from the growing working class could be articulated in a mass movement.

Figure 4.1 A young Perón.
Source: Wikimedia Commons.

In both cases, the national represents the transfer of loyalties from the local to the national community, ensuring cohesion of the "people". Neighborhood associations, trade unions, universities, sport centers and other key sites of civil society become places through which to mobilize the national imaginary of Peronism, construct loyalty to its leadership and build political consciousness of its historical project (see Figure 4.2).

There have been many variants of populism, with Peronism itself taking both right-wing and left-wing ideologies. Across the region, a new wave of populism was observed in the early 21st century as part of the left turn in governments, including the Kirchners in Argentina (2003–2015) as well as Presidents Evo Morales (Bolivia), Rafael Correa (Ecuador) and Hugo Chavez (Venezuela) (de la Torre and Arnson, 2013). Although their forms have varied – for example, they have privileged political parties, social movements and/or centralized leadership in their mobilization (Roberts, 2006) – all have tended to privilege the nation-state.

However, the strong focus on national identities risks ignoring important local differences within countries and, in order to be successful, populism must be able to articulate across geographical difference (Halvorsen and Torres, 2022). Furthermore, there is a growing tendency for populism to take root in cities worldwide, and researchers should remain alert to urban manifestations of populist movements that shift attention directly onto new scales of political identification.

Figure 4.2 A neighborhood branch of Kirchnerist organization. In Buenos Aires, populist parties build strong territorial organizations to mobilize civil and political society.

Source: Argentina Multicolor.

Decentralization and participation

If populism represented a core Latin American challenge to liberal democracy via the national-popular imaginary, then decentralization opened a different opportunity for the region to embrace participatory democracy in a post-revolutionary conjuncture. The return to democratic regimes in the 1980s brought with it an impulse to bring rule "closer" to the people, through the transfer of decision-making powers from central to local government bodies. This was a dramatic break from the highly centralized state governments that had ruled for much of the 20th century, including both the military juntas and the populist regimes.

This shift was tied into other dynamics, including the rapid expansion of neoliberal globalization that compelled governments to offload central authority to privatized and localized actors (**see Chapter 8**). It also responded to rapid urbanization and the need for city regions to have their own governance structures, including directly elected mayors (**see Chapter 7**). The latter, in turn, was often a product of political party negotiation (Falleti, 2010), as in the case of the Autonomous City of Buenos Aires that was created on the back of closed-door discussions between Argentina's two main parties in the early 1990s (Halvorsen, 2019).

The decentralization of democracy was thus part of a broad trend towards the territorialization of politics, and this opened up a new geography of participation with different outcomes on the consolidation of democracy. This can be understood through two seemingly opposing developments. On the one hand, decentralization was often seen to go hand in hand with participatory democracy and the activation of civil society that was starting to mobilize in the context of the fall of military regimes. The impulse for this was varied and responded both to internal attempts to bridge the strong tradition of collective action with newly created (liberal) democratic institutions (Pearce, 2004), as well as learning from experiments elsewhere, including the municipal governments of Spain following the fall of Franco. Local-scale institutions, and most centrally urban governments, became a strategic focus for new left organizations and parties seeking to mobilize democracy as a radical and participatory practice (Chavez and Goldfrank, 2004).

In Brazil, a coalition of trade unions and left-wing movements formed in resistance to the military dictatorship (1964–1985) establishing the Workers' Party (Partido dos Trabalhadores, or PT). They competed in the 1988 elections, winning majorities in 36 cities including São Paulo. Building on their strong linkage with civil society organizations, the PT put participatory democracy central to their platform, with local democracy providing both an end (i.e. an ideal of party activists) as well as a means towards their eventual capturing of national power in 2003 (Baiocchi, 2003).

Of the PT's local experiments with participatory democracy, none are more famous than that of Porto Alegre, capital of the Southern state of Rio Grande do Sul, that would go on to host many of the World Social Forums. Over the years Porto Alegre has become synonymous with participatory budgeting, the annual cycle of resident meetings through which deliberation would produce concrete policy outcomes (see Figure 4.3). This innovation in urban participation has been taken up by the left and right alike in cities across Europe, North America and beyond (Wampler et al., 2021), providing a global democratic export from the region, although often with limited capacity to create socio-economic transformation.

The rise in urban participation, and the corresponding development of the public sphere as a deliberative space of democracy (Avritzer, 2002), went hand in hand with a seemingly unintended outcome of decentralization: clientelism, the exchange

Figure 4.3 Participatory budgeting in Porto Alegre.
Source: Occupy.com

of political support (including votes) for material goods (e.g. welfare handouts). Clientelism took off in urban territories at a similar time during the 1980s and 1990s, tapping into informal political networks in low-income urban neighborhoods. Indeed, in some cases participatory budgeting sought to undermine clientelistic logics (Abers, 2000). The large urban peripheries of Buenos Aires provided one of the prime cases of clientelism, which responded to the dual dynamics of deindustrialization and the attempt by the historically hegemonic Peronist party to rebuild its mass basis following the collapse of state corporatism and large union structures (Levitsky, 2003).

During the 1990s, the Peronist party used local branches to create territorial networks that served a range of functions, from disturbing food and basic services to building a new activist base, as well as reproducing nation-popular identities and a sense of political belonging (Auyero, 2000). These territorial networks were key to the party's survival and success into the 1990s, allowing it to go on to withstand a major crisis at the turn of the century. For many, however, it was an inherently undemocratic process, given that it was based on informal exchanges through personalistic relations that erode collective decision-making and undercuts transparency. Nevertheless, subsequent research has indicated that clientelism may also accompany democracy (it can be competitive, inclusive and stable) or even supplement it (by providing an alternative form of governance that is not necessarily hierarchical but involves bottom-up negotiation) (Hilgers, 2012).

Plurinational democracy and the case of Bolivia

A final strand through which classical notions of liberal democracy have been contested in Latin America is attempts to expand democracy's reach to Indigenous and Afro-descendant peoples through both formal mechanisms (e.g. constitutional reform) as well as expanding the substantive realm of citizenship. Many countries have rewritten their constitutions in recent decades, often following the collapse of authoritarian regimes. On the whole, there was a concerted effort to include Indigenous and Afro-descendant peoples into new constitutions, although there was much variation. This put in motion a selective process of land-titling and, on occasions, the recognition of Afro-descendent geographical communities (maroon or quilombo). Brazil and Colombia have seen a notable presence of Afro-descendent or Black organizations in civil society demand greater democratic recognition (see Chapters 15 and 21). Even when such formal recognition was granted, however, major social, economic and political inequalities remained.

These democratic rights were consolidated in response to the growth of new social movements in civil society (see Chapter 21), many of which would institutionalize into key democratic actors. Most notably, several Indigenous movements would go on to set up political parties or highlight institutionalized social movement organizations. For example, in Ecuador, the Pachakutik (New Awakening and Revolutionary Change) was founded in 1996 as an electoral tool for promoting Indigenous candidates, while in 2000 CONAIE (the Confederation of Indigenous Nationalities of Ecuador) established a grassroots structure to promote Indigenous rights. In Bolivia, Evo Morales became the first Indigenous candidate to win the Presidential elections, on the back of his leadership of the coca growers trade union and their subsequent formation of the MAS political party. Moreover, his government oversaw an ambitious process of rewriting the constitution to explicitly recognize Bolivia as a plurinational state.

Bolivia thus presents a fascinating case of an attempt to rebuild the postcolonial state by recognizing multiple nationalities, forms of democracy and autonomous territories. Zavaleta Mercado (1986) indicates that we can understand this process in the context of what he terms a motley social formation (abigarrada), in which multiple social orders co-exists. The modern state is seen as co-existing in combination with social orders that are not modern, yet without establishing a relationship of superiority and subjection with them, and without hierarchies between the state order and the community order.

In contrast to the struggle to mobilize democracy around the hegemony of the national popular movement in Argentina, the motley social formation in Bolivia responds to the historical limitation of building a "national community". In Bolivia there are at least 30 languages or regional dialects, and close to 62% of the population identifies as Indigenous. These cultural identities are older than the republic and are carriers of different symbolic configurations, worldviews and organizational forms. For this reason, the 2009 Constitution refers to Bolivia as a plurinational state. Article 3 states: "The Bolivian nation is formed by all Bolivians, the native Indigenous nations and peoples, and the inter-cultural and Afro-Bolivian communities that, together, constitute the Bolivian people".

The 2009 constitution was a response to years of grassroots mobilization that came to surface in key moments of rupture such as the 2000 "Water War". Bolivia's fourth largest city, Cochabamba, privatized its water supply in a deal involving the North America multinational corporation Bechtel, effectively prohibiting residents from collecting rainwater to drink, provoking massive protest. This moment demonstrated the capacity for grassroots actors to resist the neoliberal policies of the (liberal) democratic state as well as the potential to constitute an alternative basis for democratic politics, led by movements rather than transnational capital. Six years later Evo Morales was sworn in as President, proposing a new bloc of economic power and a new resource distribution configuration.

Early in his Presidency, Morales supported a Pact of Unity that allowed for democratic coordination between Indigenous, peasant and other grassroots organizations, with a central objective to undertake constitutional reform. The latter was oriented around the recognition and implementation of the plurinational state that would include features such as: the election of representatives through community mechanisms; giving entity and constitutional hierarchy to community justice; communal property and control of natural assets; and recognition of different types of autonomy (regional, municipal and Indigenous) reflected in a new ordering of the state territory. Such transformations were historically unprecedented and radically reconfigured the geographies of state power.

Despite these advances, the process of putting a plurinational state into practice has come across several moments of conflict, such as the case of the TIPNIS national park discussed in **Chapter 10**, as well as hostility from local oligarchies (identified with the name of the "secessionist Media Luna") located in the eastern states of the country where the "white" population predominates. Such conflicts demonstrated the real limits to the post-colonial state's capacity to decolonize its democratic institutions (**see Chapter 12**). Bolivia thus provides a significant, yet limited, attempt to move beyond liberal ideas of democracy based on individualist and universalist principles, and towards the recognition of collective and community-based rights in a plurinational democracy that articulates the demos not as a homogenous "political nation" but a multicultural or multinational community.

Conclusion

Democracy is a contested and uneven process in Latin America. Liberal notions of state-based electoral democracy were imported through the region's colonial legacy, yet have come across a range of challenges including authoritarian rule as well as exclusions of Indigenous and Afro-descendant peoples. The chapter argues that geography provides an important tool for making sense of the multiple forms of democracy that have existed in the region, while also pointing to diverse strategies and challenges of different political actors.

The centrality of the nation-state remains the key focus of debate, and the growth of territorially based political movements are attempting to re-work taken for granted understandings of sovereignty and state power. Nevertheless, academic investigation into the geographies of democracy remain limited, both in Latin America and elsewhere, and there is a great opportunity for new and exciting lines of research. In particular, bringing together interest into grassroots social movements (and civil society more broadly) and institutional "top-down" forms of democracy will be a crucial means of generating new insights into actually existing democracy in the region.

Summary

- Democracy, as a liberal notion of periodically electing rulers, has been institutionalized across the region in a geographically and historically uneven process.
- Democracy has been contested through a range of ideas and practices that emerged through specific historical and geographical experiences in the region.
- Populism was consolidated as a political strategy in the early-to-mid-20th century, usually based on the attempt to use national identity to articulate the growing identities and demands of the popular sectors.
- Following democratization, democratic institutions have undergone significant rescaling, most notably due to decentralization.
- Indigenous and Afro-descendants' rights have been gradually incorporated into democratic models, such as the Bolivian plurinational state.

Review questions

1. What are the main geographical and historical variations of democracy in Latin America?
2. In what ways does the nation-state structure democratic strategies?
3. Has decentralization been good or bad for consolidating democracy?
4. What are the geographical implications of taking seriously Afro-descendent and Indigenous democratic rights?

Further reading

Barnett, C. and Low, M. (eds.), 2004. *Spaces of Democracy: Geographical Perspectives on Citizenship, Participation, and Representation*. Sage.
Rare attempt to provide an overview of geographical approaches to democracy, this edited collection includes specific chapters that deal with Latin American case studies.
Goldfrank, B., 2011. *Deepening Local Democracy in Latin America: Participation, Decentralization and the Left*. Penn State University.
A useful introduction to debates on local democracy and neighborhood participation based on a comparison of three South American cities.
Halvorsen, S. and Torres, F., 2022. Articulating populism in place: A relational comparison of Kirchnerism in Argentina. *Annals of the American Association of Geographers*, 112(8): 2195–2211.
A contemporary analysis of populism in Argentina that takes a geographical approach by comparison two places.
Radhuber, I. M., 2012. Indigenous struggles for a plurinational state: An analysis of Indigenous rights and competences in Bolivia. *Journal of Latin American Geography*, 167–193.
Geographical analysis of the struggles for a Pluri-national state in Bolivia.
Van Cott, D. L., 2012. *Radical Democracy in the Andes*. Cambridge University Press.
Explores a range of democratic transformations in the Andean region led by Indigenous actors.

Keywords

Decentralization: the shifting of political, fiscal and administrative power from centralized to local units.
Democracy: rule by the people, often understood in narrow terms in relation to the election of representatives.
Democratization: refers both to the end of authoritarian rule and also the processes of deepening, or extending, democratic rights and practices.
Plurinational: the coexistence of multiple ideas and practices of national belonging.
Populism: the mobilization of "the people" against an elite or oligarchical enemy, often under a charismatic leader.

References

Abers, R. N., 2000. *Inventing Local Democracy: Grassroots Politics in Brazil*. Lynne Rienner.
Auyero, J., 2000. *Poor People's Politics: Peronist Survival Networks and the Legacy of Evita*. Duke University Press.
Avritzer, L., 2002. *Democracy and the Public Space in Latin America*. Princeton University Press.
Baiocchi, G. (ed.), 2003. *Radicals in Power: The Workers' Party (PT) and Experiments in Urban Democracy in Brazil*. Zed books.
Barnett, C. and Low, M. (eds.), 2004. *Spaces of Democracy: Geographical Perspectives on Citizenship, Participation and Representation*. SAGE.

Chavez, D. and Goldfrank, B., 2004. *The Left in the City: Participatory Local Governments in Latin America.* Latin American Bureau.

Collier, R. B. and Collier, D., 1991. *Shaping the Political Arena: Critical Junctures, the Labor Movement and Regime Dynamics in Latin America.* University of Notre Dame Press.

de la Torre, C. (ed.), 2015. *The Promise and Perils of Populism: Global Perspectives.* University Press of Kentucky.

de la Torre, C. and Arnson, C. Y. (eds.), 2013. *Latin American Populism in the Twenty-First Century.* Woodrow Wilson and John Hopkins Press.

Drake, P. W., 1982. Conclusion: Réquiem for populism? In Connif, M. (ed.), *Latin American Populism in Comparative Perspective.* New México University Press.

Falleti, T. G., 2010. *Decentralization and Subnational Politics in Latin America.* Cambridge University Press.

Forment, C., 2003. *Democracy in Latin America: 1760–1900.* Univeristy of Chicago Press.

Germani, G., 1962. *Política y sociedad en una época de transición.* Paidós.

Hagopian, F. and Mainwaring, S. P., 2005. *The Third Wave of Democratization in Latin America: Advances and Setbacks.* University of Cambridge Press.

Halvorsen, S., 2019. The political opportunities of urban decentralisation: Mobilising local governance in Buenos Aires. *Political Geography,* 74: 1–10.

Halvorsen, S. and Torres, F., 2022. Articulating populism in place: A relational comparison of Kirchnerism in Argentina. *Annals of the American Association of Geographers,* 112(8): 2195–2211.

Hilgers, T. (ed.), 2012. *Clientelism in Everyday Latin American Politics.* Palgrave.

Huntington, S. P., 1991. *The Third Wave: Democratization in the Late Twentieth Century.* University of Oklahoma Press.

James, D., 1988. *Resistance and Integration: Peronism and the Working Class, 1946–1976.* Cambridge University Press.

Laclau, E., 1977. *Politics and Ideology in Marxist Theory: Capitalism, Fascism, Populism.* Verso.

Laclau, E., 2005. *On Populist Reason.* Verso.

Levitsky, S., 2003. *Transforming Labor-Bases Parties in Latin America: Argentine Peronism in Comparative Perspective.* Cambridge University Press.

Levitsky, S., Loxton, J., Van Dyck, B. and Domínguez, J. I. (eds.), 2016. *Challenges of Party Building in Latin America.* Cambridge Universtiy Press.

O'Donnell, G., 1994. Delegative democracy. *Journal of Democracy,* 5(1).

O'Donnell, G., Schmitter, P. and Whitehead, L. (eds.), 1986. *Transitions from Authoritarian Rule.* Johns Hopkins University Press.

Pearce, J., 2004. Collective action or public participation? Complementary or contradictory democratisation strategies in Latin America? *Bulletin of Latin American Research,* 23(4): 483–504.

Roberts, K. M., 2006. Populism, political conflict, and grass-roots organization in Latin America. *Comparative Politics,* 38(2): 127–148.

Romero, L. A., 2001. *Breve historia contemporánea de la Argentina.* FCE.

Sartori, G., 1976. *Parties and Party Systems: A Framework for Analysis.* Cambridge University Press.

Schumpeter, J. A., 1942. *Capitalism, Socialism and Democracy.* Harper & Brothers.

Torre, J. C. and Pastoriza, E., 1989. *La democratización del bienestar, Nueva Historia Argentina,* Tomo VIII, pp. 257–313. Buenos Aires: Editorial Sudamericana.

Wampler, B., McNulty, S. and Touchton, M., 2021. *Participatory Budgeting in Global Perspective.* Oxford University Press.

Zavaleta Mercado, R., 1986. *Lo nacional-popular en Bolivia.* Siglo XXI.

CHAPTER 5

Geopolitics

Matthew C. Benwell and Andrés Núñez

Introduction

Geopolitics is a term that is widely utilised in the news to refer to international relations between states and very often conflicts over borders, territories, or resources (**see Chapter 3**). Recent global events and disparities in media coverage across world regions mean that geopolitics is more regularly associated with wars and territorial disputes (rather than peaceful inter-state relations) in the likes of the Middle East, the Korean Peninsula and Eastern Europe, rather than Latin America. Klaus Dodds (2014) defines geopolitics as a way of looking at and engaging with the world by taking into consideration the interconnections between power, geography and knowledge. Even if we do not always hear the word geopolitics used in relation to Latin America, this chapter will illustrate why this way of looking at the world is relevant for understanding its political geographies.

Geopolitics, an academic discipline that has long had close associations with geography, has attracted attention in the context of Latin America, not all of it positive. Indeed, geopolitical thinking has a somewhat chequered regional history (particularly in the Southern Cone) given its close associations with national military academies. Repressive military dictatorships that were responsible for inflicting state terror on their civilian populations during the second half of the twentieth century in countries such as Chile, Argentina and Uruguay had close links with geographical military institutes (or *Instituto Geográfico Militar*) that drew heavily on geopolitical theory to develop and map national security strategies. These tended to utilise classical geopolitical thinking originating from Europe that positioned the natural environment, geographical location, physical geographical features and the relative strength of national militaries as central to determining the destiny and security of the state.

Although in a popular sense (in Latin America and beyond) geopolitics is still associated with the territorial struggles of states and the discourse of elites (typically male politicians or diplomats) who 'represent' them, the academic study of geopolitics has been reinvigorated through its engagements with the work of feminist political geographers

DOI: 10.4324/9781003430926-8

(see Dowler and Sharp, 2001; Massaro and Williams, 2013). At the turn of the twenty-first century, geopolitical scholarship started to develop an interest in the grounded, embodied and emotional encounters that citizens can have with the geopolitical, in stark contrast to its disciplinary traditions. These new research directions have brought different actors, spaces, objects, practices and performances into focus, encouraging a rethinking of how geopolitics and security are conceived. They have routinely critiqued the top-down, state-centric focus of geopolitical inquiry and made calls for researchers to attend to citizens' everyday lived experiences of geopolitics and (in)security – these are key sites where geopolitical power is (re)produced and negotiated. Research undertaken in Latin America and by scholars based in the region has been at the forefront of pushing these new agendas, identifying geopolitics as something felt, embodied and engaged at different sites and scales that are understood as co-constitutive.

This chapter reflects on this diversity of sites, scales, actors and objects that can be considered as co-constitutive of geopolitics. It begins by briefly outlining the geopolitical relations and dynamics influencing Latin America as a region in the twentieth and twenty-first centuries, underlining the spectre of US influence and intervention. It then introduces recent geopolitical scholarship in the region, foregrounding the work of feminist political geographers to broaden ideas about what can be thought of as geopolitical, before discussing two examples. First, it considers young people's everyday geopolitics of the home in the context of the Falklands/Malvinas sovereignty dispute – sites, scales and actors that have been rarely acknowledged as geopolitical. Second, the liveliness of geopolitical objects such as national flags are emphasised through their use and (re)appropriation by protestors during the social uprisings in Chile in 2019–20. The chapter presents geopolitics as manifest at multiple scales that are interconnected and argues that grounded and everyday perspectives are especially important, serving to unsettle narrowly conceived masculinist, militaristic and elite renderings of geopolitics and security that have predominated in Latin America.

An overview of Latin American geopolitics: past, present and future

External interference in the affairs of Latin American states by the United States and, to a lesser extent, the UK (see Cormac, 2022), have dominated regional geopolitics and (in)security for over a century. The US has at various points deployed its military forces, installed, propped up and trained murderous dictatorial regimes, supported military coups and looked to unsettle and replace progressive governments that were perceived to be hostile to their economic and geopolitical interests. Although much less widespread, recently declassified documents show the UK's programme of covert action within Latin America in the second half of the twentieth century that looked to leverage geopolitical influence and increase trade opportunities with dictatorships in Argentina and Chile, as well as interfere in electoral campaigns (see Livingstone, 2018).

The US government has long considered Latin America as within its regional sphere of influence; it has been problematically referred to as 'America's backyard', a geopolitical frame of reference revealing of the colonial attitude that continues to dominate relations. As in many parts of the globe, the US makes strategic use of military bases, airstrips and ports to preserve and further its geopolitical interests (see Dodds, 2003). This interventionist approach to Latin America's domestic affairs reached its zenith during the Cold War, when the US sought to counter the Soviet Union's regional influence by tacitly destabilising 'left-leaning' governments or directly intervening in the domestic affairs of sovereign states (e.g. Guatemala in 1954, Chile in 1973, Nicaragua in the 1980s, among many others). This was most tragically evident in Central America, where the US military launched invasions and backed death squads that led to countless human rights abuses and extensive loss of life.

The US also supported the instalment of totalitarian dictatorships in the Southern Cone (**see Chapter 4**) (encompassing Argentina, Bolivia, Chile, Paraguay and Uruguay) that guaranteed close economic and political cooperation at the expense of massive human rights violations and the 'disappearance' of thousands of citizens. This repression was consolidated in 1975 under what became known as 'Operation Condor', a secretive, transnational network that enabled the military regimes to target political opponents beyond their respective borders. Latin American states and their citizenries are still coming to terms with this notorious era of repression as human rights organisations and activists continue to seek justice for these national and transnational human rights violations (Lessa, 2022).

The spectre of US interventionism in Latin America continues to loom large and has resulted in significant suspicion and resistance from diverse sectors and governments throughout the region, most notably in Cuba, Nicaragua and Venezuela. Memory of this troubled era of US intervention and concerns about contemporary manifestations of US imperialism have seen Latin American states place an emphasis on regional cooperation and integration through multilateral organisations like the Union of South American Nations (or UNASUR) and the Community of Latin American and Caribbean States (CELAC). While the standing and efficacy of these organisations fluctuates according to the political leanings of respective national governments, they have sought to eschew US regional influence through the formation of alliances and trading agreements with other increasingly influential states and trading blocs (e.g. China, India and the EU) around the world. The engagement of Latin American states in an increasingly multi-polar geopolitical landscape is further evidenced by Brazil's current membership in BRICS (an acronym for the five leading emerging economies of Brazil, Russia, India, China and South Africa), and, up until the election of Javier Milei as President in 2023, the favourable reception towards Argentina's prospective membership.

Antarctica is another important element of the geopolitical and security strategies of several Latin American states. Argentina and Chile lay claim to almost identical Antarctic sectors that are included on official maps of their national territories. These are reinforced through everyday references to the territories via, for example, their inclusion in weather forecasts, references to Antarctica on road signs and the establishment of scientific bases and civilian settlements in polar regions (Benwell, 2014). While Argentina and Chile have committed to scientific and naval cooperation under the auspices of the Antarctic Treaty (which stipulates that the continent should be used exclusively for peaceful purposes, whilst effectively setting aside existing territorial claims), their conflicting claims (that also overlap with the UK's Antarctic sector) have the potential to cause geopolitical friction in the future (Sanchez, 2017). Many other Latin American states, including Brazil, Peru and Uruguay, retain a stake in Antarctic geopolitics by conducting annual scientific expeditions in a region that will take on increasing geopolitical significance as climate change and associated glacial retreat continue unabated.

Studying Latin American geopolitics: interconnected geographical scales

Regional geopolitical dynamics and international relations evident at the scale of the state, and sketched out above, have long attracted scholarly interest in the Latin American context. It is clearly important for any student of geopolitics to remain informed about questions of inter-state relations, security and major events occurring in the region. However, Lesley Hepple, writing in 2004, identified the tendency for regional geopolitical analyses to employ a detached and objective gaze that exaggerated the role of space and territory. This was in part due to geopolitics' close links with the military and primary concerns with state security. It resulted in geopolitical analyses that were very often simplified and partial, focusing on the scale of the state and its associated actors, in order to support particular strategies and ways of looking at the world favoured by military officials.

This 'geopolitical gaze', with its roots in the traditions and logics of classical geopolitics that dominated regional scholarship in the twentieth century, continues to influence contemporary debates. Nolte and Wehner (2015) point out the contemporary salience of these geopolitical ideas and doctrines in the Latin American context. These tend to subscribe to what they term neoclassical geopolitics, 'a policy-oriented approach, which conceptualises foreign policy challenges and the international politics of a state in light of its geographical features, or its position on the map' (Nolte and Wehner, 2015: 34). Indeed, some Latin American governments continue to construct national identities in relation to territorial, maritime or polar spaces in ways that draw on these traditions of geopolitical thinking. Examples include the Brazilian Navy's promotion of the 'Blue Amazon' in the 2000s, or more recently Argentina's 'Blue Pampa' initiative, that both focus national attention on the resource riches and strategic importance of the South Atlantic (Blair, 2022). It is also worth noting that states like Ecuador, Chile and Uruguay continue to host geographical military institutes (Argentina's was renamed the National Geographical Institute in 2009 to remove its explicit association with the military and emphasise its civilian purpose), illustrative of the persistence of these connections between the military, geography, cartography and geopolitics in Latin America.

Studies of geopolitics have routinely focused attention on the actors, spaces and institutions associated with the sovereign state in ways that have obscured other sites, subjects and protagonists from critical interrogation. Scholars of feminist geopolitics in particular have sought to directly challenge the invisibility of everyday life in this body of work, highlighting its interconnections with international relations. Sites like the body, home, school, jungle and the street are shown to be intimately connected to geopolitics, and just as worthy of study as, for instance, national congresses, geographical military institutes and the so-called corridors of geopolitical power. For instance, the intimate scale of the body has been a particular focus for work undertaken by researchers exploring abortion activism in Latin American states (Zaragocin, 2020) (**see Chapters 14 and 23**). Duffy et al. (2023) show how those involved in this activism in Peru have constructed infrastructures of abortion care by bringing together actors, technologies and strategies in the absence of effective state provision. They argue that these grounded and embodied forms of activism encountered in Latin American contexts can inform responses to increasingly restrictive state policies on abortion evident in the US and elsewhere.

Similarly, military officers, diplomats and (typically) statesmen are not the only actors of geopolitics, and research has shown how citizens (as well as those defined as non-citizens) – women, children, youth and families – can be implicated in geopolitics. Researchers interested in the geopolitical pasts and presents of Latin American countries are engaging these other geopolitical actors in their work. For example, Natale's (2021) innovative ethnographic research with military families of former officers who served during the last dictatorship in Argentina (1976–83) uncovers the often unexplored 'everyday dimensions' of military life. These emphasise perspectives that can challenge masculine narratives of the military, underlining the 'hidden' role and experiences of military wives and children, to provide alternative understandings of violence and political confrontation in the past. Research in Latin America has also shed light on the lively potential of non-human objects as part of an everyday geopolitics. The work of Juanita Sundberg (2008) at the USA-Mexico borderlands has shown how the discarded and intimate material objects of immigrants, such as identity documents, personal mementos and things essential for survival, can be (geo)politicised in different ways when US citizens encounter them. These objects are productive of geopolitical imaginings about national belonging (i.e. who does and does not belong) and can be mobilised to express indignation about unauthorised immigration or deployed to inspire more compassionate responses to migrants from Latin America and elsewhere, through the interventions of humanitarian groups and artists.

Importantly, this body of work disrupts the notion that geopolitics and its effects simply trickle down from the state to the ground level and everyday life in a uniform direction. Rather, there is a recognition of the agency of different geopolitical actors (human and non-human), although often within considerable structural constraints. An excellent example of this agency can be drawn from Sara Koopman's (2011) work with international accompaniers (people who decide to walk alongside those who are under direct threat – typically unarmed, from outside the conflict-zone and therefore less at risk as those they accompany) in the jungles of Colombia. Her work shows how grassroots peace initiatives (similar accompaniment groups have worked in Mexico and Guatemala) can push back against increasing levels of insecurity, in order to collectively build alternative nonviolent securities, in circumstances where the state is unable to quell, or is complicit in exacerbating, armed conflict. This broader, more inclusive conceptualisation of geopolitics does not disregard the actions of the state and (para)militaries but considers how other practices, objects and actors are bound up with these geopolitical events. It recognises the potential of actors, previously 'off the page' of geopolitical scholarship, to do geopolitics otherwise in ways that might help to bring about more secure, just and peaceful futures. In a similar vein, the two examples that follow look to connect geographical sites and scales of geopolitics in different contexts within the region of Latin America.

Example 1: The Falklands/Malvinas sovereignty dispute and domestic geopolitics

The intractable sovereignty dispute between Argentina and the UK over the Falkland Islands or *Islas Malvinas* has been a source of considerable geopolitical tension. It triggered a ten-week undeclared war in 1982 when Argentina, ruled by a brutal military dictatorship, invaded and occupied the islands. In the subsequent conflict, over 900 people were killed from Argentina, the Falkland Islands and the UK. The sovereignty dispute continues to overshadow contemporary Anglo-Argentine relations, as well as the UK's broader relationships with Argentina's Latin American neighbours.

The geopolitics of sovereignty disputes like the Falklands/Malvinas are often narrowly conceptualised with reference to elite actors, sites and practices. These might encompass speeches delivered by politicians at the UN or official commemorative events marking war anniversaries. Rather less well documented are the ways that geopolitics is encountered through, and influenced by, family interactions and relations in domestic spaces (Brickell, 2012). Even more marginal, when it comes to the adult, grown-up world of geopolitics, are the accounts of young people. Problematic assumptions are made about young people's lack of maturity and inability to grasp the complex world of international relations. However, research undertaken in Latin American contexts examining familial relations in domestic spaces has underlined young people's sophisticated accounts about geopolitics and the significance of the home as a site that is bound up with wider geopolitical processes (**see** Figure 5.1).

Domestic space and interactions with family are important in shaping how young people learn about, and are affected by, geopolitics, including past events like the 1982 war that took place before they were born (Benwell et al., 2020). Young people's geopolitical socialisation is not restricted to what they learn at school, and is heavily shaped by domestic settings and the intergenerational relationships and (geopolitical) objects found therein. For instance, young people from Argentina and the Falkland Islands referenced conversations with their families when the topic of the islands was raised, sometimes initiated as a result of looking at objects that their parents had collected from the 1982 war, others inspired by contemporary geopolitical developments related

Figure 5.1 A Malvinas mural painted on the wall of a house in Río Gallegos, Argentina.
Source: Author.

to the sovereignty dispute. The nature of such familial recollections in domestic spaces added an intimacy that left a lasting legacy on many of the young people.

Young people do not inevitably and passively reproduce the perspectives they hear about geopolitics from adults, and they were able to express alternative views regarding the 1982 war and the sovereignty dispute. However, the intimate and deeply personal nature of memories young people were presented with, often drawn from the direct experiences of older family members during the 1982 war, were extremely powerful and tempered the potential for the expression of alternative geopolitical perspectives. Young people are active geopolitical actors, but they are also members of communities and families that in some instances have vivid memories of violent geopolitical events in the past. These are not insignificant, and had direct implications on young people's lives (or those of close relatives) that inevitably shaped how they talked about geopolitics.

Investigating young people's negotiations of geopolitics through family relationships within domestic space can yield nuanced and sensitive understandings of how they (feel able to) express their views about geopolitical issues. This example demonstrates how the domestic sphere can be a lively setting that is imbued with geopolitics – past and present – in ways that can shape (the expression of) geopolitical subjectivities and generate knowledge about ongoing sovereignty disputes like the Falklands/Malvinas.

Example 2: The geopolitics of the national flag during the Chile uprisings, 2019–20

In October of 2019, Chile witnessed a wave of protests that erupted in the capital Santiago and swept across the country on a scale unprecedented in its democratic era. The protests were triggered by a rise in tariffs for travel on Santiago's Metro (or underground), but its roots ran far deeper, to the injustices and inequalities associated with the neoliberal Chilean state. The grievances of the protestors centred on the vast societal inequalities manifest through uneven access to healthcare and education, inadequate pension provision, as well as general disillusionment with the country's political classes and a Constitution written and ratified during the dictatorship of Augusto Pinochet. The protests attracted international attention by making creative use of social media through hashtags and street performances. Another feature of the protests was the prevalence of objects and symbols that came to signify and embody the struggle including an array of flags.

La bandera negra or black flag, which was a common sight at the protests and on social media, depicted a black and white version of the Chilean national flag with white lines and a white star drawn on a black background. Other flags on display at the protests included the *Wenufoye* flag of the Mapuche (the largest Indigenous group in Chile), as well as appropriated versions of the Chilean national flag that were adorned with political messages or bullet holes to protest against police repression unleashed against the protestors (see Figure 5.2). Many of these flags were the result of protestors actively subverting dominant representations of the nation, as well as a way to posit alternative versions and visions of Chile. Protestors were not just choosing flags and using them to protest. Instead, they were collaboratively designing and making flags in ways that expressed their feelings about the nation – a process that was, at times, cathartic, and which involved re-thinking and re-imagining the Chilean nation they wanted to be part of in the future.

The agency of non-human, material objects like flags (and their use by protestors) have not often received attention from scholars of geopolitics. In this example from

Figure 5.2 A bullet-riddled Chilean flag seen on the streets of Santiago in December 2019.
Source: Author.

Chile, they were mobilised by protestors in ways that underline their capacity to shape political practice and debate, as well as generate powerful national affective atmospheres (Benwell et al., 2021). The black flag meant different things to protestors, including anger and rejection directed at the government and police forces accused of committing human rights abuses against protestors, as well as a sense of mourning for those who had lost their lives or been injured in the clashes. The black flag bore direct resemblance to the national Chilean flag, only without its typical red, white and blue colours. In this way, it was also a provocation that looked to deconstruct the national story and encourage citizens to mourn the death of the 'old Chile', with the apathy and individualism that its neoliberal model entailed. The black flag was a subversive symbol that allowed another nation to be imagined, with an alternative 'national story'. This parallels other interventions in the streets and plazas of Chile's cities, as statues and street names evoking national 'liberators' and 'heroes' from the colonial era were replaced with, or brought into direct confrontation with, other objects such as Indigenous statues, names and imagery that constructed alternative (pluri-)national narratives.

Drawing attention to the interventions of protestors, as we have done in this example, is important because it shifts the focus from state-directed deployments of national symbols (i.e. top-down), to their (re)imagination and (re)appropriation by individuals and collectives in the everyday spaces of the street (i.e. bottom-up). Identifying the co-constitutive agencies of national citizenries and the objects they engage can reveal their politically subversive possibilities, disrupting expected performances of nation and nationhood. National flags can be (re)appropriated from the ground up by protestors, in ways that invest them with radical potential to critique, protest against and re-imagine the 'nation-state'. Contemporary geopolitical research has increasingly identified mundane material objects such as national flags as lively, with the potential to (re)produce, resist and/or disrupt predominant ideas about the nation-state and geopolitics.

Conclusion

Considerations of geopolitics in the context of Latin America require us to take on board the legacies of its close association with national militaries and the ways this continues to shape geopolitical thinking and scholarship in the region. It also demands that we consider geopolitics at multiple scales that are interconnected and co-constitutive – from the scale of inter-state relations between Latin American nations, as well as those with states outside of the region, to the everyday spaces of the street, home and university.

The work of feminist political geographers has seen geopolitical research revitalised in recent years, taking it into geographical contexts that had rarely been thought of as constitutive of geopolitics. A grounded and everyday geopolitics 'from below' foregrounds the perspectives of actors largely absent in previous scholarship (e.g. women, children, youth, families), and no longer relies exclusively on the perspectives of elite figures connected to politics, diplomacy and the military.

Taking account of other geopolitical actors, practices and objects can help unsettle masculinist, militaristic and elite renderings of geopolitics, security and the nation. It can also do far more than this, by drawing attention to how people and communities imagine and shape an alternative geopolitics (what Koopman (2011) refers to as alter-geopolitics) that can be influential in forging more peaceful relations and nonviolent securities. Contemporary geopolitical scholarship in Latin America offers exciting possibilities, then, to explore the imaginations, perspectives, agencies and actions of those previously overlooked but who nevertheless live and engage with geopolitics on a daily basis.

Summary

- Geopolitics has a notorious history in the region of Latin America, given its close historical associations with geopolitical thinking utilised by dictatorial military regimes during the twentieth century.
- Geopolitical scholarship has tended to focus overwhelming attention on the actors, spaces and institutions associated with the sovereign state, whilst overlooking how these might be intimately connected to other sites, objects and actors of geopolitics.
- A grounded and everyday geopolitics 'from below' can foreground the perspectives of other actors (e.g. including women, children, young people and families), sites and objects that have not been substantially considered in geopolitical scholarship.
- These can unsettle and push back against masculinist, militaristic and elite renderings of geopolitics, security and the nation, pointing to alternative visions that promote more peaceful, just and nonviolent societies and international relations.

Review questions

1. Why has geopolitical thinking been so closely linked to the state and state institutions in Latin America?
2. Why is it important to think about different geographical scales and sites of geopolitics as interconnected?
3. With the use of examples, discuss how grounding geopolitical research in the everyday can add to our understandings of geopolitics in the region.

Further reading

Benwell, M.C. 2020. 'Going underground: Banal nationalism and subterranean elements in Argentina's Falklands/Malvinas claim', *Geopolitics*, 25(1): 88–108.
Calls attention to the affective capacities of underground elements like earth, sand and rock in relation to geopolitics and claims to national territory.
Domínguez López, E. and Yaffe, H. 2017. 'The deep, historical roots of Cuban anti-imperialism', *Third World Quarterly*, 38(11): 2517–2535.
Highlights how imperialism and anti-imperialism have been decisive in shaping Cuban geopolitics in the past and present.
Koopman, S. 2011. 'Alter-geopolitics: Other securities are happening', *Geoforum*, 42(3): 274–284.
Exposes the ways that people do geopolitics in their everyday lives focusing on the actions of international accompaniers and peace communities in Colombia.
Lessa, F. 2022. *The Condor Trials: Transnational Repression and Human Rights in South America.* Yale University Press.
Documents the transnational networks that enabled cross-border human rights violations in the Southern Cone, as well as the efforts of transnational activists demanding justice for the victims.
Livingstone, G. 2009. *America's Backyard: The United States and Latin America from the Monroe Doctrine to the War on Terror.* Zed Books.
Outlines the long history of United States' interventionism in Latin America and its devastating implications for current levels of regional poverty and inequality.
Williams, J. and Boyce, G.A. 2013. 'Fear, loathing and the everyday geopolitics of encounter in the Arizona borderlands', *Geopolitics*, 18(4): 895–916.
Argues for an analysis of affective and emotional experience to understand geopolitics and activism at the US/Mexico border.

Keywords

Classical geopolitics: focuses attention on the influence of geographical location and features as the most important, objective variables that can inform and explain states' foreign policies.

Feminist geopolitics: connects the scales, sites and subjects that have traditionally been the focus of geopolitics to the everyday, highlighting how geopolitical power can be yielded to (re)produce inequality and exploitation.

Geopolitics: a way of looking and engaging with the world that takes into consideration the interconnections between power, geography, and knowledge.

The geopolitical gaze: a detached, transcendent and objective way of viewing the world and international politics most closely related to classical geopolitics.

References

Benwell, M.C. 2014. 'Connecting southern frontiers: Argentina, the South West Atlantic and "Argentine Antarctic territory"', in: Powell, R.C. and Dodds, K. (eds.), *Polar Geopolitics? Knowledges, Resources and Legal Regimes*. Edward Elgar Publishing. 201–218.

Benwell, M.C., Gasel, A. and Núñez, A. 2020. 'Bringing the Falklands/Malvinas home: Young people's everyday engagements with geopolitics in domestic space', *Bulletin of Latin American Research*, 39(4): 424–438.

Benwell, M.C., Núñez, A. and Amigo, C. 2021. 'Stitching together the nation's fabric during the Chile uprisings: Towards an alter-geopolitics of flags and everyday nationalism', *Geoforum*, 122: 22–31.

Blair, J.J.A. 2022. *Salvaging Empire: Sovereignty, Natural Resources, and Environmental Science in the South Atlantic*. Cornell University Press.

Brickell, K. 2012. 'Geopolitics of home', *Geography Compass*, 6(10): 575–588.

Cormac, R. 2022. 'The currency of covert action: British special political action in Latin America, 1961–64', *Journal of Strategic Studies*, 45(6–7): 893–917.

Dodds, K. 2003. 'Cold War geopolitics', in: Agnew, J., Mitchell, K. and Toal, G. (eds.), *A Companion to Political Geography*. Blackwell. 204–218.

Dodds, K. 2014. *Geopolitics: A Very Short Introduction*. Oxford University Press.

Dowler, L. and Sharp, J. 2001. 'A feminist geopolitics?' *Space & Polity*, 5(3): 165–176.

Duffy, D.N., Freeman, C. and Castañeda, S.R. 2023. 'Beyond the state: Abortion care activism in Peru', *Signs: Journal of Women in Culture and Society*, 48(3): 609–634.

Hepple, L. 2004. 'South American heartland: The Charcas, Latin American geopolitics and global strategies', *The Geographical Journal*, 170(4): 359–367.

Livingstone, G. 2018. *Britain and the Dictatorships of Argentina and Chile, 1973–82: Foreign Policy, Corporations and Social Movements*. Palgrave Macmillan.

Massaro, V.A. and Williams, J. 2013. 'Feminist geopolitics', *Geography Compass*, 7(8): 567–577.

Natale, E. 2021. 'Researching violence and everyday life in the 1970s: An ethnographic approach to the Argentine military family', *Bulletin of Latin American Research*, 41(3): 450–464.

Nolte, D. and Wehner, L.E. 2015. 'Geopolitics in Latin America, old and new', in: Mares, D.R. and Kacowicz, A.M. (eds.), *Routledge Handbook of Latin American Security*. Routledge. 33–43.

Sanchez, W.A. 2017. 'Argentina, Chile and the joint Antarctic naval patrol: A successful confidence building mechanism', *The Polar Journal*, 7(1): 181–192.

Sundberg, J. 2008. '"Trash-talk" and the production of quotidian geopolitical boundaries in the USA – Mexico borderlands', *Social & Cultural Geography*, 9(8): 871–890.

Zaragocin, S. 2020. 'La geopolítica del útero: hacia una geopolítica feminista decolonial en espacios de muerte lenta', in: Hernández, D.T.C. and Jiménez, M.B. (eds.), *Cuerpos, Territorios y Feminismos: Compilación latinoamericana de teorías, metodolgías y practices políticas*. Ediciones Abya-Yala. 83–99.

Economic and urban geographies

CHAPTER 6

Economies

Laura Sariego-Kluge

Introduction

Economies, a key aspect to all social life, are greatly researched in many disciplines. How and why do economies vary? The geographical study of economies has sought to give answers to this and other questions regarding the relationship between space and production, trade, consumption, and finance. In this sense, globalization – defined as the global integration of information and knowledge, means of production, and markets (labor, goods, services, and financial markets) – is one of the most important processes for understanding and explaining economies, their interrelations, and their patterns.

By using globalization to explain key concepts and practices of economic geography, this chapter draws on examples from Central America, a region that shares some similarities with the rest of Latin America, such as language and religion, but that differs because of its size and location. Its strategic location, which offers access to the Atlantic and Pacific oceans, proximity to both coasts of the USA (its main market for exports), and shared time zone has translated into a great portion of its economy being sustained by its linkages to the world (exports, foreign direct investment, and remittances). Moreover, the isthmus is a good example of how policy has shaped different economic pathways under similar starting conditions.

Some people, hyperglobalists, consider that a single global economy is an irreversible trend and argue that the 'end of geography' is here (O'Brien, 1992). Technological advances and reduced costs make distance and location increasingly less of a problem and, therefore, the world is getting 'smaller'. In this sense, hyperglobalists trust market forces will eventually lead to an economic convergence, narrowing the inequality gap between countries. Empirically, that is, in practice, nowadays exports amount to more than forty times their value in the early 1900s (see Our World in Data in *Further Reading*) and have increased from less than 10% of global output, to accounting for more than 32% in 2021 (UNCTAD, 2023).

DOI: 10.4324/9781003430926-10

While contemporary processes of globalization are resulting in increased economic integration around the world (Dicken, 2015), thinkers such as skeptics and transformationalists (Held et al., 1999; Martell, 2007) argue that this process of integration leads to complex, uneven geographies by creating differences between places, i.e., disparities created by free markets. Simply put, some places win, and others lose because of globalization.

An important note here is that the perspectives above are based on the role of capital in economies. While the capitalocentric economic model is pervasive in the world, it is not unique. Alternatives exist and are of increasing interest for their potential to challenge processes of commodification, the neoliberal privatization of the state, the marketization of values, and the need for accumulation of capital (Gibson-Graham, 2014). Academics are slowly but increasingly studying these alternatives and looking for diverse ways to understand economies (see, for example, **Chapter 20**). Particularly, from Latin America emerge decolonial perspectives that question how economies are organized as well as the distribution of power; emphasizing social and environmental justice rather than economic growth (**see Chapters 3 and 12**).

This introductory chapter, however, focuses on global economic disparities from a capitalocentric perspective, for it continues to be the mainstream thinking and practice. By examining the Central American case in a global context, this text shows the historical evolution of Central America's economic geography through an uneven globalization journey characterized by colonial ambitions, and the influence by the Global North's economic and political advances.

Connecting South and North America, the Central American isthmus is the sixth largest economy in Latin America (**see** Figure 6.1). It is home to the Panama Canal, one

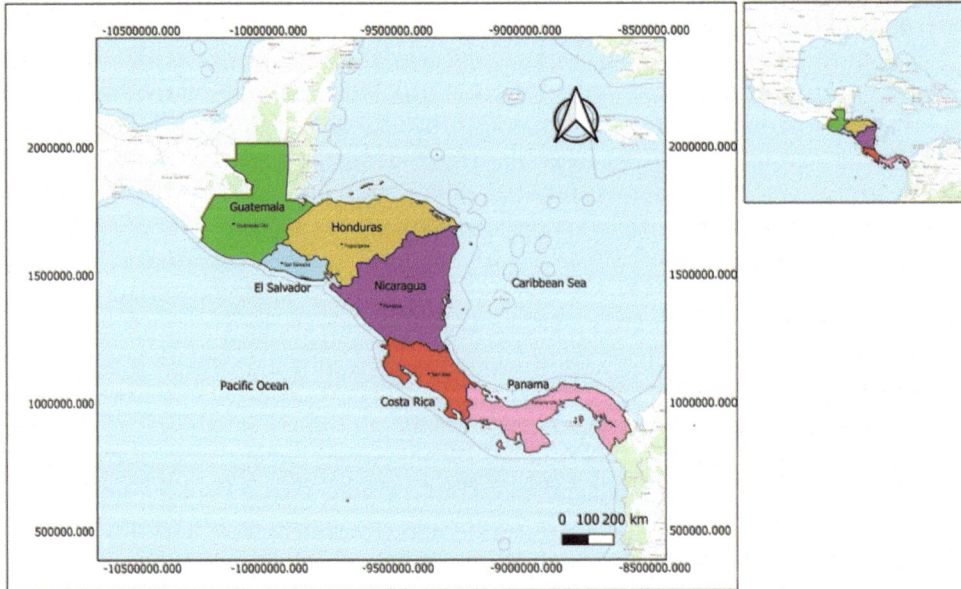

Figure 6.1 Central America: Sixth largest economy in Latin America and the Caribbean.
Source: Author. ESRI – Basemap: World Topographic Map and World Hillshade.

of the most important feats of modern times for global trade and integration, connecting the Atlantic with the Pacific oceans to allow sea shipping transit. The largest country by land area is Nicaragua, which is six times larger than the smallest, El Salvador, which is only 20,700 km². Yet, despite its size, El Salvador has a larger population than both Costa Rica and Panama combined.

These nation-states share a Spanish colonial legacy, including their language and Catholic religion. Over time, however, differences have emerged in important ways. For example, Costa Rica and Panama now rank 'very high' in the United Nations Human Development Index, while El Salvador, Nicaragua, Guatemala, and Honduras rank only 'medium', which is lower than the world average. In 2022, unemployment in Central America was on average about 6.1%, which is not too bad. Yet, the levels of informality of those employed can have alarming rates, ranging from 39% in Costa Rica to 83% in Guatemala (ILO, 2024). Informality means that people may have some sort of employment, but no social security, retirement fund, holidays, no benefits, and often little to no labor rights, nor do they pay taxes on their earnings (see Chapter 20 on urban informality). Access to public education plays a key role here; between 2009 and 2020 Costa Rica invested over 6% of their Gross Domestic Product (GDP) in education, twice as much as Guatemala over the same period (CEPAL, 2024).

In 2022, the region's productive structure was dominated by services, followed by industry, and lastly agriculture (SIECA, 2024). Guatemala is the greatest producer of goods and services in the region (highest GDP), followed by Costa Rica and Panama. Nicaragua produces the least (WB, 2024a). Despite Guatemala's high GDP, its poverty is the highest of region, with estimations that indicate 55.2 % of the population living under the national poverty line in 2022 (WB, 2024b).

In terms of its global insertion, on average, about 57.3% of Central America's GDP is based on global connections (sum of inflow of foreign direct investment, exports, and remittances) (see Table 6.1). Particularly on exports, Central America has diversified, somewhat, its traditional products of low value-added commodities such as coffee, bananas, sugarcane, and garments, to also include integrated electronic circuits, medical instruments, and processed foods, for example. Yet, by 2022, bananas, coffee, palm oil, and copper continue to be in the top five exported goods from the region (SIECA, 2024).

After this brief characterization of Central America, the chapter proceeds to unveil historical and empirical data with respect to global integration through key concepts and analysis used in economic geography. Chronologically, the chapter covers colonialism to modern days in different sections, describing global systems of production, the way the location of production and international trade are organized, and how this has changed over time as well as over space.

Central America in the periphery of production

In Latin America, colonization marked the start of a process of becoming more integrated with the world. Ever since, globalization has influenced countries of the region in different ways.

During the 300-plus-year period of colonial times, mercantilism drove the forces of integration between the subcontinent and the central powers of Spain and Portugal. With an extractive, accumulative mindset, colonialists effectively looted Latin America

Table 6.1 Central America in numbers: an isthmus of contrasts.

	Panama	Costa Rica	Nicaragua	El Salvador	Honduras	Guatemala
GDP (million USD $, 2021)	63,605.10	64,282.44	14,013.02	28,736.94	28,488.67	85,985.75
Surface area (km²)	74,180	51,060	120,340	20,720	111,890	107,160
Forest area (%, 2020)	56.8	59.4	28.3	28.4	57.0	33.0
Population (2021)	4,351,270	5,153,960	6,850,540	6,314,170	10,278,340	17,109,750
GNI per capita (current USD $) (2022)	16,750	12,670	2,090	4,720	2,740	5,350
FDI (% of GDP) (2022)	3.5	5.2	8.3	0.0	3.4	1.4
Exports (goods & services) (% of GDP) (2022)	46.3	40.6	49.8	31.2	41.9	19.0
Remittances (% of GDP) (2022)	0.81	0.8	20.6	24.4	27.4	19.0
Poverty (% at national poverty lines)	21.5 (2019)	25.6 (2022)	14.2 (2021)	26.6 (2022)	48 (2019)	55.2 (2022)
Unemployment (%, 2022)	8.8	11.5	5.6	3.8	7.1	2.6
Informal sector (% of informal employment in total employment, ILO)	55.7 (2021)	39.3 (2022)	81.8 (2012)	67.5 (2022)	82.6 (2017)	79.6 (2022)
Inequality (Gini index)	50.9 (2021)	48.7 (2021)	46.2 (2014)	39.0 (2021)	48.2 (2019)	48.3 (2014)

Source: Author, based on WB (2024a), ILO (2024), SEMCA (2024), OECD (2022), CEPAL (2024).

of their commodities to increase their economic and military power. The exploitation included precious metals and gemstones such as gold, silver, and diamonds; agricultural goods such as sugar, tobacco, and cacao; factors of production such as labor, often forced labor and slavery; and land. Few countries in Latin America were able to develop more complex economic activities, such as Brazil's shipyards for the Portuguese (this industry declined, however, after independence). Colonial power under mercantilism sought to increase their wealth, their exports, and were very selective about their imports through tariffs to make sure they were always on a favorable balance of payments.

Colonial rulers also sought to establish unity within the invaded territories and themselves in terms of government, social life, and culture. Yet, this was not always easy. For example, Central America's lack of transportation infrastructure was a major disadvantage, as most of the region was far away from Santiago de Guatemala (today Antigua), the former capital of the Kingdom of Guatemala, which comprised present-day Central American states and Mexico's Chiapas.

After independence from Spain during the first half of the 1800s, Central Americans were politically divided into two antagonist groups that would shape the economy and culture to nowadays. On the one hand, conservatives sought to keep feudal types of economic organization which favored the *status quo* class system of land-holding nobles and working *campesinos* or peasants. Under Hispanic values and institutions such as the Roman Catholic Church and a kind of paternalism over *campesinos*, conservatives feared foreigners and those who would challenge their place in the social structure. Conversely, liberals sought to favor capitalism, immigration, foreign direct investment, trade, and to break away from the church and traditional values. They sought to become more European and less Indigenous, to exploit land and rural populations to increase agricultural exports, and in doing so, threatening subsistence farmers' way of life (see Fernandes chapter). This liberal view fell under what was later called modernism and would dominate the economic and political agenda for a little over a century in Latin America.

In terms of production, the Global North continued to demand primary, low value-added, labor-intensive commodities from Latin America. The former's manufacturing industries were thriving with increased productivity introduced by new forms of production and advances in technology that led to declining communication and transport costs. One of the most important new forms of production was the Division of Labor (DoL) that Henry Ford put forward for the *Model T* car in 1913. Production was organized around an assembly line. Everyone had one repetitive task to perform and on which to specialize, thus creating speed and accuracy. It marked the origins of economies of scale and mass production of standardized goods.

This model spread across industries, each factory producing all inputs it needed (vertically integrated, full supply chain), routinizing work, increasing economic growth through a virtuous circle of demand, rising incomes, and increasingly efficient supply. In the Fordist global system of production, key regions in advanced economies were at the heart of production, such as New York or London.

Coupled with lower transport costs, this led to increased foreign direct investment (FDI), mostly in search of opportunities and resources, in less advanced economies such as Latin America (Jones and Da Silva Lopes, 2021), another form of increasing global integration. The Panama Canal, built by the US in 1904, facilitated the crossing of maritime transport coming from the Atlantic Ocean to the Pacific Ocean (see Figure 6.2). While its construction can be considered FDI (the largest public work project of the time), the negotiations for its operation were coercive and unfair (Conniff, 2008). Another example is how the United Fruit Company, which managed, often through bribes, to create large scale banana plantations in Central America (over 212,000 acres), as well as constructing infrastructure for their transportation (e.g., 180 km of railroad) (Jackiewicz and Quiquivix, 2016) while creating enclaves in the region. The company got to supply up to two-thirds of bananas to the US market in 1914.

The type of foreign investment arriving in Latin America was mostly that of extractive nature or low value added, with very few linkages to the local economy and often with damaging economic and social effects. This resulted in an economy that continued global center-periphery organization, where the periphery supplied commodities and labor-intensive manufacturing industries.

Protected industrialization in Central America during globalization

During the Cold War, globalization began to take off again in the Western world to try to secure the economic and military power of the West. Twenty-three countries signed the multinational General Agreement on Tariffs and Trade (GATT) in 1947, as a precursor of the World Trade Organization, which sought to lower tariffs and facilitate trade.

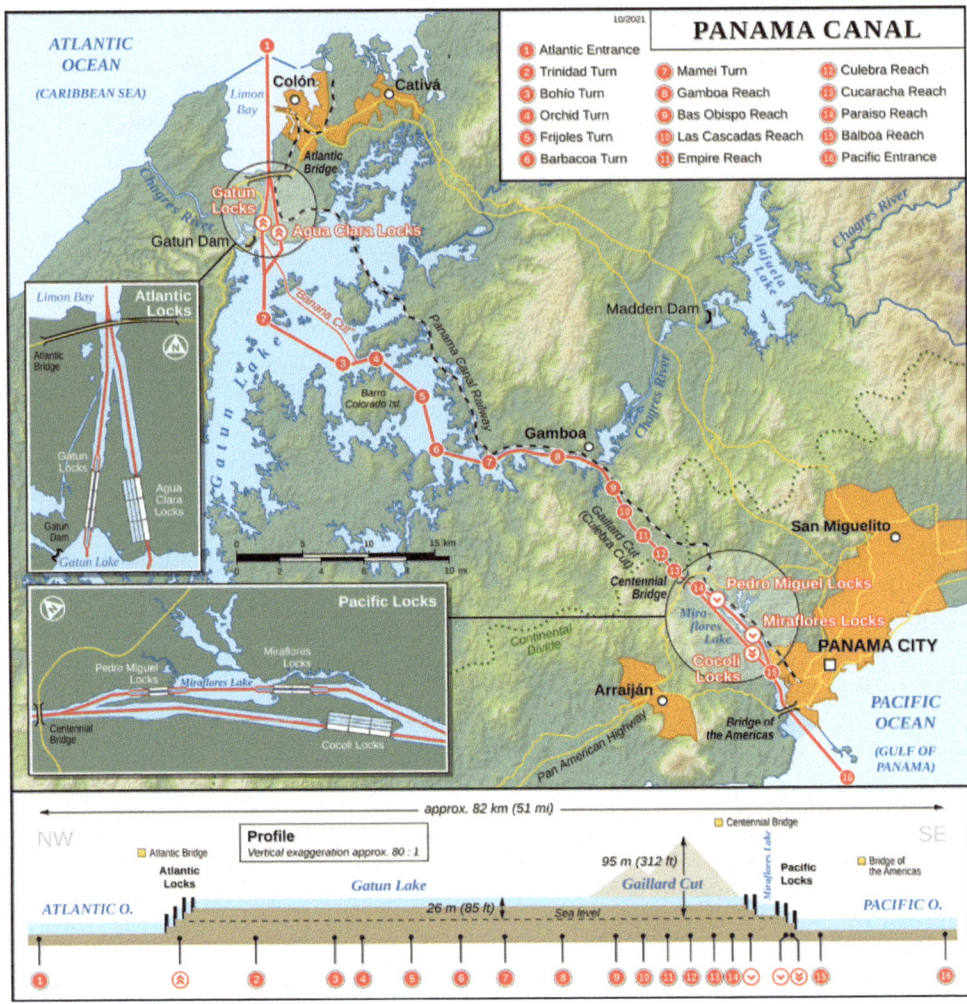

Figure 6.2 Map of the Panama Canal – The top three routes are from the US East coast towards Asia, towards South America's West coast, and from Europe to South America's West coast. The top transported goods in thousands of tons are containers with miscellaneous items and commodities such as grains, minerals, oil, chemicals, and coal (see Panama Canal website https://pancanal.com/estadisticas/).

Source: https://commons.wikimedia.org/wiki/File:Panama_Canal_Map_EN.png

Around this time, economists in Latin America began to question just how effective their agro-export-led economic model for was increasing the standard of living in a context of heightened unemployment (increased rural-urban migration). For the following two to three decades, LA policy shifted toward deglobalization (i.e., to decrease their global economic integration) and to actively encourage the development of local industries. The import-substitution industrialization (ISI) model sought to take advantage of technological advances that were finally arriving to the region. Prebisch (1959) defined it as a process to increase the proportion of goods supplied from domestic sources. The ISI model included high import tariffs, the nationalization of key industries, subsidies, investing in infrastructure, but also attracting foreign capital for the manufacturing sector to help expand the 'heavy' industries such as producing tools and machinery for production in other enterprises (through FDI, North American MNE subsidiaries got to control around one third of manufacture production in 1966).

The goal was to reach a similar technological development as other countries around the world to compete under similar conditions. As explained in **Chapter 11**, the uneven trade relationship between countries benefits those who continue to be in the developed 'center' but maintain the others in the 'periphery' by demanding commodities produced with cheap labor (labor-intensive goods) to transform adding value, to then resell them back in the form of capital-intensive manufactured goods (expensive labor). During this industrialization wave, industrial output went up six times; Mexico and Brazil manufactured over 95% of consumer goods domestically (Hoogvelt, 1997).

While the ISI model caused Latin America's economic integration with the world to slow down in terms of imports, countries sought to facilitate trade within regional blocs, concentrating in specific geographical areas. This also satisfied the need for larger markets for their newly manufactured goods. For example, Central America sought free trade within the region in most goods and established a uniform external imports tariff through the Central American Common Market, and later, the General Treaty of Central American Economic Integration. This led to a six-fold increase of inter-regional trade within seven years of signing the treaty (Cochrane, 2008). Nowadays, this regional customs union is one of their largest markets, second only to the US, concentrating about 30% of the isthmus' exports (SIECA, 2024).

The regionalization of trade was a global phenomenon. This concentrated liberalization included, in some cases, free movement of other factors of production such as labor and capital – for example, in Europe, through the European Union Treaty. In the following decades, more regional blocs emerged, such as the Caribbean Community and Common Market, the Andean Community, and the Common Market for Eastern and Southern Africa, among others. Intra-EU trade accounted for about 61% in 2021. In Asia, the value is also close to two-thirds, with 59% of trade happening intra-regionally. Yet, overall, the Global South continues to trade less with itself (around 40% of trade) than with countries from the North. The United States and China are the main trading partners not only for Central America, but for developing countries in general. Moreover, according to the World Trade Organization, the world has 356 regional trade agreements in force and about 230 under examination.

In the 1970s, in the Global North, the Fordist production system began to fall. The post-war industrialization boom had declined. Unemployment rose, consumption could not keep up, and emerging economies became increasingly competitive (note how production and consumption are closely related; this is a critical connection that is discussed in the next section). The high costs of the Fordist production system and its inflexibility to adapt led to the collapse of industries in advanced economies, and production moved en masse to locations with lower labor and production costs.

A mix of more flexible forms of production began to emerge in the new post-Fordist global production system. One example is the emergence of new industrial spaces. These are highly dense and highly flexible (i.e., with an institutional ease to change suppliers and workers as conditions change) agglomerations of firms in industries such as crafts and design, high technology, and business services (e.g., the film industry in Los Angeles, or the artisanal industry in the so-called Third Italy).

Another example is the **new international division of labor** (IDoL). Firms began dividing the different stages of the production process and locating them in different countries. A **Spatial Division of Labor** (SDoL) *within* firms was also taking off. Technological change, shifts in market demand, lower transport costs, and an increased ease for capital mobility led to a tighter global integration, and to different international divisions of labor.

With the SDoL, multinational enterprises (MNEs) had a predominant role. The cities that hosted the headquarters of these MNEs began to emerge as the core 'command points' in the global economy (e.g., Tokyo, London, New York). Others, which hosted the more elemental stages of the production such as assembly, were in the periphery of the action. An example in Central America are the *maquilas* or *maquiladoras* installed or subcontracted through North American FDI (eventually also Asian). *Maquilas* are assembly plants that typically import components produced via capital-intensive processes in advanced economies which are assembled through labor-intensive processes in developing countries (mostly women with very low wages worked here). Not quite the new industrial spaces, *maquiladoras* were also agglomerated in government set up Export Processing Zones (EPZs), areas where usually foreign firms had fewer or no restrictions to import intermediate inputs for production and often had fiscal incentivies investment.

During this time, advanced economies also began to receive FDI from newly industrialized countries, mostly from Asia. Yet, technical progress continued to spread unevenly. While economic progress advanced in northern Europe and the United States during the nineteenth century, external invasions by the US and internal civil unrest challenged most of Latin America. In these circumstances, productivity in the region became stagnant or even declined (Coatsworth, 2008).

The ISI model led to high industrial growth rates in Latin America and replaced agriculture as the leading economic sector (Dosal, 2008). Yet, in the 1970s, the model began to fail, and the early 1980s world recession which led to shortages in exchange to pay foreign debt effectively limited industrial growth in Latin America. The end of this model had arrived.

The productivity decline in Latin America during the 1980s was accelerated by the fleeing investment and private loans from the region. For example, net capital inflows decreased in 40% and international private net flows declined by 80% (Jackiewicz and Quiquivix, 2016). This led Central American countries to borrow from global lenders such as the International Monetary Fund, the World Bank, and the US Treasury.

Being financially dependent on international institutions meant having to ensure returns to their investments, and thus complying with their creditors' view about policies for increased productivity and economic growth. These policies came in the form of structural adjustment programs after the Washington consensus, which laid out some of the neoliberal principles that governments had to follow: reduce or eliminate barriers to trade (trade liberalization), deregulate FDI and movements of capital, privatization of public monopolies and enterprises, and in general making the government smaller by increasing deregulation and increasing protection of private property. In 1994, most countries around the world were signatories of the World Trade Organization treaties, which seek to facilitate global trade. Thus, once again, an export-led economic model gained prominence in Latin America.

Central America's deepening global integration

The global tendency towards increased economic integration continued throughout the 1990s, now with Latin America fully on board. An example of this deeper integration is the shift to investment in service industries, which is increasingly growing for its quicker turnover of capital. In Central America, nowadays the proportion of GDP that accounts to exports in services is about a third more than in 1995, amounting to 12.5% of the isthmus' GDP (CEPAL, 2024).

In terms of global FDI, its growth is fluctuating, going from 47% in 2015 to −20% in 2017 and 64% in 2021. Yet, overall, since the 1970s FDI has grown fast, even faster than trade (Dicken, 2015). Latin America and the Caribbean receive about 9% of the world's FDI, and for small countries such as those in Central America (between 5% and 11.5% of their GDP), this inflow has become critical for their own economies, mostly, for what it represents in terms of jobs creation.

Increased financial integration has also meant a rise of the remittances received from abroad, a crucially important revenue for most Central American countries (except Panama and Costa Rica). In 2022, net remittances amounted to about 6.67 billion dollars, representing about 12.5% of the isthmus' GDP, the same as the percentage of services exported. According to the Central American Monetary Council, Central Americans use approximately 75% of remittances for water, food, clothing, and electricity (SEMCA, 2024).

Increased globalization is also shining the spotlight onto consumption and the complex global connections that it generates with the economy and society. Consumption can be defined as the use of purchased material goods, services, and a variety of social experiences for the satisfaction of needs and desires. Production – in a capitalist system – produces commodities *and* subjects for consumption through the hiring of workers and their income. Thus, without production there is no consumption, and vice-versa.

Producers have sought to shape consumption by creating a 'consumer culture' using mass media advertising and marketing to sow false needs. Nowadays, in a context of time-space compression for production (lower costs and faster production and shipping), the role of consumption is enhanced and changing.

Countries in Central America are both producers of labor-intensive goods sold by multinational enterprises, as well as avid consumers. An example is in the rise of retail capital, such as Walmart, which entered the region in 2005 by taking over existing local super-market chains. This type of business model shifted from a producer-driven demand (having to spend time and money finding markets and persuading buyers to shop) to a buyer-driven demand. In this sense, consumption is increasingly shaping globalization. Walmart has over 850 stores in Central America (excluding Panama) (Walmart, 2024). It is the largest US Fortune 500 company, obtaining USD $570 billion in world revenue in 2022. Such powerful retailers have a massive consumer market behind them, which allows them to effectively shape production processes and be on top of the power dynamics, mostly on labor-intensive goods such as footwear, clothes, and consumer electronics, as well as primary goods such as fruits and vegetables, which are often locally sourced.

Milton Santos, a prominent Latin American geographer, would argue that this is an example of the dimension of 'verticality' in globalization (1993). That is, the trajectory of economic activities over time and space is shaped by global actors re-organizing local individuals, groups, and institutions in their favor by controlling them from a distance, for example, by homogenizing the goods that are sold (in 2019 Walmart held 72% market share in Central America). This relates to the unevenness created by globalizing processes pointed out by skeptics and transformationalists and the influence some places have over others. In this case, Walmart, a multinational from the US, has a great leverage in local production, imports, and consumption, as well as the financial market.

Research on commodity chains has helped bring awareness to consumers and raised questions about the way some of these chains perpetuate exploitation. For example, research suggests that the largest supplier of fresh pineapples to the world (Costa Rica) grows them to the social and environmental detriment of people living and working in and around its areas of production (Costa Rica is one of the top pesticide consumers per capita in the world). Yet, while consumers are increasing their awareness and demands for organic fruit, shifts in production are changing only very slowly. In the power dynamics of commodity chains, final consumers led by morals are the least powerful.

Globalization has led to high economic concentration into few players of the global economy game. While the US still dominates the arena (in commercial services and FDI), it is less dominant than before, and manufacturing production is now primordially occurring in China. Latin America continues to be in the periphery of the world economy. Uneven geographies of trade and capital movements generate imbalances that shape this increased global interconnection (Dicken, 2015). Central America is a region that receives the prices that other economies set and has no effect whatsoever on global trade.

While Central American countries are often on the receiving end of power by the world's advanced economies, this does not go unnoticed in Latin American discussions and practice by some actors (less often governments, except for Nicaragua, which has been in a dictatorship unwelcome to the US for almost 20 years). For example, the question of power relations is present in academic debates around economies, territories, and inclusive development by asking questions such as 'how do we transform power relations?' (e.g., Manzanal, 2007 in Argentina), or through participatory research which involves direct collaboration with local actors and economies in the form of resistance alternatives to global pressures to conform or simply because there is not even an option to conform (e.g., see Saquet and Cichoski, 2022 for an explanation in the case of Brazil).

In this sense, of individuals not having the possibility to partake in the benefits of globalization, the social economy in the form of cooperatives or similar types of associations have created important opportunities in the region (see Figure 6.3). Some *campesinos* in vulnerable situations have been able to improve their livelihoods by voluntarily becoming members of these social democratic organizations to help each other in their labor, consumption, commercialization, and financial needs and aspirations. These organizations may sell goods (such as coffee cooperatives) or services (e.g.,

Figure 6.3 Home of an associate at the Union of Cooperatives, UCA San Ramón, in Matagalpa, Nicaragua. This cooperative specializes in agriculture and rural development to offer economic opportunities to families in rural areas.

Source: Author.

financial services cooperatives), and with the revenues seek to create social development, economic and environmental opportunities for their associates (see Infocoop for the case of Costa Rica: www.infocoop.go.cr/estadisticas).

Overall, contemporary processes of globalization show an increased instability of the aggregate growth of world merchandise production and trade (Dicken, 2015). Because of globalization, events such as the 2008 financial crisis caused by the crash of the subprime mortgage market in the US; the recent COVID-19 virus pandemic; and the Ukrainian invasion by Russia have had remarkable contagion effects in the global economy. Increased transport prices, lower tourist visitations, and stricter national and global regulations have affected the entire world, including Central America.

Conclusion

Processes of globalization help us understand economies and the uneven patterns of production, trade flows, consumption, and finance that are created. The post-Fordist global production system has led to a more interconnected world than ever before, although this is recently beginning to contract because of current events such as the global pandemic, the climate crisis, the shipping crisis, and geopolitical instability and wars. Paradoxically, this heightened global connectivity coexists with a simultaneous geographical concentration of economic activities. The concepts of global and local interplay with one another, emphasizing that geography still matters in shaping the dynamics of the global economy. Both economic specialization and the reworkings of the division of labor have shaped new patterns of global production and trade, reflecting social-spatial inequalities. Central American nations have witnessed these complexities first-hand, as they navigate the tension between global integration and local contexts.

The examination of Central America within the context of uneven globalization processes reveals significant insights into the evolving dynamics of the global economy. The 1960s marked a clear qualitative change in the region's economy. Key elements driving this transformation were FDI and MNEs, whose influence and presence became increasingly prominent. Central American countries, like many others, have experienced both winners and losers in the process of globalization, with some nations (and cities) benefiting from increased integration, while others face challenges in adapting to the changing economic landscape.

Moreover, the relationship between consumption and production emerges as a dialectical one, underscoring the importance of consumption in understanding the broader dynamics of the global economy. Consumption extends beyond mere economic transactions; it encompasses social and cultural dimensions as well. Central American cases illuminate the significance of sensitivity to geographical difference, revealing hidden geographies of connections between the worlds of consumption and production.

In summary, the chapter demonstrated the multifaceted nature of the global economy, with Central America serving as case study. By delving into these complexities, we gain valuable insights into the ever-evolving landscape of the global economy and its implications for the periphery and beyond.

Summary

- The study of globalization from an economic geography perspective shows the uneven patterns of production, trade, and consumption in space.
- Colonialism and neocolonialism have had great influence in the economies of Central America, which are also diverse and unequal.
- Globalization is increasing over time but has also faced obstacles in the form of global financial crises, pandemics, or wars.
- Some actors in Central American and Latin American countries in general have sought alternatives to globalization for their economies, mostly because they are outright excluded from the benefits.

Review questions

1. What patterns of international production, FDI, and trade can you observe? How have these changed over time?
2. What have been the implications of globalization for Central American economies?
3. Which perspective do you agree with, hyperglobalists or skeptics/transformationalists? Discuss.
4. Should globalization be regulated? How? At which scale?

Further reading

Bebbington, A. (2009). Latin America: Contesting Extraction, Producing Geographies. *Singapore Journal of Tropical Geography*, 30. 7–12. https://doi.org/10.1111/j.1467-9493.2008.00349.x
Short critical essay on continued extraction and contestation forces in Latin America.
Estado de la Nación is funded by the Council of National Universities (CONARE) in Costa Rica.
They produce high value research on Costa Rica and Central America. Some of their reports or their summaries can also be found in English. https://estadonacion.or.cr/english-docs/
Keeling, D. J. (2008). Latin America's Transportation Conundrum. *Journal of Latin American Geography*, 7(2). 133–154. www.jstor.org/stable/25765222
Shows how lack of adequate transportation (one of the earliest issues in the region during and after colonization), continues to be one of the most important challenges for integration.
Ioannou, S. and Wójcik, D. (2022). The Limits to FinTech Unveiled by the Financial Geography of Latin America. *Geoforum*, 128. 57–67. https://doi.org/10.1016/j.geoforum.2021.11.020
Offers insight into financial integration and its unevenness within and without Latin America.

Keywords

Conservatives (colonial times): uphold traditional structures and values (church), hierarchies, and elite dominance, resisting change.

Consumption: the act of using or purchasing goods and services to satisfy human wants and needs.

Division of labor: the specialization of tasks and roles among individuals or groups in the production process, leading to increased efficiency and productivity.

Enclave: concentrated specialized economic activity disconnected from the local economy.

Foreign Direct Investment (FDI): investment made by a company or individual from one country into a business or enterprise located in another country.

Globalization: global integration of information and knowledge, means of production, and markets (labor, goods, services, and financial).

Hyperglobalists: supporters of unrestricted global trade, capital flows, and integration.

Liberals (colonial times): supporters of open markets, free trade and FDI flows.

Maquila: manufacturing or assembly plants where imported materials are processed and exported back, typically operating under favorable tax and trade conditions.

Remittances: financial transfers made by migrant workers to their families or home countries, typically sent as a form of support or contribution to their households' well-being.

Skeptical internationalists: critical approach to globalization, considering potential negative impacts.

Spatial division of labor: the distribution of economic activities and specialized tasks across different geographical locations or regions, reflecting the advantages, resources, and capabilities of each location.

Transformationalists: scholars emphasizing the transformative nature of globalization and dynamic interactions between global, national, and local forces.

References

CEPAL. (2024). *CEPALSTAT Bases de Datos y Publicaciones Estadísticas*. United Nations Economic Commission for Latin America and the Caribbean. https://statistics.cepal.org/portal/cepalstat

Coatsworth, J. (2008). Economic Development. In *Encyclopedia of Latin American History and Culture* (2nd ed.). (3). 11–32. Gale Cengage Learning.

Cochrane, J. (2008). Central American Common Market. In *Encyclopedia of Latin American History and Culture* (2nd ed.). (2). 250–251. Gale Cengage Learning.

Conniff, M. L. (2008). Panama Canal. In *Encyclopedia of Latin American History and Culture* (2nd ed.). (5). 29–32. Gale Cengage Learning.

Dicken, P. (2015). *Global Shift: Mapping the Changing Contours of the World Economy* (7th ed.). The Guildford Press.

Dosal, P. J. (2008). Industrialization. In *Encyclopedia of Latin American History and Culture* (2nd ed.). (3). 830–835. Gale Cengage Learning.

Gibson-Graham, J. K. (2014). Rethinking the Economy with Thick Description and Weak Theory. *Current Anthropology*, 55(S9). S147–S153.

Held, D., McGrew, A., Goldblatt, D. and Perraton, J. (1999). *Global Transformations*. Polity Press.

Hoogvelt, A. (1997). *Globalisation and the Postcolonial World. The New Political Economy of Development*. Red Globe Press.

ILO. (2024). *Informality*. International Labour Organization. https://ilostat.ilo.org/topics/informality/

Jackiewicz, E. and Quiquivix, L. (2016). Ch. 3 Cycles of Economic Change. In Jackiewicz, E. and Bosco, F. (eds.), *Placing Latin America – Contemporary Themes in Geography* (3rd ed.). Rowman & Littlefield.

Jones, G. and Da Silva Lopes, T. (2021). International Business History and the Strategy of Multinational Enterprises. How History Matters. In Mellahi, K. et al. (eds.), *The Oxford Handbook of International Business Strategy*. Oxford University Press.

Manzanal, M. (2007). Territorio, poder e instituciones. Una perspectiva Crítica. In Manzanal, M., Arqueros, M. and Nussbaumer, B. (eds.), *Territorios en construcción. Actores, tramas y gobiernos, entre la cooperación y el conflicto*. 15–50. CICCUS.

Martell, L. (2007). The Third Wave in Globalization Theory. *International Studies Review*, 9(2). 173–196.

O'Brien, R. (1992). *Global Financial Integration: The End of Geography*. Council on Foreign Relations Press.

Prebisch, R. (1959). International Trade and Payments in an Era of Coexistence. Commercial Policy in the Underdeveloped Countries. *The American Economic Review*, 49(2). 251–273.

Santos, M. (1993). Los espacios de la globalización. *Anales de Geografía de la Universidad Complutense*, 13. 69–77.

Saquet, M. and Cichoski, P. (2022). Capítulo 6. Territorios y (des)arrollo raíz: contribuciones para una perspectiva de investigación y cooperación popular, decolonial y contrahegemónica. In Morales, D. et al. (eds.). *Territorios y Desarrollo. Teorías, Debates y Casos desde América Latina*. Universidad de Costa Rica.

SEMCA. (2024). *Estadísticas*. Secretaría Ejecutiva del Consejo Monetario Centroamericano. www.secmca.org/secmcadatos/

SIECA. (2024). *Estadísticas*. Secretaría de Integración Económica Centroamericana. www.sieca.int/estadisticas/

UNCTAD. (2023). *Key Statistics and Trends in International Trade 2022*. United Nations Conference on Trade and Development. https://unctad.org/system/files/official-document/ditctab2023d1_en.pdf

Walmart. (2024). *Nuestra operación*. Walmart México y Centroamérica. www.walmartcentroamerica.com/conocenos/quienes-somos

WB. (2024a). *DataBank World Development Indicators*. World Bank. https://databank.worldbank.org/

WB. (2024b). *The World Bank in Guatemala – Overview*. World Bank. www.worldbank.org/en/country/guatemala/overview.

CHAPTER 7

Urbanization

Juan Miguel Kanai

Introduction

Urbanization processes tell us a powerful story about Latin American geographies. With one of the highest levels of urbanization in the world – i.e. the proportion of people living in metropolitan areas, intermediate cities, and smaller urban settlements – the region also exhibits the starkest social and spatial inequalities. Whilst most countries qualify as middle-income countries, and some are considered fast-growing markets, income distribution is extremely uneven. This manifests both in the most internationalized cities, where conspicuous wealth coexists with the squalor of unmet basic needs, and in the 'left behind' regions, where structural poverty is entrenched, even in those areas with booming agricultural or extractive industries.

This chapter explores the idea that the drivers of urbanization – the reasons why cities add population, grow in area, and acquire economic and administrative functions – are intrinsically linked to the inequality that plagues the region and inspired the moniker of Latin America as a 'lopsided continent'. The chapter has three main sections. The first section begins with a review of the period of high economic growth spanning the mid-twentieth century decades when populous metropolitan regions expanded as the result of the import substitution industrialization applied in the largest countries. The second section takes stock of the neoliberal restructuring that has taken place since the 'lost decade' of the 1980s with the collapse of the previous model of economic development based on relatively autonomous national industries. As a consequence, cities and regions found new roles but also became more exposed to the instability of world markets. The third section highlights the complex relation that has emerged in the twenty-first century between the current drivers of growth, of an increasingly extractivist and speculative nature, and new urbanization processes. The section reviews the development patterns associated with primary/extractive industries and their infrastructures, and points to the precarity that people experience in their everyday lives across the diverse range of conditions that characterize what is now a thoroughly yet unevenly and fragmentarily urbanized continent.

DOI: 10.4324/9781003430926-11

The analysis herein highlights the weaker-than-expected correlation between urban expansion, economic growth, and social development; the importance of deficit-ridden urban contexts to understand multidimensional poverty; and the challenges for state-led and community-driven strategies to make Latin American urbanisation more inclusive, sustainable, and liveable. These themes have important implications for how we understand Latin American geographies overall, but are far from exhaustive, even within the urbanization research agenda. Concluding remarks highlight changes and continuities through the historical period covered. An acknowledgment is made of important historical dimensions and questions of embodiment not covered within the chapter's primarily political economy approach. Avenues for integrative further research are presented.

Rise of the metropolis in Latin America

The region experienced its fastest and longest continuous period of urbanization and metropolitan expansion in the middle decades of the twentieth century. Roughly between 1930 and 1970, the urban population ratio more than tripled as Latin America became one of the most urbanized world regions, and especially after World War II the growth of large cities paralleled economic, social, and political transformations that had proven previously elusive (Almandoz, 2014). The dimensions of such modernization included large-scale industrialization; the growth of mass societies articulated around consumption and expectations of progress (with modernist high-rises and ambitious public works as the most conspicuous material evidence of this ethos in cities); and the rise of new political actors demanding a share of power and the reform of local and national governing institutions (Roberts, 2020). The rise of the modern industrial metropolis was accompanied with changing relations between Latin American nations and the outside world, particularly the advanced capitalist economies of the North Atlantic Basin. Whilst imports were restricted, the region was open to foreign direct investment (such as e.g. North American and European automotive companies), technology transfers, and international cooperation to develop its infrastructure and strengthen what was known at the time as national economic spaces (Ocampo, 2013).

Even amid the high-growth era, it became evident that urban development was uneven and would not benefit all peoples and places across the continent. Path dependencies stretching back to the colonial era meant that cities with entrenched political and economic functions, such as Mexico City, Buenos Aires, Bogotá, and Santiago de Chile, received the brunt of new investments and grew disproportionately vis-à-vis the overall size of their national urban systems whilst historically relegated peripheral territories fell further behind (Chase-Dunn, 1985). Compounding the complexity of socio-spatial inequalities, the expanding metropolis became a magnet for internal migration from poorer parts of the country seeking work opportunities, as demonstrated by the large numbers of arrivals from the Northeast of Brazil in Rio de Janeiro and São Paulo (Valladares, 2019). Newcomers would encounter low-paying jobs, precarious housing located largely in informal settlements without adequate public services (**see Chapter 19**), and various forms of discrimination related to their geographical origin, racial and ethnic backgrounds, and a perceived lack of urbanity. If the emergence of slums predated the rise of the metropolis, precarious, informal settlements became more pervasive as cities grew and urban planning either ignored their reality or found itself unable to provide comprehensive solutions beyond piecemeal removals and later upgrades, which nevertheless remained controversial for their exclusionary implications and negative implications for vast numbers of the urban poor.

The rise and subsequent decline of the metropolis was accompanied by a rich body of urbanization research in and from Latin America. In early stages, the literature focused on how to optimize growth through physical interventions during the optimist earlier stages, first through rational or scientific city planning and from the 1950s on through developmentalist regional planning. Yet, a more critical social science and political economy approach gained momentum in the interpretation of why metropolitan pathways in the region differed from those in North America and Europe (Gorelik, 2017). Architecture and planning followed external models (Almandoz, 2006). The figure of Swiss-French architect Le Corbusier loomed large in Brazilian, Argentine, and Uruguayan circles from the 1920s onward. Monumental works from this period include the opening of Buenos Aires' *Avenida 9 de Julio* grand boulevard conceived by Paris-trained Carlos Maria Della Paolera (see Figure 7.1) and the various public space projects that landscape architect Roberto Burle Marx produced for Rio de Janeiro, including the iconic sidewalks alongside Copacabana Beach completed in 1970. Burle Marx's early collaborators Lucio Costa and Oscar Niemeyer would go on to build Brasilia alongside similar modernist principles. As developmental ideas took hold, large-scale projects, such as the construction of *Ciudad Guayana* in Venezuela, required infrastructure and economic planning and were largely influenced by Francois Perroux's 'growth pole' theory and the operations of the Tennessee Valley Authority created during the United States' New Deal period.

Figure 7.1 Buenos Aires' 9 de Julio Avenue.
Source: Creative Commons.

Critiques of the 'modern industrial metropolis' model emerged from a general discontent with imported notions of progress and development. Heterodox economists introduced an analysis of the region's peripherality caused by external factors, which from the 1960s neo-Marxist social scientists built into a systemic interpretation of Latin America's dependency (Connel, 2007: 139–164) (see **Chapters 11 and 19**). Dependent development was taken up by Peruvian scholar Anibal Quijano to analyse uneven urbanisation and the impact of urbanization processes on rural societies (Vegliò, 2021). Brazilian geographer Milton Santos critiqued the role that both planning and regional science played in exacerbating inequalities and poverty in peripheral regions (1977). Writing on Mexico City, anthropologist Nestor Garcia Canclini pointed out that the lack of public cultural spaces was one of the major deficits from the massive metropolitan expansion that took place since 1950 (Canclini and Liffman, 2000). This correlated with a shift towards cultural consumption in private spaces, particularly through television, which in turn reshaped imaginations of the city and nation as cohesive spaces.

Neoliberal restructuring: urbanization under market rule

Since the late twentieth century, Latin American countries have been immersed in a series of economic, social, political, and more recently environmental crises (see **Chapter 17**) that have reshaped urbanization processes in the region. These regional crises have been linked with changes in the global capitalist system in terms of both the drivers of economic profit – increasingly linked to technology, financial speculation, and specialized consumption – and the policies undertaken to pursue and sustain the growth paradigm under such shifts. Whilst the model of import substitution industrialization (see **Chapter 6**) had already shown signs of stagnation since the late 1960s, and the region was impacted by the oil and dollar crises of the 1970s, systemic collapse occurred in the 1980s, when national governments throughout Latin America were forced to default on their foreign debts and found themselves unable to provide their citizens with even basic services.

During this 'lost decade' for development, economic growth halted, industries closed, and food riots took place. Influenced by international financial institutions, such as the International Monetary Fund, the World Bank, and the United States Department of Treasury, governments throughout the region embarked in a neoliberal programme of reform that became known as the 'Washington Consensus'. Policies focused on liberalizing the economy, especially allowing more industrial imports; reducing regulations to save on production costs; pursuing balanced budgets through fiscal discipline; attracting foreign investment by privatizing state enterprises and offering concessions on natural resource exploitation; and promoting exports in sectors with an apparent global comparative advantage, especially in the primary sector (see **Chapter 8**).

Market rule brought mixed results for national economies and spearheaded more unequal, fragmentary, and contentious urbanization dynamics (Portes and Roberts, 2005; Kanai, 2011; Boano and Vergara-Perucich, 2017). Under the new open economy regime, growth objectives were met sporadically during boom periods, especially immediately after the rollout of high-profile initiatives such as export-processing zones and other measures to attract foreign investment, but these were interspersed with periods of sharp contractions (Schindler and Kanai, 2021; Schrank, 2008; Cypher and Delgado

Wise, 2010). The income distribution bifurcated, with some growth at the top fuelled by high corporate incomes, and a much larger expansion of low-income groups composed of former industrial workers and middle-class occupations associated with state services. This inequality was soon reflected in changing cities as gleaming office districts, spaces of ostentatious consumption, and luxurious gated communities emerged as bubbles of wealth across seas of urban decay and impoverishment. City planning became entrepreneurial and focused on attracting investment, international visitors, and increasingly high-income residents, be they 'creative' expats or digital nomads. Regional planning was largely abandoned and fewer new infrastructure projects were built; the privatization drive emphasized transferring existing assets to bidders committing to upgrades and service improvements. Discontent with market-oriented reforms and their social costs soon translated into oppositional movements as those excluded from new growth dynamics began organizing both in metropolitan areas and peripheral regions. Whilst some of these politics aimed to reverse neoliberal policies through state power, and hence focused on electoral strategies and partisan coalition building, others sought to create spaces of social solidarity with higher autonomy from both state and market institutions.

Research on urbanization under hegemonic neoliberalism problematized the social and spatial implications of the rise of 'global city' functions in erstwhile more state-controlled and endogenously oriented urban systems (Sassen, 2002). In the areas most favoured by domestic and foreign capital, this not only resulted in new business and employment clusters, but also a revitalization of the built environment, particularly through property-led growth. From the example of Managua, Rodgers (2012) argues that investments in transport infrastructure focused on serving a few elite sectors whilst sidelining or even inflicting 'infrastructural violence' on disadvantaged areas. Focusing on Asunción, one of the least studied cities in the region, Boschmann (2020) shows that improvements in the urban centre were also selective, related to the expansion of tourism and with a new focus on river-front spaces (see Figure 7.2). Whereas higher exposure to global tourism flows created new business opportunities, it also produced thorny debates on who local cultures and physical assets should serve, and has been linked to exclusionary redevelopment, gentrification, and displacement (Kanai, 2014a; Janoschka and Sequera, 2016). More broadly, the inequality-enhancing effects of neoliberal urbanism spearheaded various discontents, including with the rise of urban violence, fear of crime, and the stigmatization of the most precarious social-economic groups, youth in particular (Portes and Roberts, 2005; Moser and McIlwaine, 2004; Caldeira, 2000). Urban resistance to the privatization of public services resulted in popular protest, and at times violent repression, such as the episodes remembered as Bolivia's 'water wars' with an epicentre in the city of Cochabamba (Perreault, 2006).

Contemporary challenges: precarity and contestation in a thoroughly urbanized continent

Extension and heterogeneity characterise contemporary urbanization processes, with various forms of precarity emerging across the transnationally interconnected urban system. Political contestation to neoliberal reforms and policy reversals continue shaping urban dynamics in the early decades of the twenty-first century as critical research

Figure 7.2 Asunción's Avenida Costanera.
Source: Creative Commons.

points out that neoliberalism has mutated rather than ended (Gago and Mezzadra, 2017). Whereas the privatization agenda and state rollback have been largely contested, such as with the case of water companies and other utilities, most Latin American nations remain firmly committed to global market integration (Kanai, 2016). In the context of a multipolar yet unevenly developed world, the region's governments pursue various agendas to integrate country resources and productive forces to global value chains – with rapid growth in the primary sector of agricultural exports and natural resource extraction. Adding a layer of complexity, urban and regional policies are now influenced not only by the US-dominated multilateral system of international finance and development assistance, but are also increasingly dependent on Asian investments and bilateral negotiations with the People's Republic of China, which has become a major trading partner and source of investment capital for the region (Schindler and Kanai, 2021).

Regional planning has experienced a resurgence accompanied with the aim of reducing logistic costs and pursuing cross-border integration of transport, energy and communication grids (ibid.). Pursuing this connectivity imperative has resulted in the expansion of infrastructure networks and new forms of urbanisation. Growth has occurred in secondary cities of erstwhile remote regions where new forms of export-oriented agriculture and extractive industries have developed, including paradigmatic soy and oil palm mega-plantations and large-scale oil and gas extraction, alongside mining for old and new resources such as gold, copper, coal, and lithium (Arboleda, 2016). Sparser, peri-urban settlements have emerged along newly built transport routes at ever longer distances from metropolitan cores. These new urban spaces are not only surrounded by the environmental decay that extractivism causes, but they also lack social amenities, as their planning is solely focused on economic functions (Kanai and Schindler, 2022).

In more consolidated metropolitan regions, older neighbourhoods and suburbs evince highly varied development pathways. Some entrepreneurial regimes experiment with new technologies and 'smart city' approaches to urban management (Irazábal and Jirón, 2021). Whilst exclusionary forms of regeneration continue triggering displacement in select areas (Garmany and Richmond, 2020), community involvement supports local transformations in others (Schindler et al., 2023). The impacts of the global climate emergency are increasingly felt throughout urban Latin America, but consequences are particularly salient for poorer, informal settlements (Hardoy and Pandiella, 2009; Ferrari et al., 2021).

Urban infrastructure provision is a key concern within contemporary urbanization research. Expanding roadway networks have an influence on metropolitan development. These interconnections promote the formation of new satellite nodes and ever wider peri-urban zones around older cores, but also spearhead more diffuse and fragmented yet also functionally articulated urban growth (Brenner, 2019). Whilst much of the expansion has taken place in the Brazilian Amazon (Becker, 1995; Monte-Mor, 2014; Kanai, 2014b), extensive, networked urbanization is also being studied in Colombia and Ecuador, where riverine functional articulation is also attempted for remote regions (Wilson and Bayón, 2017; Uribe, 2019). Studies from northern Chile and Bolivia show the intrinsic relation between infrastructure provision and growth of extractivism (Arboleda, 2016; Perreault, 2018), and concerns with the resulting precarity in newly urbanized locations (which are often not recognised as such within policy discourse) add to the critique of the social and environmental consequences of this mode of development (Kanai and Schindler, 2022). The occurrence of extreme climatic events, including torrential rains, megadroughts, and heat waves, affects both the older metropolitan and newly urbanized regions, and various infrastructure solutions are attempted for disaster mitigation. Whilst the bias for foreign investment and large firms is clear in large-scale engineering works (Silva, 2016), even 'green' projects with a more innovative ecosystems-based approach follow market-based approaches and may result in displacement and the sidelining of vulnerable groups (Millington, 2018).

Conclusion

This introduction to urbanization in Latin America reviewed the constitutive role that urbanization processes play in shaping the region's geographies, from mediating between political economic shifts and social experiences of growth and decline to exhibiting and accelerating the environmental change that is inherent to such shifts. With a scope over a century long, the review showed that key to the political economy of urbanization is the position of Latin America within the world economic system and largely unequal and fragmented value chains tying the region to global capitalism. Equally importantly, it was pointed out, an endogenous critical mass of interpretive urban scholarship emerged with the aim of not only explaining outcomes confounding theories coming from the North, but also providing practical insights on how to redress the deficits of a thoroughly yet unevenly urbanized continent. Whilst such focus points to key avenues for further research, it also left some important issues out of the narrative. These need to be acknowledged before concluding.

A thorough understanding of Latin American urbanization must not be limited to the region's pathways since its metropolitan transformation and instead adopt a far-reaching historical scope. Rich literatures show that some settlements, and particularly in the Andean region larger cities such as Quito and Cuzco, date back millennia (Schaedel et al., 2011); the institutions, built environments and idiosyncrasies of older urban cores are sedimented over by centuries of European colonial rule led by the Spanish and Portuguese crowns (Almandoz, 2014); and the conflict-ridden consolidation of nation-states over the nineteenth century both determined the power invested in national capitals and is key to understanding contradictions between democratic discourses of equality and progress and the geographies of uneven development relegating numerous peripheral regions to long-term backwardness (Almandoz, 2006).

It is also necessary to centre research on the people whose bodies and identities have been directly exposed to the dynamics of dispossession and exclusion through urban development, both historically and in the present day. A growing body of research pays attention to the urbanisms of Indigenous peoples, Afro-descendants, and other historically relegated groups (see Chapters 13 and 15). Gender and sexuality are recognised as dimensions of exclusion in the urban arena but also as forms of difference that can animate transformative politics for cities otherwise (see Chapter 23). Valuable studies adopt an intersectional approach to understand how men and women negotiate, both individually and collectively, the differences and hierarchies that they embody (Ulloa, 2016). Such struggles and everyday tactics generate a welter of knowledges that political economy approaches cannot ignore (Oliveira, 2021).

In the face of a global climate crisis, the agenda for urbanization research includes emerging challenges concerning threatened built and natural environments alongside longstanding concerns with justice and inclusion (see Chapters 16 and 18). Urban and regional political economy has contributed important insights on Latin American geographies since the rise of the metropolis, pointing both to asymmetries of power with global cores and geographically uneven development within countries. Yet for this approach to remain relevant it must be open to understanding the multiple locations and typologies of urban growth and transformation that result and interact with the workings of capital. Continuing effort is required to meet the challenge of interdisciplinarity and integration of perspectives whereby urban political economy analysis synergizes with deep histories and cultural, embodied, and identity-based understandings of cities and regions.

Summary

- Over the past century, Latin American geographies have been transformed by large-scale urbanization processes.
- The rise of the modern metropolis in Latin America did not fit pathways identified in North America and Europe. The industrial base was smaller, more people were concentrated in fewer cities, and socio-spatial inequalities were starker.
- Neoliberal reforms since the late twentieth century have reshaped urbanization in Latin America. Cities and regions are now more exposed to global forces. Inequality and exclusion have deepened.

Review questions

1. Is it accurate to speak of Latin American urbanization? Or is there too much variation between and within countries to articulate general arguments about the process? How could we defend each position?
2. What were the distinctive spatial and social features within each period reviewed in the chapter? Can we also identify continuities?
3. Who are the key state, economic, and social actors shaping urbanization processes across Latin America? What are their interests and how are these negotiated and/or contested?

Further reading

Arboleda, M. (2020). *Planetary mine: Territories of extraction under late capitalism*. Verso Books.

A timely book that brings together debates on the changing shape of capitalism, extractive industries, and contemporary urbanization in Latin America.

Perlman, J. (2010). *Favela: Four decades of living on the edge in Rio de Janeiro*. Oxford University Press.

An update on seminal work on informal settlements and exclusion in metropolitan Brazil. It traces live stories over several decades and analyses shifts leaving some feeling most marginalized than ever.

Sá, L. (2014). *Life in the Megalopolis: Mexico city and São Paulo*. Routledge.

A multidisciplinary study of the region's two largest cities. Introduces their key dynamics and iconic places through a reading of fictional works, film, and performance art.

Keywords

Socio-spatial inequalities: the multiple inequalities that condition people's life chances in terms of not only their income but also physical access to employment, housing, and social services.

Urban infrastructure: the material bases of urbanization processes, including but not limited to surface transport and telecommunications, water and energy provision, and public open and green space.

Urbanisms: the ideas, images, practices, and actions that various actors such as governments, communities, and city-builders adopt to influence urbanization processes.

Urbanization: the demographic, physical, and economic transformation of places whereby people come to live in closer proximity to one another and diversify their activities.

References

Almandoz, A. (2006). Urban planning and historiography in Latin America. *Progress in Planning*, 65, 81–123.

Almandoz, A. (2014). *Modernization, urbanization and development in Latin America, 1900s-2000s*. Routledge.

Arboleda, M. (2016). Spaces of extraction, metropolitan explosions: Planetary urbanization and the commodity boom in Latin America. *International Journal of Urban and Regional Research*, 40(1), 96–112.

Becker, B. K. (1995). Undoing myths: The Amazon – an urbanized forest. In: Clüsener-Godt, M. & Sachs, I. (eds.), *Brazilian perspectives on sustainable development of the Amazon region*. Paris: UNESCO and Parthenon Publish Group Limited, 53–89.

Boano, C. & Vergara-Perucich, F. (2017). *Neoliberalism and urban development in Latin America*. Routledge.

Boschmann, E. E. (2020). Historical evolution and neoliberal urbanism in asunción. *Journal of Latin American Geography*, 19(4), 140–169.

Brenner, N. (2019). *New urban spaces: Urban theory and the scale question*. Oxford University Press.

Caldeira, T. (2000). *City of walls: Crime, segregation, and citizenship in São Paulo*. University of California Press.

Canclini, N. G. & Liffman, P. (2000). From national capital to global capital: Urban change in Mexico city. *Public Culture*, 12(1), 207–213.

Chase-Dunn, C. (1985). The coming of urban primacy in Latin America. *Comparative Urban Research*, 11(1–2), 14–31.

Connel, R. W. (2007). *Southern theory: Social science and the global dynamics of knowledge*. Polity.

Cypher, J. M. & Delgado Wise, R. (2010). *Mexico's economic dilemma: The developmental failure of neoliberalism*. Rowman & Littlefield Publishers.

Ferrari, S. G., Kaeshage, K., Narváez, S. C. D. & Bain, A. A. (2021). Adaptation strategies for people: Mitigating climate-change-related risks in low-income and informal urban communities through co-production. *Journal of the British Academy*, 9(s9), 7–37.

Gago, V. & Mezzadra, S. (2017). A critique of the extractive operations of capital: Toward an expanded concept of extractivism. *Rethinking Marxism*, 29(4), 574–591.

Garmany, J. & Richmond, M. A. (2020). Hygienisation, gentrification, and urban displacement in Brazil. *Antipode*, 52(1), 124–144.

Gorelik, A. (2017). Pan-American routes: A continental planning journey between reformism and the cultural Cold War. *Planning Perspectives*, 32(1), 47–66.

Hardoy, J., & Pandiella, G. (2009). Urban poverty and vulnerability to climate change in Latin America. *Environment and Urbanization*, 21(1), 203–224.

Irazábal, C. & Jirón, P. (2021). Latin American smart cities: Between worlding infatuation and crawling provincialising. *Urban Studies*, 58(3), 507–534.

Janoschka, M. & Sequera, J. (2016). Gentrification in Latin America: Addressing the politics and geographies of displacement. *Urban Geography*, 37(8), 1175–1194.

Kanai, J. M. (2011). Barrio resurgence in Buenos Aires: Local autonomy claims amid state-sponsored transnationalism. *Political Geography*, 30(4), 225–235.

Kanai, J. M. (2014a). Buenos Aires, capital of tango: Tourism, redevelopment and the cultural politics of neoliberal urbanism. *Urban Geography*, 35(8), 1111–1117.

Kanai, J. M. (2014b). On the peripheries of planetary urbanization: Globalizing Manaus and its expanding impact. *Environment and Planning D: Society and Space*, 32(6), 1071–1087.

Kanai, J. M. (2016). The pervasiveness of neoliberal territorial design: Cross-border infrastructure planning in South America since the introduction of IIRSA. *Geoforum*, 69, 160–170.

Kanai, J. M. & Schindler, S. (2022). Infrastructure-led development and the peri-urban question: Furthering crossover comparisons. *Urban Studies*, 59(8), 1597–1617.

Millington, N. (2018). Linear parks and the political ecologies of permeability: Environmental displacement in Sao Paulo, Brazil. *International Journal of Urban and Regional Research*, 42(5), 864–881.

Monte-Mor, R. L. (2014). Extended urbanization and settlement patterns in Brazil: An environmental approach. In: Brenner, N. (ed.), *Implosions/explosions: Towards a study of planetary urbanization*. Jovis, 109–120.

Moser, C. & McIlwaine, C. (2004). *Encounters with violence in Latin America: Urban poor perceptions from Columbia and Guatemala*. Psychology Press.

Ocampo, J. A. (2013). *The history and challenges of Latin American development*. Santiago: Economic Commission for Latin America and the Caribbean.

Oliveira, F. A. D. (2021). Who are the super-exploited? Gender, race, and the intersectional potentialities of dependency theory. In: Madariaga, A. & Palestini, S. (eds.), *Dependent capitalisms in contemporary Latin America and Europe*. Palgrave Macmillan, 101–128.

Perreault, T. (2006). From the Guerra Del Agua to the Guerra Del Gas: Resource governance, neoliberalism and popular protest in Bolivia. *Antipode*, 38(1), 150–172.

Perreault, T. (2018). Energy, extractivism and hydrocarbon geographies in contemporary Latin America. *Journal of Latin American Geography*, 17(3), 235–252.

Portes, A. & Roberts, B. R. (2005). The free-market city: Latin American urbanization in the years of the neoliberal experiment. *Studies in Comparative International Development*, 40, 43–82.

Roberts, B. (2020). *The making of citizens: Cities of peasants revisited*. Routledge.

Rodgers, D. (2012). Haussmannization in the tropics: Abject urbanism and infrastructural violence in Nicaragua. *Ethnography*, 13(4), 413–438.

Santos, M. (1977). Planning underdevelopment. *Antipode*, 9(3), 86–98.

Sassen, S. (2002). *The Global City: New York, London, Tokyo*. Princeton Univeristy Press.

Schaedel, R. P., Hardoy, J. E. & Scott-Kinzer, N. (eds.). (2011). *Urbanization in the Americas from its beginning to the present*. Walter de Gruyter.

Schindler, S. & Kanai, J. M. (2021). Getting the territory right: Infrastructure-led development and the re-emergence of spatial planning strategies. *Regional Studies*, 55(1), 40–51.

Schindler, S., Kanai, J. M. & Bay, J. D. (2023). Deindustrialisation and the politics of subordinate degrowth: The case of Greater Buenos Aires, Argentina. *Urban Studies*, 60(7), 1212–1230.

Schrank, A. (2008). Export processing zones in the Dominican Republic: Schools or stopgaps? *World Development*, 36(8), 1381–1397.

Silva, E. (2016). Patagonia, without Dams! Lessons of a David vs. Goliath campaign. *The Extractive Industries and Society*, 3(4), 947–957.

Ulloa, A. (2016). Territory feminism in Latin America: Defense of life against extractivism. *Nómadas*, 45, 123–139.

Uribe, S. (2019). Illegible infrastructures: Road building and the making of state-spaces in the Colombian Amazon. *Environment and Planning D: Society and Space*, 37(5), 886–904.

Valladares, L. D. P. (2019). *The Invention of the Favela*. University of North Carolina Press.

Vegliò, S. (2021). Postcolonizing planetary urbanization: Aníbal Quijano and an alternative genealogy of the urban. *International Journal of Urban and Regional Research*, 45(4), 663–678.

Wilson, J. & Bayón, M. (2017). Fantastical materializations: Interoceanic infrastructures in the Ecuadorian Amazon. *Environment and Planning D: Society and Space*, 35(5), 836–854.

Development and environmental geographies

CHAPTER 8

(Post)Neoliberalism

Jean Grugel and Pía Riggirozzi

Introduction

In the first two decades of the twenty-first century, the electoral pendulum in Latin America swung to the left, giving rise to a phenomenon known as the 'Pink Tide' or 'postneoliberalism'.[1] Most 'Pink Tide' governments sought to modify neoliberal approaches to growth that had taken root across the region in the 1980s and 1990s and to use the state to actively reduce socio-economic inequalities, though the reform impetus was less visible in Central America, where democracy and civil society were also much weaker than in South America (Cannon and Hume 2012; Grugel and Riggirozzi 2018). Instead of deregulation and privatisation, governments where the left was elected into office, especially in Argentina, Brazil, Bolivia, Ecuador and Uruguay, offered a mix of interventionism in the economy, an expansion of welfare, higher rates of taxation (especially over agricultural and mining exports), labour protection and social investments. Though electorally popular, at least initially, Pink Tide governments also provoked hostility, sometimes from business, sometimes from civil society movements. The wave of Pink Tide governments had largely been brought to a close by 2019, generally via the ballot box (for example in Argentina, Peru, Ecuador, El Salvador and Uruguay) but sometimes via a coup, as in in Honduras, or a 'soft' coup, as in Brazil, Bolivia and Paraguay, where there was no direct overthrow but leftist governments were forced from office (Farthing 2023). The legacies of this period continue to reverberate, since the right-wing governments that replaced the Pink Tide also proved unpopular, as inequalities rose once again. Consequently, the left began to return once again to office, for example in Chile in 2022 under Gabriel Boric, and in Brazil, where Lula, leftist President from 2003 to 2010, achieved a third presidential term in in 2023 in a closely run contest against incumbent Jair Bolsonaro. The ambitions of this new left, though influenced by the postneoliberal era, are, however, considerably more moderate than their predecessors. This chapter explores some of the main contours of postneoliberalism, indicating some important implications for the geographies of Latin America,

DOI: 10.4324/9781003430926-13

What was/is '(post)neoliberalism'?

Following democratisation in the 1980s and 1990s (**see Chapter 4**), Latin America adopted neoliberal prescriptions for economic growth in the midst of economic crisis. The shared script of deregulation, labour market reform, privatisation and trade liberalisation opened up regional economies and caused rapid and enormous social and economic hardship. In response, in the early years of the twenty-first century, leftist governments won elections in Venezuela (1998), Brazil (2002), Argentina (2003), Uruguay (2004), Bolivia (2005), Ecuador (2006), Nicaragua (2007) and, for shorter periods, in Paraguay (2008), El Salvador (2009) and Peru (2011) (**see Figure 8.1**). Of course, these governments were characterised by significant differences. Postneoliberalism reflected national political tensions, traditions and political economies. In Argentina, the postneoliberal Peronist governments of Nestor Kirchner and Cristina Fernandez de Kirchner, who were Presidents from 2003–2015, reflected long-standing traditions of Argentine populism (**see Chapter 4**), whilst in Venezuela, first under Hugo Chavez and later Nicolas Maduro, governance has been shaped above all by rentier politics, that is, extreme dependence of state institutions on external income derived from natural resources and, as such, the impact on the state of the rise and fall of the country's nationalised oil industry. Nevertheless, despite differences, all postneoliberal governments sought to revive elements of the 1950s/1960s idea of Latin American developmentalism, namely the idea that states should develop autonomy and capacity to promote growth and steer the country towards a mixed economy (**see Chapter 11**). As such, Therborn (2011) describes neodevelopmentalism as *sui generis* variants of welfare capitalism, while Roberts (2008: 217) views Pink Tide governments as largely 'social democratic' (excepting Venezuela) because they echoed European welfare state commitments to redistribute wealth and income to lower-income groups.

All postneoliberal governments – once again with the exception of Venezuela, which has to be considered as a case apart – introduced reforms to expand social citizenship, reduce poverty and restore dignity to the many citizens who had not prospered under

Figure 8.1 Leaders of the Pink Tide governments at a meeting in Argentina, 2007.
Source: Wikimedia.

marketized governance, often through the expansion of cash transfer schemes. There were also new approaches to housing and attempts to reduce the social and economic exclusion of disabled people in Ecuador, the expansion of childcare facilities in Uruguay and the introduction of the country's first old age pension scheme in Bolivia. Governments strengthened the role of the state in key economic activities by nationalising some public services or expanding investment in strategic industries, such as mining industries. Gudynas (2012) describes this spending as a form of 'compensation' to poor people rather than wholesale reform; despite investments, national economies remained highly dependent on external markets. In fact, however, the scale of income redistribution was far more extensive than merely 'compensation'. Feierherd et al. (2023) show that the social policies of postneoliberal governments were highly effective and that they 'lowered income inequalities' faster than other governments in the region, with the very poorest experiencing the biggest income improvements.

Finally, all postneoliberal governments sought to re-make the region of South and Central America in the direction of more autonomous foreign policy making. There was an attempt to build regional relationships in ways that would counter the influence of the United States and its Free Trade Area of the Americas (FTAA) project. The emergence of new regional bodies such as the Bolivarian Alliance of the Americas (ALBA) in 2004, the Union of South American Nations (UNASUR) in 2008 and the Community of Latin American and Caribbean States (CELAC) in 2011 are all manifestations of this trend (Riggirozzi and Tussie 2012). As Riggirozzi and Tussie (2023: 13) argue, the region became a political space to try and rework the normative frameworks and practices of regional and national governance.

Why is postneoliberalism controversial?

If postneoliberal governments reduced poverty, why was the postneoliberal period contentious and controversial? Three important critiques can be made: the ecological critique, the democratic critique and the not-radical-enough critique. Let's start with the latter. Postneoliberalism represents a break with the dominant neoliberal model of political economy by rejecting the inevitability of inequality and the view that inequality is functional for a dynamic economy. But postneoliberal governments in practice offered only relatively moderate reforms, always within the context of capitalism. All governments maintained some core aspects of the Washington Consensus, including openness to foreign investment and a focus on growth through the region's comparative advantage in mining and natural resources. This led to the criticism that these governments were not, in fact, 'post' neoliberalism but were simply another variant of neoliberalism. Yates and Bakker (2014) suggest that they should be regarded instead as examples of 'variegated neoliberalism'.

Secondly, there is the critique that postneoliberalism damaged regional democracy. All postneoliberal governments relied on strong executives, sometimes called hyper-presidentialism, and some critics took the view that this represented a populist challenge to still-fragile regional democracies (de la Torre and Peruzzoti 2018). Certainly some postneoliberal governments sometimes pushed at the democratic boundaries, as did their political opponents. For example, in Honduras, even moderate reforms met with the overthrow of left-wing President Manuel Zelaya in 2009, whilst in Bolivia, the long-time leader of the left, Evo Morales, was overthrown in a palace coup in 2019, after thirteen years in the Presidency a reflection of what Dunkerley (2007) refers to as the existence of 'two Bolivias', or a country so divided that political crisis is almost inevitable. Postneoliberal governments did not only face hostility from right-wing critics, however. At times, they also incurred the wrath of social movements and communities who felt left out, for whom the pace of change was slow or who lost land and livelihoods through the expansion of the extractive frontier on which postneoliberal governments came to depend. Civil society critiques of postneoliberal governments were particularly marked in Bolivia and Ecuador (see **Chapter 10**). Insofar as respect for human rights and civil liberties were concerned, postneoliberal governments (with the usual exception of Venezuela) generally performed no worse than their predecessors, and occasionally better than them, though greater attention was paid to reducing socio-economic inequalities than in addressing identity-based or gender-based forms of discrimination (Fontana and Grugel, 2018).

The ecological critique that can be made against postneoliberalism resonates most strongly. Postneoliberalism did not seek to challenge the 'commodity imaginary' that has been dominant in the region since its colonisation (Giraudo and Grugel 2022). Instead, governments took decisions to continue to rely economically on mining, natural resources and agricultural exports, which intensified the trend of devastating and destroying environments and habitats across the region, whilst at the same time failing to halt the long-established trend towards de-industrialisation (Bresser-Pereira 2011). Postneoliberal governments neglected the need to transform economies in ways that would produce stability, decent work and environmentally sustainable growth over the longer term (see **Chapter 17**).

The role of extractivism

The social spending that characterised postneoliberalism means that governments had to increase revenue and taxation. To do so, governments chose not to remake the economy root-and-branch, but to work within the political economies of extraction inherited from the past, expanding mining, agricultural production and foreign ownership of land. Acosta and Machado (2012) refer to this as 'neoextractivism'. Dependence on extractive industries also accentuated the political authority of extractive companies and national elites tied to agriculture and mining.

Postneoliberalism thus came to depend on a 'commodity consensus' (Svampa 2015) or the large-scale expansion of natural resources exports, including gas, oil, new or 'critical' minerals such as lithium, and primary products including foodstuffs such as soy beans (Purdy and Castillo 2022). World Bank data suggest that exports as a percentage of GDP rose from 13.4% in 1960 to 25.5% in 2005, driven by a combination of expansion of production and rising prices. In Argentina, for example, there was the expansion of soy bean production, which came to represent 16% of global production and over 6% of all soy bean exports by 2016 (World Bank 2013).

The tensions that were generated by the rapid expansion of extractive and agricultural frontiers were evident across the region, perhaps particularly in Brazil, which under Lula became the second-largest exporter of food and raw materials in the world (see **Chapter 22**). By 2015, Brazil was exporting more soy, beef, coffee, poultry and sugar than any other country in the world. Export taxes provided the fiscal resources to take millions of Brazilians out of poverty, but the policies also led directly to deforestation, environmental destruction and the violent land dispossession of peasant and Indigenous communities. Eventually, they brought the government into a *de facto* alliance with agricultural business groups. Under Dilma Rousseff, the left-wing President who replaced Lula in 2011, the key government department of Agriculture, Livestock, and Provision, fell under the control of the powerful Confederation for Agriculture and Livestock Production (Giraudo, 2021) and the government even pardoned the fines that the state environmental protection agency had tried to impose on large landowners for destruction of land.

Social inclusion and economic citizenship

The victory of the postneoliberal left drew on popular demands for inclusionary rather than commodified neoliberal citizenship. This was understood chiefly in terms of the need to remedy inequality, informality and poverty. In 2002, 44% of the total population of the region were poor. Social expenditures rose in all Pink Tide countries (most especially in Argentina, Uruguay and Brazil) between 2004 and 2015 along with spending on primary education, health and pensions. There were rapid falls in infant and maternal mortality, wages increased, and security of employment improved – with the partial exception, in Andean countries especially, of the extractive industries where labour conflicts were tightly regulated (Santilla Ortiz and Webber, 2015). Cash transfer programmes were extended to include families in rural areas, with school-age children, pregnant women and women with care responsibilities for disabled people. The number of individuals living in recipient households increased across the region from fewer than 1 million in 1996 to 131.8 million in 2015, or 20.9% of the total population (**see** Figure 8.2). Poverty and income inequality fell (**see** Figure 8.3), most dramatically in Argentina, Uruguay and Brazil. As well as social expenditure and welfare, postneoliberal governments reduced the heavy burden of indirect taxes on consumption (Barlow and Peña 2022).

The limits of inclusion: neglecting the 'entangled' inequalities of gender, race and ethnicity

Addressing socio-economic inequalities in Latin America is absolutely critical to produce fairer societies, but there are other drivers of inequality that postneoliberal governments were less effective in reducing, in particular the 'entangled inequalities' (Costa 2018) of land, racial and ethnic biases and manifestations of patriarchal power. These 'entangled inequalities' meant that the constitutional reforms characteristic of many postneoliberal governments, especially in the Andes, though important, had only limited traction. Policies to address ethnic and racial inequalities were always partial. Indigenous land titling, which is central to the principle of recognition and reparation with Indigenous and peasant communities, was slow and partial, and resource governance strategies, even in Bolivia and Ecuador, excluded Indigenous communities from

Figure 8.2 Latin America and the Caribbean: Population of households participating in CCT programs, 1996–2016.

Source: Cecchini and Atuesta (2017: 22).

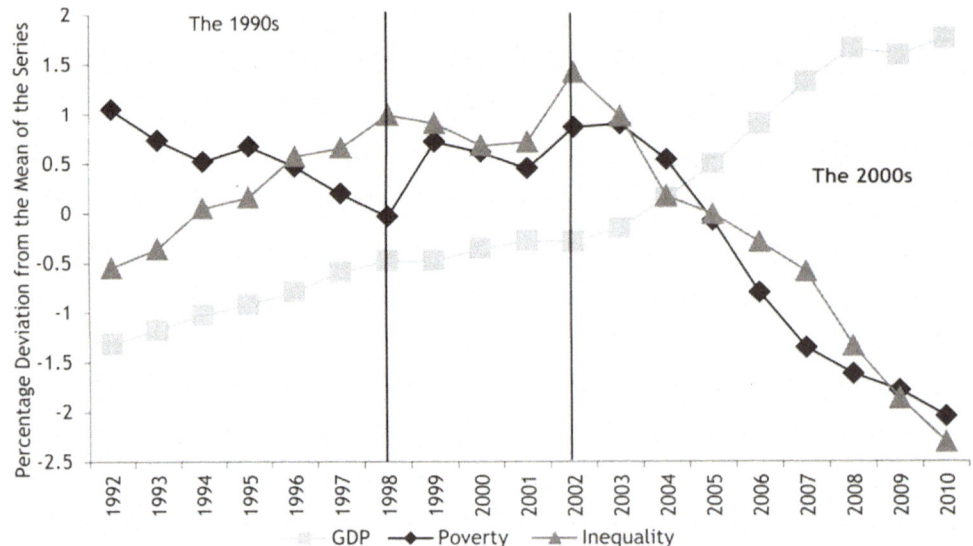

Figure 8.3 Poverty, inequality and per capita GDP in Latin America, 1992–2010.

Source: Gasparini and Cruces (2013: 54).

decision making, despite the promises of plurinationalism and *Buen Vivir* (Radhuber and Radcliffe 2023) (**see Chapter 13**). In Brazil, meanwhile, where Indigenous communities own around 25% of the Amazon basin, both Lula in his first two presidencies and Dilma Rousseff failed to prevent incursions on Indigenous land, including logging, agriculture, mining and hydrocarbon and infrastructure projects, whilst moving slowly on land titling of Indigenous land, thereby contributing to deforestation and threatening the very existence of Indigenous communities (Bebbington et al. 2018). Ultimately, the search for growth, modernity and the capture of rents from the Amazon meant that, even under the left, Brazil's Indigenous peoples were positioned as barriers to growth. Across Latin America as a whole, Breton et al. (2022) estimate that 2.1 million hectares, or a third of all land purchased 2000 and 2012 alone, passed into the hands of foreign companies, mainly for agricultural or mining exploitation. These failures were not only due to governments' fiscal dependence on exports or extractivism. More fundamentally, they are rooted in the persistence of racist and colonial ideas of who has worth and whose knowledge matters.

Additionally, there was limited progress regarding women's equality. Although some governments discursively acknowledged the need for gender reform, change was slow. Blofield et al. (2017: 345) show how Pink Tide governments did not develop 'clearly articulated gender equality initiatives' and improvements in the social and economic rights of poor women, especially poor Indigenous women, lagged considerably behind those of men (Fontana and Grugel 2016). The socio-economic wellbeing of poor women improved as a result of cash transfers, but opportunities for women to find decent work improved only marginally. Governments did not legislate to recognise women's reproductive rights (with the noticeable exception of Uruguay) and they generally failed to challenge or reform the restrictive reproductive laws and policies, particularly with regard to abortion, that shape and limit the everyday lives of women and girls cross the region. Wilkinson (2019) goes as far as describing Ecuador under Rafael Correa as 'a lost decade for women' while Friedman (2019) notes the contrast between the recognition of LGBTIQ+ rights in Argentina under the government of Nestor Kirchner and government refusal to legalise abortion, which was only decriminalised in 2021 (Riggirozzi and Grugel 2021). Pink Tide governments also ignored the impact of the care burden imposed on women and girls, especially poor women and girls, and the low wages and lack of regulation poor women who worked as paid carers, cleaners and maids. In fact, the reliance of postneoliberal governments on cash transfer programmes to address socio-economic inequality tended to intensify women's social reproductive role. Only in Chile was there a major expansion of state provision of childcare in order to enable women to take on formal employment under Michelle Bachelet (President 2006–2010, 2014–2018). As a consequence, women and girls were left exceptionally vulnerable to violence, impoverishment and insecurity during the COVID-19 crisis, just a few years after most Pink Tide governments had left office (World Bank 2021; UN Women 2020).

Conclusion: resurgence or accommodation? The post-postneoliberal left

The transformations of the state, society and the economy promised by postneoliberalism were partial and once in office, postneoliberal governments were pragmatic, contradictory and sometimes concentrated power narrowly. At the same time, there was an accommodation, not a break, with neoliberalism. Even the expansion of social and economic citizenship was tied to the cash transfer model, but there was a power nonetheless in its critique of neoliberalism and the hope it represented for a different kind of development.

In recent years, in particular since 2018, there has been something of a resurgence of a left as part of the cycle of politics. However, the left after 2018 is not really a second coming of postneoliberalism. While leftist leaders won the presidency in Mexico and Argentina in 2019, Bolivia and Peru in 2020, Chile in 2021, Colombia in 2022 and Brazil in 2023, these governments are less ambitious about redistribution and the economic climate has been more challenging. On the one hand, there have been efforts at regional cooperation in support of more heterodox political economies, including efforts by Brazil and Argentina to launch a regional currency as a counterbalance to the dollar. On the other, there have been fewer socio-economic initiatives. Low productivity, weak growth, limited fiscal capacity and rising inflation restrict how far governments can afford to extend rights and citizenship (ECLAC/ILO 2020). At the same time, the region faces an unprecedented humanitarian crisis as a result of mass forced migration from Venezuela and Central America to other countries in the region (UNHCR 2022) (**see Chapter 9**). Offsetting these challenges is proving exceptionally difficult. Nevertheless, other rights, which cost less financially, are being recognised, including identity rights which are beginning to rise up the agenda across the region, at least partially. There is now a marked trend, for example, to recognition of the rights LGBTQ+ communities.

On balance, then, there is a greater accommodation with neoliberalism than radical reform or further constitutional changes. Chile, where President Gabriel Boric first emerged as a student leader in the 2011 protest movements against neoliberalism, is unlikely to endorse dramatic changes to the Chilean model of development. His capacity for moving left or introducing measures to protect Indigenous communities and limit the expansion of foreign ownership of Indigenous land was severely limited by the electorate's rejection of a new constitution in a referendum in September 2022. And Lula, who was re-elected in Brazil in 2022 with the smallest of margins, has proposed a relatively limited social reform agenda, although he has endorsed the need for more protection of the environment and greater protection for Indigenous communities. Indeed, the most significant promise he made in the presidential campaign was to halt deforestation and protect the Amazon from further exploitation, echoing that of Colombia's new president to freeze new oil and gas exploration (Garcia 2022). This certainly provides the basis for a greater recognition of the rights of Indigenous communities, although delivering these promises is not proving easy. A major lesson from past postneoliberal experiences is that any model of growth and social development that relies on the extension of income derived the exploitation of natural resources for social programs will produce a destructive 'battle of rights' and generate tensions between advancing socioeconomic claims versus ecological rights (Riggirozzi 2020), with postneoliberal governance becoming a divisive and truncated project of change based on trade-offs between the poor and nature. Learning the lessons from postneoliberalism and finding a route to equitable and sustainable development and a just and inclusive society in Latin America remains a challenge.

Summary

- Pink Tide governments won widespread support from electorates across Latin America in the early years of the twenty-first century after widespread protests against social inequality, austerity measures and the privatisation of public resources, in response to decades of neoliberalism.

- Postneoliberalism put in place interventionalist states, expanded welfare systems, and social investment as a response to the (socio-economic) inequalities that resulted from neoliberal policies in the 1990s.

- Postneoliberalism cemented agriculture, mining and extractivism at the heart of national development strategies, but failed to halt the established trend towards de-industrialisation, and neglected the need to transform economies in ways that would produce stability, decent work and environmentally sustainable growth over the longer term.

- Postneoliberal governments performed better in addressing socio-economic inequalities than in addressing identity-based discrimination.

- Postneoliberalism did not deal well with the need to strengthen democratic governance and relied on hyper-presidential forms of governance.

- The tension between the rights of people to human development and the rights of nature were not resolved.

- Despite the mixed record of postneoliberal governance, the left remains vibrant in Latin America, but the path to sustainable and equitable development and how to deal with the legacies of colonialism remain challenging.

Review questions

1. How far did the Latin American left establish an alternative model of development, going beyond neoliberalism? Reflect on whether – and if so, how – postneoliberalism comprised a change in the norms of development and democracy in the region.
2. Why was postneoliberalism so contentious?
3. Did postneoliberal governments perform better in addressing socio-economic inequalities than in tackling identity-based inequalities?
4. What explains the failure of postneoliberal governments to challenge extractivism in the region?

Further reading

Grugel, J., and P. Riggirozzi (2012) 'Post Neoliberalism: Rebuilding and Reclaiming the State in Latin America', *Development and Change*, 43(1), 1–21.
Outlines and provides an overview of the post neoliberal debate and assesses government strategies and social responses to neoliberalism that led to the emergence of postneoliberal projects across Latin America

Nehring, D., G. Gómez Michel, and M. López (eds) (2019) *A Post-Neoliberal Era in Latin America? Revisiting Cultural Paradigms*. Bristol University Press.

Explores cultural forms, such as literature, underground cinema and street fairs that support social and political relations and expressions of anti-neoliberal resistance and the contribution of these to postneoliberalism

Ruckert, A., L. Macdonald, and K. R. Proulx (2017) 'Postneoliberalism in Latin America: A Conceptual Review', *Third World Quarterly*, 38(7), 1583–1602.

Offers a critical literature review of the concept and models of postneoliberalism across Latin America

Siegel, K. M. (2016) 'Fulfilling Promises of More Substantive Democracy? Post-Neoliberalism and Natural Resource Governance in South America', *Development and Change*, 47(3), 495–516.

Examines the relationship between development and democracy, focusing on South America, arguing that reliance on neo-extractivism acts as the main development strategy in post-neoliberal democracies

Keywords

Neoliberalism: development model that gives primacy to the free market, privatisation and a reduction of the welfare state.

Pink Tide: Anglophone catch-all term for progressive governments elected across Latin American in the early twenty-first century.

Postneoliberalism: an aspiration to break with neoliberal development models.

Note

1 In this chapter, we tend to use 'postneoliberalism' and 'Pink Tide' interchangeably. Along with 'neodevelopmentalism', both terms seek to capture the attempt to break with 'neoliberal' development strategies. In fact, the term 'postneoliberalism' captures an aspiration rather than a reality. Some scholars have suggested that neoliberalism continued almost without modification whilst others suggest that the left failed to establish a truly developmental/neodevelopmental state in the sense of building autonomy to shape the national development process, since they remained dependent on primary commodity exports.

References

Acosta, A., and D. Machado (2012) 'Ambientalismos y conflictos actuales en América Latina', *Revista OSAL (Observatorio Social de América Latina – CLACSO)* – 32, ISSN 1515–3282. http://biblioteca.clacso.edu.ar/clacso/osal/20120927103642/OSAL32.pdf

Barlow, M., and A. Peña (2022) 'The Politics of Fiscal Legitimacy in Developmental States: Emergency Taxes in Argentina Under Kirchnerism', *New Political Economy*, 27(3), 403–425.

Bebbington, D. H., R. Verdum, C. Gamboa, and A. J. Bebbington (2018). 'The Infrastructure-Extractives-Resource Governance Complex in the Pan-Amazon: Roll Backs and Contestations', *European Review of Latin American and Caribbean Studies/Revista Europea de Estudios Latinoamericanos y Del Caribe*, 106(2018), 183–208.

Blofield, M., C. Ewig, and J. M. Piscopo (2017, Winter) 'The Reactive Left: Gender Equality and the Latin American Pink Tide', *Social Politics*, 345–336.

Bresser-Pereira, L. (2011) 'An Account of New Developmentalism and Its Structuralist Macro-economics', *Revista de economia política*, 31(3), 493–502.

Breton, V., M. González, B. Rubio, and L. Vergara-Camus (2022) 'Peasant and Indigenous Autonomy Before and After the Pink Tide in Latin America', *Journal of Agrarian Change*, 22(1), 547–575.

Cannon, B., and M. Hume (2012) 'Central America, Civil Society and the 'Pink Tide': Democratization or De-democratization?', *Democratization*, (19), 6.

Cecchini, S., and B. Atuesta (2017) *Programas De Transferencias Condicionadas en América Latina y el Caribe: Tendencias de Cobertura e Inversión*. CEPAL.

Costa, S. (2018) 'Entangled Inequalities, State, and Social Policies in Contemporary Brazil', in Ystanes, M., and Strønen, I. Å. (eds), *The Social Life of Economic Inequalities in Contemporary Latin America*, 59–80. Palgrave Macmillan.

de la Torre, C., and E. Peruzzoti (2018) 'Populism in Power: Between Inclusion and Autocracy', *Populism*, 1(1), 38–59.

Dunkerley, J. (2007) 'Evo Morales, the 'Two Bolivias' and the Third Bolivian Revolution', *Journal of Latin American Studies*, 39(1), 133–166.

ECLAC/International Labour Organization (ILO) (2020) 'Work in Times of Pandemic: The Challenges of the Coronavirus Disease (COVID-19)', Employment Situation in Latin America and the Caribbean, No. 22 (LC/TS.2020/46), Santiago.

Farthing, L. (2023) 'Latin America's New Left Surge', *NACLA*. https://nacla.org/latin-america-new-left-surge

Feierherd, G., P. Larroulet, W. Long, and N. Lustig (2023) 'The Pink Tide and Income Inequality in Latin America', *Latin American Politics and Society*, 1–35. https://doi.org/10.1017/lap.2022.47

Fontana, L., and J. Grugel (2016) 'The Politics of Indigenous Participation through Free Prior Informed Consent: Evidence from Bolivia', *World Development*, 77(1), 249–261.

Friedman, E. (ed) (2019) *Seeking Rights from the Left; Gender, Sexuality, and the Latin American Pink Tide*. Duke University Press.

Garcia, D. (2022) 'How the New Latin America Left is Seeking a Greener Future', *Reuters*. www.reuters.com/world/americas/how-new-latin-america-left-is-seeking-greener-future-2022-05-16/

Gasparini, L., and G. Cruces (2013) 'Poverty and Inequality in Latin America: A Story of Two Decades', *Journal of International Affairs*, 66(2), 51–63.

Giraudo, M. E. (2021) 'Taxing 'the Crop of the Century' The Role of INstitutions in Governing the Soy Boom in South America', *Globalizations*, 18(4), 516–532. https://doi.org/10.1080/14747731.2020.1795426

Giraudo, M. E., and J. Grugel (2022) 'Imaginaries of Soy and the Costs of Commodity-Led Development: Reflections from Argentina', *Development and Change*, 53(4), 796–826. https://doi.org/10.1111/dech.12714

Grugel, J., and P. Riggirozzi (2018) 'New Directions in Welfare: Rights-Based Social Policies in Post-Neoliberal Latin America', *Third World Quarterly*, 39(3), 527–543.

Gudynas, E. (2012) 'Estado Compensador y Nuevos Extractivismos', *Nueva Sociedad*, (237), 128–146.

Purdy, C., and R. Castillo (2022) 'The Future of Mining in Latin America: Critical Minerals and the Global Energy Transition', *R4D Report*. https://r4d.org/resources/the-future-of-mining-in-latin-america/

Radhuber, I. M., and S. A. Radcliffe (2023) 'Contested Sovereignties: Indigenous Disputes Over Plurinational Resource Governance', *Environment and Planning E: Nature and Space*, 6(1), 556–577. https://doi.org/10.1177/25148486211068476

Riggirozzi, P. (2020) 'Social Policy, Inequalities and the Battle of Rights in Latin America', *Development and Change*, 51, 506–522. https://doi.org/10.1111/dech.12571

Riggirozzi. P., and J. Grugel (2021, January 27) 'Argentina's Legalisation of Abortion is Only the Beginning of the Battle for Reproductive Rights in Latin America', *LSE Blogs*. https:// blogs.lse.ac.uk/latamcaribbean/2021/01/27/argentinas-legalisation-of-abortion-is-only-the-beginning-of-the-battle-for-reproductive-rights-in-latin-america

Riggirozzi, P., and D. Tussie (2012) *The Rise of Post-Hegemonic Regionalism in Latin America.* Springer.

Riggirozzi, P., and D. Tussie (2023) *Post-Hegemonic Regionalism. Oxford Research Encyclopaedia of International Studies*, 29. Oxford University Press. https://oxfordre.com/ internationalstudies/view/10.1093/acrefore/9780190846626.001.0001/acrefore-9780190846626-e-657

Roberts, K. (2008, September–October) '¿Es Posible una Socialdemocracia en América Latina?', *Nueva Sociedad*, 217, 86–98.

Santilla Ortiz, A., and J. R. Webber (2015, August 14) 'Cracks in Correrismo?', *Jacobin*. www. jacobinmag.com/2015/08/correa-ecuador-pink-tide-protests-general-strike/

Svampa, M. (2015) 'Commodities Consensus: Neoextractivism and Enclosure of the Commons in Latin America', *South Atlantic Quarterly*, 14(1), 65–82.

Therborn, G. (2011) *Inequalities and Latin America: From the Enlightenment to the Twenty First Century.* https://refubium.fu-berlin.de/bitstream/handle/fub188/19695/1_WP_Therborn_Online.pdf?sequence=1

UNHCR (2022) *Global Trends in Forced Displacement in 2021.* Geneva, Switzerland: UNHCR. www.unhcr.org/62a9d1494/global-trends-report-2021

UN Women (2020) 'The Gendered Impacts of COVID-19 on Labor Markets in Latin America and the Caribbean', *Report.* https://lac.unwomen.org/en/noticias-y-eventos/articulos/2020/11/ impacto-cconomico-covid-19-mujeres-america-latina-y-el-caribe

Wilkinson, A. (2019) 'Ecuador's Citizen Revolution 2007–17: A Lost Decade for Women's Rights and Gender Equality', in Friedman, E. J. (ed), *Seeking Rights from the Left: Gender, Sexuality, and the Latin American Pink Tide.* Duke University Press.

World Bank (2013) *Exports of Goods and Services, Online Data.* https://data.worldbank.org/ indicator/NE.EXP.GNFS.ZS

World Bank (2021, May 5) *The Economic Impact of COVID-19 on Women in Latin America and the Caribbean. Results Brief.* www.worldbank.org/en/results/2021/05/05/ the-gendered-impacts-of-covid-19-on-labor-markets-in-latin-america-and-the-caribbean

Yates, J., and K. Bakker (2014) 'Debating the 'Post-Neoliberal Turn' in Latin America', *Progress in Human Geography*, 38(1), 62–90.

CHAPTER 9

International migration and displacement

Marcia A. Vera Espinoza and Vania Reyes Muñoz

Introduction

Migration, the movement of individuals or groups from their usual place of residence across an international border or within a country, has been central to Latin American geography and also to its socio-economic dynamics and political debates. While there are different types of mobility encompassed within the dynamics of migration, in this chapter we will focus on international migration. From transoceanic migrations and the transatlantic slave trade that shaped the period of colonisation (in Latin America and the Caribbean), to the displacement of over seven million Venezuelans since 2014 to date, and many other mobilities in between, migration has been a key feature in the development of the region.

International migration has been analysed across different dimensions, including the reasons for the move, the distance people move and the duration of it (McDowell and Sharp, 1999). In many cases, the analysis has not been able to contextualise and historicise contemporary migrations. This has been evidenced, more broadly, in the increasing political salience of migration and the way that has been portrayed in media and political discourses in the United States and Europe, but also in Latin America. Mitchell et al. (2019) have stated that critical geography has contributed to mapping out the geography of current migration and to situating it in the context of larger geopolitical dynamics shaped by specific geoeconomics. Central to this debate is also the analysis of how the management of migration 'to contain people on the move' provides a necessary context for the understanding of mobility, but also of migrants' experiences and subjectivities (Weima and Hyndman, 2019, 31; see also Vera Espinoza et al., 2021).

The study of international migration has also traditionally shown a geographical narrowness, as its focus has been predominantly South to North and characterised by an alarmist approach (i.e. the myth of invasion, as Hein de Hass has called it). However, we know that international migration goes in all directions, with people also moving North-South and also South-South. The latter has received increasing attention in the last few years with a growing body of scholarship published in Anglophone geography, but also in Spanish and in Portuguese (Crawley and Teye, 2024; see also the Revista Interdisciplinar da Mobilidade Humana – REMHU).

Latin American countries are not only important within South-South migration debates as a region of emigration. The countries of the region are also sites of destination, transit, return and further mobility (Jubilut et al., 2021). Within these dynamics there are countless challenges for migrants, including refugees, and also for the societies in receiving and transit places (at the local and national level). Migration is complex, normal, divisive, collective and intimate, with levels of analysis that go from the embodied experiences of migration to the global management of it. This chapter aims to look at the intersection of the migrants' experiences and the governance structures that determine their (im)mobility. By doing so, we map some of the geographies of migration in Latin America that include transnational spatialities, power negotiations and processes of resistance.

The chapter begins with a brief overview of migration patterns in the region, then identifies some key characteristics of migration in Latin America, such as the feminisation of migration and the transnational practices of care. We then move to discuss bordering practices, both within and outside the nation-states, and the extent to which migrants, including refugees, challenge and resist migration management.

Migration and displacement flows in Latin America

Since 2015, there has been an increasing salience of international migration, both in political discourses and in media coverage across the globe. This has been more evident in relation to the so called 'migration crisis' in Europe and to the former US president Donald Trump's attempts to build a wall along the United States' southern border. Since then, other 'crises' of migration and displacement continue to take up media and political attention, with a particular focus on migrants going to the Global North. Within these debates, migration patterns in Latin America have received limited attention internationally, except when coupled with the images of migrant caravans heading towards the US. However, within Latin American countries, migration has also become a daily feature of national and regional debates. These debates are mostly framed around discourses of 'crisis' and 'exceptionality' (Vera Espinoza, 2024), which obscure the fact that migration has been intrinsically linked to territorial formation (see Chapter 3) and the development of the region.

Literature exploring migration dynamics in Latin America and the Caribbean has broadly identified four broad phases of migration in the continent (see Aruj, 2008; Cerrutti and Parrado, 2015). We expand on these four, to broadly speak of seven moments or phases of migratory flows. The first is marked by transoceanic migrations and the transatlantic slave trade. It is estimated that the transatlantic slave trade exceeded 12 million people between 1500 and 1865, when the enslavement of people was banned in several colonies (see http://slavevoyages.org). The second one is linked to internal migration during the struggles for independence of Latin American colonies in the early 1800s (see Chapter 2). The third phase considers Latin America's immigration from the late nineteenth century to 1930, a period in which around 13 million Europeans went to the region. The same period is also known for registering the first wave of Arab immigrants to Latin America. The fourth phase is linked to the economic crises and internal migrations in countries driven by the urban growth model – farm to city – in the middle of the last century (1930–1960) (see Chapter 7). A fifth moment relates to border migrations with neighbouring countries and intra-regional migration, particularly during the period of military dictatorships in the 1960s and 1970s. The sixth phase is linked to the promotion of global market economies and the circulation of capital and labour, including South-North and South-South flows (2000s) (see Chapter 6). We seem to now be in a seventh stage of migration, where the idea of capital movement is liberalised and labour circulation is severely restricted by the securitisation of borders and visa regimes.

Contemporary trends in migration, since the middle of the last century (1960–1970), saw Argentina and Venezuela as the main destination countries, due to their industrial boom. The oil industry in Venezuela attracted border migrants from Colombia and the Andean sub-region (Peru, Bolivia and Chile). Latin America then witnessed the large-scale displacement caused by the military dictatorships of the 1970s as well as the mobility flows following the return to democracy in countries like Chile (see Chapter 4), and the economic growth of Brazil. Cerrutti and Parrado (2015, 400) point out that in the first decade of the 2000s, more than three million South Americans lived in a South American country other than the country in which they were born. If we add to this the displacement of the Venezuelan population in the last ten years, the figure triples.

In the Caribbean, there are two main trends. The first, and most prominent, is emigration outside the region, mainly to North America (with the United States as the main destination country) and Europe. The second is intraregional migration, characterized largely by migration from Haiti to the Dominican Republic, and to a lesser extent by emigration to other island territories and countries in the region, including the Bahamas (Mcauliffe and Oucho, 2024).

The reasons for both intra- and extra-regional migration are complex and varied. A key perspective is understanding migration as a form of social hope (Kleist and Thorsen, 2016). Social hope, which relates to the simultaneous potential and uncertainty in anticipation for something better, has been used to problematise migration within the region and from the South, including the migration of Africans to Latin America (Jung, 2021). This implies that due to the null possibility of societal hope at the countries of origin, those who migrate direct their hopes to the market. Any market, any country that appears to have a market that absorbs migrant labour, emerges then as an alternative.

It is also important to consider that in recent years Latin America has emerged as an attractive destination when opportunities to enter Europe and the United States are reduced. In recent decades, the literature has also focused on individual decisions, social networks, family trajectories and economic and political instability as key drivers of migration. There are also forced displacements, driven by internal conflict, political persecution and the persistence of extractivist logics in the region. Climate migration cannot be thought of separately from ongoing climate change and power asymmetries in the production of unequal economies. Latin American countries, together with the countries of the so-called Global South, have fewer resources to adapt to and mitigate the effects of climate change. Moreover, the absence of food security for rural, Indigenous and Afro-descendant communities (see Chapter 15) have also been a reason for the migration of youth and young adults. Migration that has happened internally at first is now a continuous movement for better living conditions across borders and throughout the region.

Key characteristics of migration in Latin America

Migration in Latin America has key multidimensional characteristics. Here, we focus on two of them: the feminisation of migration and care chains. In other words, who is in charge of social reproduction in times of global mobility? The first one implies the increase of women in the composition of the migration flows, which also suggests their employment in low-paying and less socially prestigious jobs. The second refers to the management of care that falls to women, due to their gender role, and to the provision of care in countries of origin once the caregivers have migrated to provide care in the countries of destination.

We draw here from our own research in Chile to show these dynamics. Let's consider the experience of Ana, a young Peruvian woman who came to Chile to work as a teacher, a degree she obtained in her country of origin. While in Santiago, the difficulties in finding work in her profession and the high cost of rent led her to take a job as a domestic worker living in the house where she was employed. Working as a 'nanny' meant a loss of fixed working hours and the expectation of full-time availability to meet the demands of the care imposed by the employer. In her job Ana prepares meals, looks after the children, plays with them, takes them to the park, does the shopping, looks after the grandparents, gives them their medicines, and does the daily cleaning. She has one day off a week, where she goes to the city centre to meet other Peruvians.

The case of Chile is emblematic with respect to Peruvian women doing domestic work at their destination, regardless of their professional qualifications. At the beginning of the millennium, there was an evident feminisation of Peruvian migration (65% of the migratory flow), unlike in other neighbouring countries where the percentage of men and women was similar (Stefoni, 2002). Peruvian women occupied the jobs that the local labour force rejected. These jobs were mainly associated with care and cleaning in private homes in high-income sectors. The feminisation of migration also accounts for the assignment of specific tasks to particular ethnic communities across the region, including Nicaraguan women in Costa Rica (Lerussi, 2008) or Bolivian women in Argentina (Magliano, 2007). The contractual relationship of women in temporary jobs, both in domestic services and in agribusiness, is based on the naturalisation of female skills or abilities.

The feminisation of migration in the last three decades, based on the incorporation of women from the South as a new labour force in global production regimes (Sassen, 2003), has emphasised the idea of cross-border dynamics where the receiving and sending countries earn foreign currency at the expense of the precariousness of irregular/regular labour in border areas. The growing female presence in these territories cannot be considered as a particular individual experience, but as a gender dynamic (ibid.). In the case of Latin America, it is argued that the feminisation of migration patterns in the last decade is part of a family strategy of economic survival. This strategy would be based on the geographical proximity of the countries of the South, which allows stable family ties to be maintained in a continuous transit determined by the temporality of the economic work performed (Mora, 2008).

While the increasing feminisation of migration entails a positive articulation with respect to economic transformations that could lead to emancipatory actions by women, it also deepens intergenerational and gender asymmetries that affect family and care relations (Zapata Martínez, 2016). That means that when one woman migrates, other women, generally older adults or girls and adolescents, have to take care of the family group that remained in the place of destination. This is understood under the concept of care chains, which problematises female migrant labour participation in the modern model of care. Global care chains refer to the transfer of care that involves links between households in the country of origin and in the receiving country (Cerrutti and Maguid, 2010, 13). The dual role of women in domestic reproductive and public productive tasks generates a space in which care tasks are mostly solved by resorting to other women, instead of the incorporation of men into the private reproductive sphere.

Migration governance and the production of (im)mobility

The control of mobility is probably one of the most visible aspects of international migration. Every time we travel, we need a passport or a visa that 'validates' our mobility, depending on where are we from or where are we travelling to. When we enter another country, we are expected to do so by passing through an immigration control. These documents and border infrastructures not only determine entrance to another territory, but also the length of our stay. They are also key materialities of borders, those complex and changing spaces that as Samers and Collyer (2017, 161) argue 'symbolize the material and exclusionary power of national states'.

What is less visible is the wide range of norms, rules and decision-making processes that produce these infrastructures, as well as the creation of different migrant categories that determine who enters a territory, for how long and under which conditions. It is then imperative to remember that migration categories are not 'natural', they are social constructs that change contingent to context and time (Carens, 2013). However, they have a huge impact on the ability of people to migrate, or can even determine their immobility. These categories also play a role on how we understand migration and the binaries we create about migrants, including refugees (see Vera Espinoza, 2019).

What is relevant here is to think about the different levels involved in migration dynamics. In many cases we focus only on the movement of people, and not on the decisions and rules that are taken at the national, regional and global level that also impact migration. Let's then zoom out to those levels by discussing what is understood by migration management and the governance of migration. Migration management tends to refer to a set of policies and norms that are designed to regulate and control the number and the type of migrants (see Samers and Collyers, 2017). Migration and refugee laws as well as temporary measures, such as migration bans or humanitarian visas, are also part of this migration management.

Linked to migration management is the concept of migration governance. The latter refers to the 'combined frameworks of legal norms, laws and regulations, policies and principles as well as organizational structures (subnational, national, regional and international) and the relevant processes that shape and regulate States' approaches with regard to migration in all its forms, addressing rights and responsibilities and promoting international cooperation' (Mcauliffe and Oucho, 2024; Betts, 2011). Here the focus is on the states' actions in relation to migration. It has been generally understood in the literature that the global governance of migration is less accomplished than the global governance of refugees, as the definition of refugee is stamped in the 1951 Convention Relating to the Status of Refugees and its 1967 protocol, which determines a series of states' responsibilities, including that a person cannot be sent back to a country where they may be facing persecution (see Jubilut et al., 2021). Global non-binding agreements such as the 2018 Global Compact for Migration and the Global Compact on Refugees are also part of this migration governance. These compacts, conventions and declarations don't happen in a vacuum and are negotiated and agreed in relation to wider social and political processes (Geddes et al., 2019).

At the regional level, Latin America has also developed regional and sub-regional frameworks and agreements to manage migration. Across the region there are soft law instruments such as the Cartagena Declaration of 1984, which is the cornerstone of the current refugee protection regime in the region, enlarging the refugee definition to be applied in Latin America. Other agreements, organisations and spaces have been developed in the sub-regions (South America, Central America and North America, and also across them) to agree on cooperation managing diverse forms of migration. These negotiations between states are in many cases related to other wider socio-economic agendas that are of interest for the countries taking part. That is to say that the negotiations, and associated conditionalities, may vary greatly between countries of the Southern Cone and the ones that take place between Mexico, the United States and Canada.

And while the decision-making process may take place at the global and regional levels, its influence cascades down to the legislations and policies each country adopts and also impacts the daily lives of migrants, including refugees, both outside and inside the destination or transit countries. Some scholars have invited us to think about these impacts as forms of 'bordering'. Bordering, as coined by Yuval-Davis et al. (2019), constitutes a principal organising mechanism in 'constructing, maintaining and controlling social and political order' that through these rules, norms and frameworks, determines who is and who is not entitled to enter a territory, who can stay and acquire social rights. These bordering practices happen at different spaces: one of them is of course at the border, but bordering can also take place even before a migrant moves (through practices of externalisation) and other bordering processes determines migrants' social lives when they are inside a territory.

Case study: Venezuela displacement

A clear example to reflect this discussion is related to the Venezuelan displacement (see Figure 9.1). With more than seven million Venezuelans across the world, more than six million of them across Latin America, the countries of the region have been trying to deal with the consequences and opportunities of this mobility. Countries such as Colombia have received more than 2.5 million Venezuelans, followed by Peru (1.5 million), Ecuador (502,000) and Chile (440,000) (R4V, 2023), with many of them transformed from being transit countries to destinations. Since the start of this displacement back in 2015, the countries of the region have opted to implement a series of ad-hoc measures to manage these flows (Gandini et al., 2019). For example, countries such as Brazil and Mexico are the only two countries of the region that have applied the Cartagena refugee definition to specific national groups, including Venezuelans (Blouin et al.,

Figure 9.1 Colombia-Venezuela border, 2022.

Source: Wikimedia Commons https://commons.wikimedia.org/wiki/File:Colombia_-_Venezuela_border_re-open_(52417656550).jpg

2020), recognising them as refugees. Other countries such as Chile, Peru and Colombia have developed temporary protection measures that do not recognise them as refugees (refugee status determination may lead to permanent residency in many of these countries) and also comes with temporal limitations (Zapata et al., 2023).

Let's take the example of two individuals of the same family.[1] Pedro left Venezuela by the end of 2016 and made his way to Chile. He did not need a visa to travel to Chile. When he arrived, he applied for asylum and his refugee status was recognised a year later. According to the Refugee Law in Chile, once Pedro obtained the refugee status determination, he then had access to permanent residency which granted him the same rights and responsibilities as any other Chilean. He managed to find a job and rented accommodation soon after arrival. His sister Carla, however, couldn't travel to Chile that year as she was taking care of her parents. When Carla decided to migrate to Chile in late 2018, the Chilean government of Sebastian Piñera had issued an executive decree, alongside other reforms, to change visa procedures for Venezuelan migrants. This decree created the Visa of Democratic Responsibility for Venezuelans. This consular visa could be issued in any Chilean consulate abroad subject to specific requirements such as a passport (or ID national card) and proof of non-criminal record. That meant that Carla couldn't travel to Chile, as her brother did, without a visa, but instead had to apply for one from Venezuela. The requirements for this visa increased and changed over time, and Carla could not afford to pay for some of the documents requested. The consular visa that was promoted in the media as a special visa to help Venezuelans fleeing the Maduro regime ended up being a de facto barrier to legal entry for a targeted nationality. Similar barriers have been imposed in other countries, such as Ecuador and Peru (Freier and Luzes, 2021). Carla resorted to a dangerous and long journey to arrive in Chile, throughout which she had to deal with both people smugglers and state agents. She entered through a non-regulated point (known as *trochas*). Three years after her arrival she still was not able to regularise her migratory status. She did not have a stable job and has not been able to rent her own place, living sporadically with her brother or with friends, while also trying to overcome the trauma of the journey.

There are two interconnected academic debates to be taken from this discussion. First, migration management and migrant categorisation not only determine entry and staying requirements, but its effects also have material, symbolic and embodied impacts on peoples' lives (Vera Espinoza et al., 2021). Second, we are witnessing an increased securitisation of migration across the region (Brumat and Vera Espinoza, 2023) as well as weakening of the regional refugee protection regime (Zapata et al., 2023). This shows the diffusion of ideas around control, externalisation and crisis, as the region emulates practices similar to countries in the Global North (see Mountz, 2020).

'Luchas migrantes': migrants' resistance

It has been widely stated, both by migrants themselves and in an interdisciplinary body of literature, that migrants, including refugees, are not passive within their experiences of mobility. Far from it, independently of the reasons of their migration, mobile people exercise agency as part of their journey and settlement experiences. The bordering practices explored in the previous section provide an overview of the multiple exclusionary spaces that migrants have to navigate across different territories. They also provide the context to understand the emergence of different spaces and practices of resistance as part of migrants' political agency and solidarity (Rojas, 2021).

This resistance has been understood as part of 'luchas migrantes' (migrant struggles). Amarela Varela (2020) is a leading Latin American scholar exploring what she calls the 'sociology of migrant struggles'. This body of work analyses migrants' collective actions and their 'different expressions and modalities, strategies, actors, alliances and contexts' (Varela, 2020, 657–658). These actions remind us Rojas (2021, 39) challenge both the State and society at large. While they take different forms and enact diverse aims, they tend to develop a critique of racism and xenophobia while contesting border regimes but also redefining the normative frameworks of social recognition.

Joane Florvil was a Haitian woman who migrated to Chile in 2016 and died while in police custody in 2017 (see Figure 9.2). Florvil had been arrested a month earlier for allegedly abandoning her two-month-old baby in front of a municipal building, where she had gone to report a robbery. Unable to communicate in Spanish, she left her baby to a guard's watch, as she looked for a translator. While Joane considered this a safe practice, the guard interpreted this action as abandonment. Joane was arrested the day after and died a month later. In November 2017, the justice system determined that there was no abandonment. The entanglement of irregularities and negligence that led to Joane's death has been described by migrant and human rights organizations as a reflection of the institutional and structural racism existing in Chile.

The migrant caravans that have departed from Central America across Mexico toward the US, at least since 2011 but that reached higher visibility since 2018, have been understood as a part of this migrant struggles. For Frank-Vitale and Núñez Chaim, this group of people walking together across borders and territories 'combine

Figure 9.2 Commemoration of the death of Joane Florvil by Migrants Organizations, September 30, 2023, Santiago de Chile.

Source: Vania Reyes.

humanitarian accompaniment with political protests to protect people on their walk and demand respect for their rights' (2020, 37). Caravans are then a strategy to collectively address the risks on the migrants' journey to their destinations. Among these risks are the violence of criminal groups, the violence of the police in different states and even the risks of the natural environment (Oswald, 2023), for example, in the crossing of the Darien Gap (Tapón del Darien) in Panama. In this context, Varela (2020, 256) has described the migrant caravans as 'a form of migrant self-defence, an innovative form of transmigration in the region, and exercise of collective self-care, a rebellion against the global government of borders', particularly against the Plan Frontera Sur (Southern Border Plan), the Mexican attempt to manage borders through externalisation. Then, the caravans represent, not without political and social contestation, a collective action that rebels against border imperialism and the governance of migration.

We have also seen how migrants, both as individuals and collectively, came together to support other migrants – as well as other local communities – during the COVID-19 pandemic. When most countries of the region were closing borders and lockdown measures had specific impacts on migrant population, particularly those who were undocumented, migrant collectives and organisations came together to disseminate information, to cover basic needs, and to create awareness campaigns, among other actions. Elsewhere, one of the authors has described how these actions not only resisted exclusionary governance practices, but also emerged as examples of 'inclusive resistance' (Vera Espinoza, 2022). Exploring the case of migrant organisations in Chile, Vera Espinoza shows how actions at the local level such as 'ollas comunes' (common pots) to transnational regularisation campaigns (#RegularizacionYa), not only challenged bordering practices, but also generated spaces of inclusion and social cohesion within communities. These collective actions (**see Chapter 21**) took place both within specific territories and in transnational digital spaces. Migrant struggles emerge then as the collective manifestation of migrants' political lives, as a challenge to the regimes and norms that contain people's mobility and as multiple spaces of organisation and care.

Conclusion

Migration, both internal and international, has been pivotal to territory formation and to the development of Latin America. While contemporary international migration has generated media and political salience, mobility has been a constant in the region. Countries of the region are not only sites of emigration, but also of transit, destination and return. While the drivers of migration are varied, most people move as a form of social hope in response to diverse socio-economic inequalities.

We briefly discussed two key characteristics of migration in Latin America: the feminisation of migration and care chains. While these are not the only dimensions that tell us about the lives of migrants in the region, and are not exclusive to this region alone, both are relevant to understanding the dynamics and experiences of international migration in Latin America. These experiences are further shaped by the works of migration management and border control, which impact the lives of migrants inside and outside the territory, while they are also resisted through collective action.

Summary

- Migratory flows are a fundamental character of the growth and consolidation of the regional territory, even before the creation of national states. The patterns of international migration are dynamic and are driven by economic and labour asymmetries in the region, development expectations and also by the increase of social and natural disasters.
- The feminisation of migration implies the increase of women in the composition of the migration flows, which also suggests their employment in low-pay and less socially prestigious jobs. Care chains problematise the maintenance of gender asymmetries and the provision of care by migrant women in the region.
- Migration management and migrant categorisation not only determine entry and staying requirements for mobile people, they also have material, symbolic and embodied effects that impact migrants' lives.
- Migrants, including refugees, resist bordering practices through collective actions and strategies across spaces. These 'luchas migrantes', migrant struggles that derive in political actions, not only challenge the governance of migration and normative social order, they also emerge as spaces of inclusion and solidarity across and within communities.

Review questions

1. A key characteristic of migration in the region is the development of care chains within and outside the territory. Critically reflect on who bears the responsibility of care.
2. Using examples, discuss how bordering practices are shaping migration dynamics and with what consequences.
3. In what ways can migrant struggles be understood as practices of 'inclusive resistance'? How do they challenge migration management?

Further reading

Guizardi, M. (2021). *The Migration Crisis in the American Southern Cone. Hate Speech and its Social Consequences.* Springer.
This edited book presents theoretical and empirical contributions to migration studies from a Latin American perspective.
Samers, M., and Collyer, M. (2017). *Migration.* Routledge.
The book provides a critical and conceptually informed introduction of migration and immigration.
Vergara-Figueroa, A. (2018). *Afrodescendant Resistance to Deracination in Colombia: Massacre at Bellavista-Bojayá-Chocó.* Palgrave.
The book examines how the concepts of forced displacement and migration could be formulas for historical erasure.

Keywords

Forced displacement: the movement of persons who have been forced or obliged to flee or to leave their homes or places of habitual residence. This may be as a result of armed conflict, situations of generalised violence, violations of human rights or natural or human-made disasters (Mcauliffe and Oucho, 2024).

Migrant: there is not precise definition of 'migrant', although is mostly used to refer to individuals that reside in another country for a determined period of time. In some cases, it is associated with specific legal categories. In this chapter we use the term 'migrant' as interchangeable with 'immigrant', but also inclusive of other people on the move, such as refugees.

Migration: the movement of individuals or groups from their usual place of residence across an international border or within a country.

Refugees: the 1951 Refugee Convention defines a refugee as 'someone who is unable or unwilling to return to their country of origin owing to a well-founded fear of being persecuted for reasons of race, religion, nationality, membership of a particular social group, or political opinion'. The Cartagena Declaration of 1984 broadened the definition of refugee to include 'persons who have fled their countries because their lives, safety or freedom have been threatened by generalised violence, foreign aggression, internal conflicts, massive violation of human rights or other circumstances that have seriously disturbed public order'.

Note

1 This is a fictional account based on interviews developed for different projects in the last four years.

References

Aruj, R. (2008). Causas, consecuencias, efectos e impacto de las migraciones en Latinoamérica. *Papeles de Población*, 55, 95–116.

Betts, A. (ed) (2011). *Global Migration Governance*. Oxford University Press.

Blouin, C., Berganza, I., and Freier, L. F. (2020). The Spirit of Cartagena? Applying the Extended Refugee Definition to Venezuelans in Latin America. *Forced Migration Review*, 63, 64–66.

Brumat, L., and Vera Espinoza, M. (2023). Actors, Ideas, and International Influence: Understanding Migration Policy Change in South America. *International Migration Review*, 0(0). https://doi.org/10.1177/01979183221142776

Carens, J. (2013). *The Ethics of Immigration*. Oxford University Press.

Cerrutti, M., and Maguid, A. (2010). *Familias divididas y cadenas globales de cuidado: la migración sudamericana a España. Serie políticas sociales* (Vol. 163). Cepal. http://repository.eclac.org/bitstream/handle/11362/6168/lcl3239e.pdf?sequence=1&isAllowed=y

Cerrutti, M., and Parrado, E. (2015). Intraregional Migration in South America: Trends and a Research Agenda. *Annual Review of Sociology*, 41, 399–421. https://doi.org/10.1146/annurev-soc-073014-112249

Crawley, H., and Teye, J. K. (eds) (2024). *The Palgrave Handbook of South–South Migration and Inequality*. Palgrave Macmillan.

Frank-Vitale, A., and Núñez Chaim, M. (2020). "Lady Frijoles": las caravanas centroamericanas y el poder de la hipervisibilidad de la migración indocumentada. EntreDiversidades. *Revista de Ciencias Sociales y Humanidades*, 14(1), 37–61. https://doi.org/10.31644/ed.v7.n1.2020.a02

Freier L. F., and Luzes, M. (2021). How Humanitarian are Humanitarian Visas? An Analysis of Theory and Practice in South America. In L. Jubilut, G. Mezzanotti, and M. Vera Espinoza (eds) *Latin America and Refugee Protection: Regimes, Logics and Challenges*. Berghahn.

Gandini, L., Prieto Rosas, V., and Lozano-Ascencio, F. (2019). El éxodo venezolano: migración en contexto de crisis y respuestas de los países latinoamericanos. In L. Gandini, F. Lozano-Ascencio, and V. Prieto Rosas (eds) *Crisis y migración de población venezolana. Entre la desprotección y la seguridad jurídica en Latinoamérica* (pp. 9–32). Universidad Nacional Autónoma de México.

Geddes, A., Vera Espinoza, M., Hadj-Abdou, L., and Brumat L. (eds) (2019). *The Dynamics of Regional Migration Governance*. Edward Elgar Publishing.

Jubilut, L., Vera Espinoza, M., and Mezzanotti, G. (2021). *Refugee Protection in Latin America. Logics, Regimes and Challenges* (pp. 1–29). www.berghahnbooks.com/title/JubilutLatin

Jung, P. R. (2021). Hope, Disillusion and Coincidence in Migratory Decisions by Senegalese Migrants in Brazil. *Social Inclusion*, 9(1), 268–277. https://doi.org/10.17645/si.v9i1.3721

Kleist, N., and Thorsen, D. (eds) (2016). *Hope and Uncertainty in Contemporary African Migration. Routledge Studies in Anthropology*. Taylor & Francis. https://doi.org/10.4324/9781315659916

Lerussi, R. (2008). Trabajo doméstico y migraciones de mujeres en Latinoamérica. El caso de las nicaragüenses en Costa Rica. Punteo para un enfoque de reflexión y acción feministas. *Anuario de Estudios Centroamericanos*, 34, 183–203.

Magliano, M. J. (2007). Migración de mujeres bolivianas hacia Argentina: cambios y continuidades en las relaciones de género. *Amérique Latine Histoire et Mémoire*, 14, 1–16. https://doi.org/10.4000/alhim.2102

Mcauliffe, M., and Oucho, L. A. (eds) (2024). *World Migration Report 2024*. International Organization for Migration (OIM).

McDowell, L., and Sharp, J. (1999). *A Feminist Glossary of Human Geography*. Arnold, a Member of the Hodder Headline Group.

Mitchell, K., Jones, R., and Fluri, J. (2019). Introduction to Critical Geographies of Migration. In K. Mitchell, R. Jones, and J. Fluri (eds) *Handbook on Critical Geographies of Migration* (pp. 1–17). Edward Elgar.

Mora, C. (2008). Globalización, género y migraciones. *Polis Revista Latinoamericana*, (20), 1–11. https://scielo.conicyt.cl/scielo.php?script=sci_arttext&pid=S0718-65682008000100015

Mountz, A. (2020). *The Death of Asylum Hidden Geographies of the Enforcement Archipelago*. University of Minnesota Press

Oswald, Ú. (2023). Migración climática y fronteras militarizadas: seguridad humana, de género y ambiental. *Frontera Norte*, 35(01), 1–29.

R4V. (2023). Refugees and Migrants from Venezuela. Available at: https://www.r4v.info/en

Rojas, B. (2021). Romper las fronteras. Las luchas de las organizaciones de migrantes y la emergencia de una reivindicación antirracista (1998–2018, Santiago). *Palimpsesto*, 11(18), 34–61.

Samers, M., and Collyer, M. (2017). *Migration* (Second Edition). Routledge.

Sassen, S. (2003). *Contrageografías de la Globalización. Género y ciudadanía en los circuitos transfronterizos*. Ed. Traficantes de Sueños.

Stefoni, C. (2002). Mujeres inmigrantes peruanas en Chile. *Papeles de Población*, 33, 117–144.

Varela, A. (2020). Notes from an Anti-racist Feminism in the Wake of the Migrant Caravans. *The South Atlantic Quarterly*, 119(3), 655–663. https://doi.org/10.1215/00382876-8601506

Vera Espinoza, M. (2019, November 6). The Human Behind the Refugee Category and the Industry Behind Refugees' Representation. *Discover Society*.

Vera Espinoza, M. (2022). Gobernanza excluyente vs. resistencia inclusiva: el manejo de las migraciones durante la pandemia en Chile. In G. Zapata, M. Vera Espinoza, and L. Gandini (eds) *Movilidades y covid-19 en América Latina: inclusiones y exclusiones en tiempos de "crisis"* (pp. 87–109). UNAM.

Vera Espinoza, M. (2024). Migration Governance in South America: Change and Continuity in Times of "Crisis". In H. Crawley, and J. K. Teye (eds) *The Palgrave Handbook of South – South Migration and Inequality*. Palgrave Macmillan. https://doi.org/10.1007/978-3-031-39814-8_29

Vera Espinoza, M., Prieto Rosas, V., Zapata, G. P., Gandini, L., Fernández de la Reguera, A., Herrera, G., López Villamil, S., Zamora Gómez, C. M., Blouin, C., Montiel, C., Cabezas Gálvez, G., and Palla, I. (2021). Towards a Typology of Social Protection for Migrants and Refugees in Latin America during the COVID-19 Pandemic. *Comparative Migration Studies*, 9(1). https://doi.org/10.1186/s40878-021-00265-x

Weima, Y., and Hyndman, J. (2019). Managing Displacement: Negotiating Transnationalism, Encampment and Return. In K. Mitchell, R. Jones, and J. Fluri (eds) *Handbook on Critical Geographies of Migration* (pp. 30–44). Edward Elgar.

Yuval-Davis, N., Wemyss, G., and Cassidy, K. (2019). *Bordering*. Polity Press.

Zapata, G. P., Gandini, L., Vera Espinoza, M., and Prieto, R. (2023). Weakening Practices Amidst Progressive Laws: Refugee Governance in Latin America during COVID-19. *Journal of Immigrant & Refugee Studies*, 21(4), 547–565. https://doi.org/10.1080/15562948.2022.2163521

Zapata Martínez, A. (2016). Madres y padres en contextos transnacionales: el cuidado desde el género y la familia. *Desacatos. Revista de Ciencias Sociales*, 0(52), 14–31. https://doi.org/10.29340/52.1631

CHAPTER 10

Sustainable development

Jessica Hope

Introduction

Sustainable development is a dominant agenda for securing social, economic, and environmental wellbeing in response to intensifying climate change. Originally coined to resolve conflicts and contradictions between development goals and conservation priorities during global conferences on environment and development during the 1970s and 1980s (Adams, 2009:59–86), it sets out an alluring, normative agenda to secure human development whilst protecting nature. However, sustainable development has been much critiqued by Anglophone critical geographers, primarily as a vague and ill-defined term that is easily appropriated by those with financial and political power.

I structure this chapter by first analysing sustainable development as broadly and critically defined. Taken as a normative agenda to change how we live, sustainable development encompasses questions about what development and progress are, what nature is and how different definitions of both development and nature change the trajectories of sustainability. Latin American thinking and activism grapples with many of these questions. I introduce two key ideas from the region: alternatives to development and pluriversality. Second, I outline the contemporary global sustainable development project, using political economy and political ecology to analyse sustainable development as a set of global development goals (the Sustainable Development Goals, or SDGs/ODS in Spanish) and their wider agenda (Agenda 2030). I present some of my own research in Bolivia, which analysed the take-up of the SDGs and how they impacted Indigenous-led anti-extractive territorial movements – those who are defending their land in the face of intensifying resource and fossil fuel extraction. Finally, I propose that Latin American geography helps us to rethink sustainability in three key ways – extending imaginaries of sustainability, changing those who are identified as experts and prioritising an environmentally just future.

DOI: 10.4324/9781003430926-15

Sustainable development

Sustainable development is a normative agenda that can be hard to succinctly define. The term was formulated over a series of documents (*The World Conservation Strategy, Our Common Future*, and *Caring for the Earth*) and global conferences (starting with the UN Conference on the Human Environment in Stockholm, 1972) (Adams, 2009:59–85). It has its roots in a range of ideas, including Northern American and European environmentalism, conservation, international environmental organisations, the emergence of ecology science, concerns about population growth and developing global scientific networks (Adams, 2009:56). However, there is no specific theory or metric that defines what sustainable development actually is; rather, it is the Brundtland Report definition that is most often quoted. This defines sustainable development as 'meeting the needs of the present without compromising the ability of future generations to meet their own needs' (Brundtland, 1987).

In Latin America, 'development' is a contested term. Post-development scholars, for example, criticised 'development' as hinging on the largely unquestioned assumption that North America and Europe provide a model for countries and societies within Asia, Africa and Latin America. Arturo Escobar, a Colombian anthropologist, led post-development debates in the 1980s and 1990s. He argued that rather than being a humanitarian project, development was a powerful discourse, backed by imbalances of power, that categorised the world into the hierarchy of First World, Second World and Third World. This justified intentional interventions and projects in Latin America that were grounded in modernisation theory – the idea that development is a linear progression from pre-modern to modern, dependent on expert knowledge, technology transfers and industrialisation (see Chapter 11). For Escobar, development discourse recast Latin America as poverty stricken and in desperate need of intervention (Escobar, 1995). In the 1980s and 1990s, post-development scholars and activists called for an end to the development project because of the failings of these interventions in Latin America. This was partly because large-scale technology transfers, mega-infrastructure projects and structural adjustment policies were having devastating outcomes on societies (and poverty) (see Kohl and Farthing, 2006).

In the early 2000s, echoes of post-development thinking can be found in Latin American alternatives to development (see Chapter 17). A starting point is that development is a 'zombie concept, dead and alive at the same time' (Gudynas, 2011:441) and that neoliberal structural policy adjustments have not worked. They are explicit in critiquing and rejecting the entire concept of development (as linear progress), especially definitions of development that centre economic growth. Development's promotion of modernity is understood to be entangled with coloniality, namely 'a matrix of global power that has hierarchically classified populations, their knowledge, and cosmologic life systems according to a Eurocentric standard' (Walsh, 2010:52). In response, social movements and Indigenous scholars developed alternatives, most notably Buen Vivir/Vivir Bien/sumak kawsay/suma qamana (from here, written as Buen Vivir).

Buen Vivir is loosely translated into English as collective wellbeing and the 'good life'. It constitutes a 'critical reaction to classical Western development theory, (as well as) alternatives to development emerging from Indigenous traditions' (Gudynas, 2011:441). Definitions of wellbeing go beyond material wellbeing and instead involve ensuring harmonious communities of humans but also of humans and the non-human.

Community is thus both social and ecological and thus, the 'classical Western dualism that separates society from Nature vanishes . . . as one contains the other, and they are not separable' (Gudynas, 2011:444). Whereas sustainable development centres around individuality and economic growth, Buen Vivir is the 'result largely of the social, political, and epistemic agency of the Indigenous movement over the last two decades . . . (and) . . . responds to the urgency of a radically different social contract that presents alternatives to capitalism and the 'culture of death' of its neoliberal. . . development project' (Walsh, 2010:18).

Although there are overlaps, definitions of Buen Vivir are geographically diverse. One of the most well-known approaches to Buen Vivir is the Ecuadorian concept of sumak kawsay, the Kichwa wording for a fullness of life in a community, together with other persons and Nature. This was written into the Ecuadorian constitution in 2007. In Bolivia, the Aymara concept of suma qamana was developed from the critique of Aymara sociologist Simon Yampara. Rather than being a term long used in rural Indigenous communities (which are diverse), suma qamana and Vivir Bien was a pointed response to the failings of classical development – one rooted in traditional Indigenous knowledges. Vivir Bien was written into the new 2009 constitution as the guiding ideology of the MAS government. The inclusion of Indigenous knowledges into the constitutions of both countries marked a crucial shift in how Indigenous concepts and social movement politics were included in the fabric of both states. However, inclusion has been a contested and conflictive process, especially as Buen Vivir has been used to justify expanding extractive frontiers. Buen Vivir thus also entails a contested politics of authorship and implementation, elaborated later in the analysis of the take-up of the SDGs in Bolivia.

The second concept introduced here is pluriversality, which is a concept that also engages with questions of how we live and what nature is, partly through exploring what is known, how and by whom. The Pluriverse draws on Zapatista thinking to promote (and create) a 'world where many worlds fit'. This means acknowledging that there is a diversity of ways that the world is known and thus a multiplicity of worlds. Again, this term encompasses academic work but also the demands of Indigenous territorial movements, some of which understand their worlds as intimately tied to their territory (see Chapter 12). Pluriverse scholarship advocates for territorial, cultural and political autonomy and includes claims for land rights. For Blaser, a 'pluriverse' enacts multiple, distinct ontologies or worlds, which 'bring themselves into being and sustain themselves even as they interact, interfere, and mingle with each other' under asymmetrical circumstances (2013:32). This recognises the politics inherent in whose knowledges are acknowledged, taken seriously and valued, as well as the colonial histories that subjugated Indigenous knowledges to Enlightenment thinking.

In terms of sustainability and environmental geography, the Pluriverse 'repoliticises nature for the Anthropocene' by responding to the 'ruinations' of colonialism and extractive capitalism' (De la Cadena and Blaser, 2018) with alternative ways to know, value and live with the non-human. In other words, it is both a response and an alternative. Partly, pluriversal scholarship does this by analysing how claims for territory challenge state power, for example to enact ways of living beyond capital and the state (Aguilar, 2014; see Radcliffe and Radhuber, 2020), secure access to land, for example in contemporary concerns for environmental justice (see Mena et al., 2020), and enact place-based political projects and knowledges (see De la Cadena and Blaser, 2018).

Pluriversality is decolonising in taking seriously multiple (including Indigenous) knowledges and worlds and in promoting (and enacting) a world where these many worlds fit (following Zapatista thinking).

In summary, though Buen Vivir and the Pluriverse engage with questions about how we live, they depart from sustainable development in foundational ways. First, they tackle development very differently. Whilst sustainable development aims to make existing trajectories of development (and growth) more sustainable, both Buen Vivir and pluriversality abandon 'development' as a failed project. Rather, they reconceptualise the kinds of collective lives we should value and enact. This is most prominent in Buen Vivir. Second, whilst sustainable development sets out a broad term for all, Buen Vivir and pluriversality start by recognising diversity and difference. Most prominent with pluriversality, this offers a very different starting point for thinking about sustainability (and unsustainability).

Sustainable development, a global project

Departing from sustainable development as a broad normative concept, a productive way to understand contemporary sustainable development is by analysing how the term is being operationalised. This means analysing how it is being remade by the United Nations Agenda 2030 and Sustainable Development Goals (SDGs). The 2015 UN SDGs combine a response to climate change with development targets, comprised of 17 goals, quantified by 169 targets, and measured by 230 indicators (UN, 2015). Agenda 2030 is focused on building consensus between development banks, multilateral and bilateral institutions, states, international non-government organisations (INGOs), civil society organisations, and, increasingly, private sector actors. It aims to balance social, economic, and environmental arenas within a 'plan of action for people, planet and prosperity' (UN, 2015) and a term adaptable enough to reach across borders and bank accounts.

Agenda 2030 and the SDGs are a flagship project of 'global development', meaning a reframing of development that casts (sustainable) development as for both the Global North and South (see Figure 10.1). Unlike past development goals, the SDGs are for the UK, North America and Europe, as much as countries in the Global South. This marks a significant shift from 'International Development', which has a historic focus on particular regions of the world (the 'third world', the 'developing world' and the Global South). For the UK academic institutions that have renamed themselves as global development institutions, 'global development' 'changes the implicit (and much critiqued) binary logic of international development – that the Global North is "developed," while the Global South needs to be so' (Hope et al., 2022:155). For scholars at the Global Development Institute at the University of Manchester, for example, the COVID-19 pandemic consolidated the need for their renaming by highlighting 'the falsity of any assumption that the Global North has all the expertise and solutions to tackle global challenges' instead revealing 'the need for multi-directional learning and transformation in all countries towards a more sustainable and equitable world' (Oldekop et al., 2020:105044). There are divergencies and tensions, however, between how global development is understood by critical development scholars and by those implementing Agenda 2030.

Figure 10.1 SDGs on display at a public park in La Paz, Bolivia.
Source: Author.

Using political economy to analyse sustainable development as a flagship project of global development foregrounds how sustainable development is related to global economic governance, as well as private sector actors and logics. Partly, this is about funding. Emma Mawdsley, a UK based development geographer, has argued that a key conversation regarding Agenda 2030 has been about how to finance the SDGs. Private sector representatives have been 'invited to drive and shape global development governance and policy by the United Nations and other multilaterals, and national development agencies' (Mawdsley, 2018:192). This may seem a logical step in ensuring the ambitions of Agenda 2030, but it has significant consequences for *how* sustainable development is envisioned and implemented. Private sector representations, for example, promote private sector logics and strategies, such as 'blended finance', debt and equity finance for public – private partnerships (PPPs), as well as a focus on infrastructure, land and digital financial technologies (Gabor and Brooks, 2016; Mawdsley, 2018). These are pitched as essential drivers of economic growth, which is understood as trickling down into poverty reduction (Mawdsley, 2018:193). For critical development scholars, such as post-development scholars, trickle-down economics is a proven fallacy.

I have combined this approach with political ecology, a broad inter-disciplinary field researching how natures are made by politics, history and policy, to sharpen our environmental focus. Being more critical of sustainable development as a normative agenda, political ecology focuses analysis on the political actors and processes that enact sustainable development policies and projects, revealing the environments and natures being created in this process. In my own research, I have used political ecology to frame analysis of the early take-up of the SDGs in Bolivia – a country on the frontlines of negotiations between resource extraction, different models of development (including Vivir Bien), Indigenous territorial autonomy and land rights. To be clear, Bolivia is not a case study example of a 'developing country', nor a generic example of Latin American geography, but rather a very particular case where political activists and scholars are proving crucial in determining wider trajectories of sustainability, particularly those concerning fossil fuel extraction, energy and land rights.

Bolivia and the SDGs

In 2005, the election of Evo Morales, himself an Indigenous social-movement leader, was celebrated as marking the collapse of a neoliberal, elite-led government. Within critical social science (and more broadly), his election was seen as a victory for the social movement politics that had campaigned to defend the interests of society and nature (see Kohl and Farthing, 2006; Harten, 2011). Bolivia has since been much researched as a site of emergent (and contentious) politics of post-neoliberalism (see Chapter 8) and pluri-nationalism. The former refers to contested efforts to re-work the state, economy and civil society in response to the failings of neoliberalism and the demands of social movement mobilisation (Grugel and Riggirozzi, 2012) and the latter, to a process for 'territorial resignification and demographic occupation of state territory by multiple social movements' (Mamani, 2011:32), engaging ongoing debates about Indigenous territorial rights, citizenship and control over natural resources (Gustafson, 2009; Radcliffe, 2012). In subsequent years, however, it has become clear that these radical agendas have been much constrained by the state's neo-extractivist development model (see Chapter 17). Bolivian commitments to neo-extractivism form a part of an unprecedented rise in resource extraction across Latin America since the early 2000s, when a resource boom was exploited by both neoliberal and post-neoliberal states alike (Bebbington and Humphreys Bebbington, 2011). Extractivism is now understood as a defining feature of post-neoliberalism.

In Bolivia, despite new welfare projects that reduced poverty significantly, the vast majority of public investment was allocated to energy and hydrocarbon development (40%) and transportation infrastructure (29%), accounting for nearly 70% of the 2016 public investment budget for 2016 (MEFP, 2015 in McKay, 2017:414). This revealed that social capital investments prioritised further accumulation, particularly in the extractive sectors (McKay, 2017). Redistributive policies also failed to reflect a serious commitment to diversifying the economy and the control and distribution of royalties remained highly centralised, with minimal contributions made to activities that directly supported Indigenous groups (Radhuber, 2012). For the MAS and their supporters, neoextractivism is a key way to retain (and share) benefits in Bolivia. For some of the opposition, neoextractivism is undermining existing land rights and environmental protections, as well as commitments to Vivir Bien. They are not benefitting. This tension between extractive-led models of development and anti-extractive territorial movements continues, almost 20 years later (Riofrancos, 2020) and it is in this context that the SDGs have been implemented.

Whilst global development goals are often studied as a global policy framework (Hulme, 2009) or evaluated in terms of their effectiveness (Hulme, 2009), data capture (Jerven, 2013), and North-South development politics (Bond, 2006), I analysed how an assemblage of institutions, development discourses, landscapes and energy infrastructures were disciplined and held together in Bolivia to determine the scope and content of the SDGs, as they were taken up (see Hope, 2021). I investigated how anti-extractive territorial movements were treated and included by the SDG agenda, focusing on two cases studies – a long-running conflict over road building to access hydrocarbon pools in the Territorio Indígena y Parque Nacional Isiboro Sécure (TIPNIS; Isiboro Sécure Indigenous Territory and National Park) (see Figure 10.2) and resistance to hydropower dams in Chepete and Bala. From this research, I first argued for the anti-politics of the SDGs, meaning that they render neutral and apolitical the conflictive politics of extractivism and sustainability. Second, I argued that the SDG assemblage reveals the 'lost geographies' of Agenda 2030, namely the Bolivian territories and political movements that are fighting against extractive frontiers.

I reached this conclusion partly from how the SDGs were accepted by the Bolivian government. In-country, the implementation of the SDGs is authorised by an agreement between the UN and the central government. Despite the creation of Agenda 2030 being newly participatory, in practice the UN needs to partner with governments to adopt the SDGs. In Bolivia, this process was complicated by the reluctance of government ministers to work with the UN or adopt 'foreign' development agendas. Although early in the Morales Presidency the UN was used as a platform, his administration simultaneously

Figure 10.2 Community meeting in TIPNIS, 2012.
Source: Author.

rejected dominant development paradigms and institutions as the imperialism of the Global North. Bolivia's relationship with global development has long been conflictive because of the devastation caused by past policies. Although partnership was secured, the SDGs were matched to pre-existing national development targets.

The vagueness of 'sustainable development' and the new flexibility for states has enabled partnership between the Bolivian government and the UN – significant given Bolivia's recent history with international development institutions. However, the merging of Vivir Bien and the SDGs by the UN and the Bolivian central government is contradictory, when viewed from the perspective of wider activist and academic debates. Sustainable development's explicit commitments to growth are at odds with how Vivir Bien/Buen Vivir has been conceptualised and advocated by activists and scholars. Further, deference to state-led development is particularly relevant to the environmental remit of the goals because post-neoliberal extractivism depends on contractual partnership with transnational firms and extractive capital. Included in Bolivia's sustainable development assemblage are the 2006 contract renegotiations with global hydrocarbon firms, when the Morales administration renationalised the sector. This gave Bolivia a higher percentage of royalties from private gas companies (amounting to $5.5 billion USD in 2014 and $3.7 billion USD in 2015; Fabricant and Gustafson, 2016:273). The state contracts transnational firms within a nationalised hydrocarbon framework. Bolivia contracts work, for example, from Brazil's state-oil giant Petrobras, Spain's Repsol, the UK's British Gas, and France's Total (Fabricant and Gustafson, 2016). Shell returned to Bolivia in 2015 after eight years' absence. In summary, the partnerships of the SDGs emerge from existing development agendas, networks and politics, which pre-date the environmental goals for Agenda 2030 and discipline its uptake.

My argument was also based on the discipling of non-governmental organisations (NGOs), who are key actors in working within Agenda 2030. Disciplining is here examined for the more explicit and repressive ways that power is felt in the SDGs assemblage. It is used to both analyse how the work of NGOs is modified to suit extractive interests and to explain the wider holding-together of the sustainable development assemblage. Globally, the relationship between Indigenous groups and international development actors has been seen as crucial in helping minority groups negotiate with states, but the uptake of the SDGs in Bolivia reveals a drastic reworking of the 'boomerang effect' (Keck and Sikkink, 1999), as development infrastructure is being disconnected from anti-extractive Indigenous territorial politics (Hope, 2021). In Bolivia, it was evident that NGOs were disciplined to ensure that the contentious politics of extractivism was not a part of their work. International NGOs were clear that they could not get involved in disputes between the state and civil society, despite previously supporting those leading Bolivia's eco-territorial turn. This was explained as undermining the sovereignty of Bolivia and the government's right to determine the country's national development agenda. However, reticence to publicly engage with conflicts over extractivism was also explained as a response to the legislation and bureaucracies of government. NGOs described a climate of 'fear and mistrust' when working in the country – one that challenged their work and created a sense of insecurity. NGOs referred to government attempts to pass new legislative frameworks for their work and to the ever-changing administrative process for gaining authorisation to work within the country, with several NGOs experiencing unexpected U-turns. Several bureaucratic and administrative shifts were cited as creating this new work culture and one NGO worker said they 'spent most of (their) time doing bureaucracy'. It was also widely cited that the government ejected an international NGO from Bolivia in 2010 for their support of the TIPNIS opposition. The MAS government accused them of political meddling ('injerencia política'), though it published no specific allegations. In this context, NGOs expressed difficulty in knowing how to best respond to rising social conflict, in ways sensitive to the responsibilities they had to their Bolivian partners and staff. One small Bolivian NGO worked in the immediate vicinity of one of the case studies. Yet they were clear that their work, though explicitly environmental in its remit, could not include or address the conflict.

In summary, researching the SDGs in Bolivia meant researching how powerful, extractivist development logics are maintained and reworked, despite increasing momentum to combat climate change and environmental crisis. I examined how the particular partnerships and modes of consensus-building across the SDG assemblage secure complementary interpretations and practices of global sustainable development, revealing how particular issues (extractivism) and geographies (Indigenous territories) are excluded.

Conclusion

In this chapter, I propose that Latin American geography helps us to rethink sustainable development in three key ways. First, Latin American geography extends imaginaries of sustainability. There is a large, important and still relevant body of critical Anglophone geographical work on sustainable development. Summarizing this imperfectly, a repeated critique within this work is that sustainable development remains centred on growth-led development and is thus an oxymoron. Within Agenda 2030, for example, building consensus between the private sector, states and civil society is a priority. For those challenging states or industry on environmental grounds, as shown in the Bolivian example and in wider environmental justice campaigns, this is a problem. However, in Latin America a significant body of scholars and activists are developing alternative conceptualisations of sustainability – ones that are more radical in reconceptualising relationships between land, non-humans and societies. Work on Buen Vivir and the Pluriverse, as introduced in this chapter, reveal (and enact) a multiplicity of worlds, as well as foreground the post-colonial and decolonial politics of how knowledges are recognised, erased and/or shared (see **Chapter 13**). This starting point of multiplicity marks a significant difference to sustainable development as it grounds imaginaries of sustainability in attentiveness to diversity, power and justice.

Second, and relatedly, Latin American geography presses us to rethink expertise. Conflicts over land, development and resources are proving crucial in revealing both the pitfalls and possibilities for wider trajectories of sustainability: for example, revealing that the SDGs are not proving useful for those challenging extractive frontiers in Bolivia. Those defending land rights in the face of extractive capital are negotiating first-hand the multiple claims being made on their land in the name of development, as well as simultaneously developing alternatives (see **Chapter 22**). They encounter and negotiate global capital and state development visions in very immediate ways and their experiences and ideas are crucial for understanding both unsustainable practices and future trajectories of sustainability.

Third, and finally, Latin American geography foregrounds agendas for decoloniality and justice. Decoloniality is premised on recognition of the historical violence of coloniality and modernity, including the environmental 'ruinations' caused. Particular agendas relevant to questions of sustainability, for example Buen Vivir and the Pluriverse, include recognition that many societies have unequally contributed to our current environmental crisis and that the frontlines of environmental degradation in Latin America link to wider processes of development and capitalism. In this context, Buen Vivir and the Pluriverse offer diverse possibilities for a more sustainable future. For those of us in the Global North, these challenge us to more carefully question our own positionality and complicity, as well as to act as better allies. Latin American geography both proposes and evidences that sustainability is a matter of justice.

Summary

- Sustainable development remains a core, global response for securing social, economic and environmental wellbeing in response to intensifying climate change, despite being much critiqued and contested. The contemporary global project is hugely influenced by global economic governance, as well as private sector actors and logics.
- In Latin America, 'development' is a contested term and project. Alternatives have been developed which challenge linear ideas of progress, growth-based measurements of wellbeing and the Nature/Culture binary.
- Recent agendas for pluri-nationalism and post-neoliberalism in Bolivia and Ecuador have been underpinned by established and experienced Indigenous political organisations and leaders.
- In Bolivia, the SDGs acted as anti-politics – neutralising and depoliticising the politics of land and resource use. Specifically, this excluded the anti-extractive activism of Indigenous territorial movements, despite increased global momentum to combat climate change.
- Latin American geography helps us to rethink sustainable development in three key ways. First, Latin American geography extends imaginaries of sustainability. Second, and relatedly, Latin American geography presses us to rethink expertise, foregrounding those defending land rights in the face of extractivism. Third, and finally, Latin American geography foregrounds agendas for decoloniality and justice.

Review questions

1. What is the difference between 'alternatives to development' and 'development alternatives'?
2. In what ways do Latin American alternatives to development challenge mainstream sustainable development? Use examples.
3. When trying to combat climate change, when/why is the consensus building of the SDGs useful and when/why does it hinder transformation?

Further reading

Adams, B., 2020. *Green development: Environment and sustainability in a developing world*. Routledge.

Green Development is a seminal text that analyses sustainable development in both theory and practice. The fourth edition gives more attention to the political ecology of development, market-based and neoliberal environmentalism, and degrowth.

Aguilar, R.G., 2014. *Rhythms of the Pachakuti: Indigenous uprising and state power in Bolivia*. Duke University Press.

Between 2000 and 2005, Bolivia was radically transformed by a series of Indigenous uprisings against neoliberal and antidemocratic policies. In this book, Raquel Gutiérrez Aguilar documents these mass collective actions and their radical agendas.

https://ejatlas.org

This online atlas and work-in-progress, documents environmental justice conflicts worldwide and helps you to identify trends in what is unsustainable.

Oslender, U., 2016. *The geographies of social movements: Afro-Colombian mobilization and the aquatic space*. Duke University Press.

In this wonderful book, Oslender explores how the social relationships of residents of Colombian lowland territories are entangled with the region's rivers, streams, swamps, rain, and tides/Oslender argues that this "aquatic space" shapes local knowledge and related political processes.

Riofrancos, T., 2020. *Resource radicals: From petro-nationalism to post-extractivism in Ecuador*. Duke University Press.

Riofrancos analyses the political struggle in Ecuador between resource nationalists who want the profits of extraction to benefit more Ecuadorians and anti-extractive grassroots activists who prioritise land rights and nature.

Keywords

Buen Vivir: loosely translated into English as collective wellbeing and an agenda for a 'good life', Buen Vivir constitutes an alternative to development that moves away from linear ideas of progress and economic growth. Instead, it foregrounds harmony within communities of humans but also between humans and the non-human. However, there are some major differences between interpretations of Buen Vivir, especially between the Buen Vivir implemented by states and the Buen Vivir claimed by communities.

Pluriversality: the term 'Pluriverse' comes from the Zapatista agenda to 'make a world where many worlds fit'. It means starting from recognition that there are multiple ways of knowing, experiencing and living in the world, thus multiple worlds. This is a very different starting point to a universal approach as it foregrounds the need to equitably negotiate differences and multiplicity.

Political ecology: this is a broad, interdisciplinary field that examines how natures are made through politics, policies, history and culture. Political ecology is not a theory, nor tied to a certain set of theories, though often involves a commitment to social justice, ethnographic research and attention to the local scale. Political ecology spans academic research and activism.

> **Post-development:** post-development theory emerged in the 1980s in response to failures of the development project in the Global South (when development was exclusively orientated to so-called 'developing countries'). Lead by scholars from the Global South, post-development theory argues that divisions of 'developed' and 'developing' are linked more to the hegemony of Western ideals and the justification of interventions, than ambitions to meaningfully respond to the particular and diverse needs of countries in the Global South.

References

Adams, B., 2009. *Green development: Environment and sustainability in a developing world.* Routledge.

Aguilar, R.G., 2014. *Rhythms of the Pachakuti: Indigenous uprising and state power in Bolivia.* Duke University Press.

Bebbington, A. and Humphreys Bebbington, D., 2011. An Andean avatar: Post-neoliberal and neoliberal strategies for securing the unobtainable. *New Political Economy, 16*(1), pp. 131–145.

Blaser, M., 2013. Ontological conflicts and the stories of peoples in spite of Europe: Toward a conversation on political ontology. *Current Anthropology, 54*(5), pp. 547–568.

Bond, P., 2006. Global governance campaigning and MDGs: From top-down to bottom-up anti-poverty work. *Third World Quarterly, 27*(2), pp. 339–354.

Brundtland, G.H., 1987. Our common future: Call for action. *Environmental Conservation, 14*(4), pp. 291–294.

De la Cadena, M. and Blaser, M. eds., 2018. *A world of many worlds.* Duke University Press.

Escobar, A., 1995. *Encountering development: The making and unmaking of the third world.* Princeton University Press.

Fabricant, N. and Gustafson, B., 2016. Revolutionary extraction? Mapping the political economy of gas, soy, and mineral production in Evo Morales's Bolivia. *NACLA Report on the Americas, 48*(3), pp. 271–279.

Gabor, D. and Brooks, S., 2016. The digital revolution in financial inclusion: International development in the fintech era. *New Political Economy, 22*(4), pp. 423–436.

Grugel, J. and Riggirozzi, P., 2012. Post-neoliberalism in Latin America: Rebuilding and reclaiming the state after crisis. *Development and Change, 43*(1), pp. 1–21.

Gudynas, E., 2011. Buen Vivir: Today's tomorrow. *Development, 54*(4), pp. 441–447.

Gustafson, B., 2009. New languages of the state: Indigenous resurgence and the politics of knowledge in Bolivia. Duke University Press.

Harten, S., 2011. *The rise of Evo Morales and the MAS.* Bloomsbury Publishing.

Hope, J., 2021. The anti-politics of sustainable development: Environmental critique from assemblage thinking in Bolivia. *Transactions of the Institute of British Geographers, 46*(1), pp. 208–222.

Hope, J., Freeman, C., Maclean, K., Pande, R. and Sou, G., 2022. Shifts to global development: Is this a reframing of power, agency, and progress? *Area, 54*(2), pp. 154–158.

Hulme, D., 2009. *The millennium development goals (MDGs): A short history of the world's biggest promise.* DiFD. Available at https://www.gov.uk/research-for-development-outputs/the-millennium-development-goals-mdgs-a-short-history-of-the-world-s-biggest-promise (last accessed 18/11/2024).

Jerven, M., 2013. *Poor numbers: How we are misled by African development statistics and what to do about it.* Cornell University Press.

Keck, M.E. and Sikkink, K., 1999. Transnational advocacy networks in international and regional politics. *International Social Science Journal, 51*(159), pp. 89–101.

Kohl, B. and Farthing, L.C., 2006. *Impasse in Bolivia: Neoliberal hegemony and popular resistance*. Zed Books.

Mamani, P., 2011. Cartographies of indigenous power: Identity and territoriality in Bolivia. In N. Fabricant and B. Gustafson (Eds.), *Remapping Bolivia: Resources, territory, and indigeneity in a plurinational state* (p. 32). NMSAR Press.

Mawdsley, E., 2018. Development geography II: Financialization. *Progress in Human Geography*, 42(2), pp. 264–274.

McKay, B.M., 2017. Agrarian extractivism in Bolivia. *World Development*, 97, pp. 199–211.

Mena, C.F., Arsel, M., Pellegrini, L., Orta-Martinez, M., Fajardo, P., Chavez, E., Guevara, A. and Espín, P., 2020. Community-based monitoring of oil extraction: Lessons learned in the Ecuadorian Amazon. *Society & Natural Resources*, 33(3), pp. 406–417.

Oldekop, J.A., Horner, R., Hulme, D., Adhikari, R., Agarwal, B., Alford, M., Bakewell, O., Banks, N., Barrientos, S., Bastia, T. and Bebbington, A.J., 2020. COVID-19 and the case for global development. *World development*, 134, p. 105044.

Radcliffe, S.A., 2012. Development for a postneoliberal era? Sumak kawsay, living well and the limits to decolonisation in Ecuador. *Geoforum*, 43, p. 240.

Radcliffe, S.A. and Radhuber, I.M., 2020. The political geographies of D/decolonization: Variegation and decolonial challenges of/in geography. *Political Geography*, 78, p. 102128.

Radhuber, I.M., 2012. Indigenous struggles for a plurinational state: An analysis of Indigenous rights and competences in Bolivia. *Journal of Latin American Geography*, pp. 167–193.

Riofrancos, T., 2020. *Resource radicals: From petro-nationalism to post-extractivism in Ecuador*. Duke University Press.

Walsh, C., 2010. Development as Buen Vivir: Institutional arrangements and (de) colonial entanglements. *Development*, 53(1), pp. 15–21.

PART II

Key perspectives

CHAPTER 11

Dependency and capitalism in Latin America

Chris Hesketh

Introduction

Dependency theory is one of the most important radical theoretical contributions to emerge from Latin America with key implications for how to understand its geographies of development. Dependency theory fundamentally goes beyond a superficial analysis of the global order that sees it as a world of formally independent nation-states. Instead, dependency theory provides a framework to analyse the relations of subordination between different geographical areas and how this relationship is produced and reproduced, through an international division of labour. The wealth of some countries is intrinsically linked to the poverty of other countries, and vice versa. Dependency perspectives therefore constitute a vital project of 'counter-representation' of the world, as well as a distinctive set of theoretical ideas and concepts (Slater 2004). This chapter provides the reader with an introduction to dependency theory and the geographies of capitalist development. A useful starting point is to examine what the ideas and practices of development were that dependency theory sought to challenge.

Development

After the Second World War, the dominant view of development was known as modernisation theory. Developed by scholars such as Walt Rostow (1960), this was a linear theory of development that posited that all countries would go through the same stages of development on their evolutionary path. This path went as follows: 1) traditional forms of society, 2) transitional stage, 3) take off, 4) drive to maturity, and 5) high consumption.

The obvious implication of this analysis for the poorer countries of the world was simply to attempt to mimic the development patterns found within Western capitalism. Prior to the phrase being popularised by UK Prime Minister Margaret Thatcher in the 1980s, the implication for developing countries was clear: 'There is no alternative.'

DOI: 10.4324/9781003430926-17

Latin American societies needed to further embrace capitalism and market integration to overcome economic backwardness.

Linked to this idea about the need for further market integration was an assertion that Latin America suffered from a dualistic form of society. That is to say, there were sectors of society on the one hand characterised by modern capitalist industry, but on the other hand there were sectors characterised by backwards, traditional, and feudal forms of production. The task of development was therefore to transform these latter areas into more capitalist-oriented ones. Modernisation theory demonstrates a Euro-centric approach in that it denies any geographical specificity and history to other nations and assumes that the particular history of Europe (and the United States of America) is a universal model that all must follow. Here is a good juncture to introduce the importance of dependency theory. Not only did the dependency theorists challenge prevailing ideas about development, they also challenged contemporary political practice.

In the 1950s and 1960s, modernisation theory was the dominant perspective on how countries individually developed. Furthermore, in the Latin American context, the broad idea that countries' development moved in fixed stages was echoed by orthodox Communist parties in the region, who followed the line set by Stalin and the Communist International (Comintern). The argument put forward was that socialism in Latin America was only possible on the basis of the productive forces developing (thus creating wealth to be distributed). The task for erstwhile revolutionaries was therefore to back capitalist development to speed up the process of transition. Such assertions had clearly been upended by the reality of the Cuban Revolution (1959). The Revolution sent shockwaves through both the continent and the world. It challenged the reformist, gradualist politics of the time as well as the hegemony of the United States of America in the regional affairs of Latin America. It was in this febrile atmosphere of the 1960s, which included wider debates about decolonisation (see Chapter 2) and insurgent movements for national liberation, that dependency theory came to prominence.

Debating development

Dependency theory is not a homogenous theoretical perspective. There remain divisions between its more radical and reformist elements as well as a different level of emphasis placed on economic issues versus a more philosophical approach to the question of what dependency means (see Kay, 2010; Slater 2004). However, despite these different forms, what unites all dependency approaches is a critique of unilinear, stadial approaches to development and a critique of geographical understandings that are state-centric.

Dependency theorists first and foremost take a long-term historical view of Latin American development. This begins with the fundamental impact that colonisation had on the continent. The colonisation of Latin America is fundamental in a double sense. First, it destroyed the basis of local manufacturing by bringing in foreign imports whilst banning things like refineries and destroying looms and local spinning mills. Economic activity was also reoriented to favour export-oriented **laitifundia** (Galeano 2008: 56, Gunder-Frank 1966: 12). Second, not only did colonisation drain wealth from the region, thus hindering the possibility of development, but at the same time, the export of precious metals served as a form of primitive accumulation that expanded the means of payment necessary for the genesis of European capitalism (Galeano 2008; Marini 2022). From the outset, then, a key focus of dependency theory was to explore capitalism as a world-historical process characterised by uneven development. Moreover, not only is development uneven, it is *purposefully constructed* as uneven owing to the subservient relationships established by central/core countries over peripheral/ satellite countries. The above points are vital to affirming that development is both an uneven and combined process. The combined element is important to stress, as it means that as a result, the economic preconditions that countries in Western Europe may have enjoyed for the emergence of capitalism simply cannot be replicated in Latin America (contrary to the understanding of modernisation theory). Indeed, it was the very establishment of a global division of labour that made these conditions impossible (Taylor 1979: 43). Capitalism must therefore be conceptualised as a global whole, but with different countries and regions occupying different structural positions according to the dependency perspective.

Scholars such as Theotonio Dos Santos (1970) and Andre Gunder-Frank (2000) rigorously challenged the dominant assumptions shared by neo-classical economists (in the guise of modernisation theory) and the orthodox Communist parties of the time. As mentioned above, both these schools of thought argued that Latin America had a dualistic economy whereby a modern, capitalist sector co-existed with a backward, feudal sector. Despite the seeming difference in positions of these traditions, both argued that capitalism was a necessary stage of development. The political task, therefore, was to transform space, meaning the backward, feudal sectors of the economy would be transformed into capitalist ones. More market integration was therefore necessary. For the neo-classical thinkers, this represented the pinnacle of economic evolution, whereas for the Communist parties it represented a regrettable but necessary stage in the move towards socialism (Castañeda 1993: 72; Laclau 1971: 19; Ruccio 2011: 113). In other words, history unfolded in every individual country in a linear succession of modes of production. It was this argument that dependency theory took issue with (Henfrey 1981: 18).

Rejecting the prevailing wisdom, dependency theorists argued that capitalism had to be conceptualised as a singular global process (Dos Santos 1970: 231). Rejecting the so-called 'dual-society' thesis whereby there existed pristine, untouched feudal forms of social relations, Gunder-Frank (2000: 4–5) argued that the capitalist mode of production dominated the spatial totality. What appeared at first glance to be backwards, feudal relations pertaining to Latin America were not original, untouched spaces, but rather were spaces that had been produced through their interaction with capitalism. In Gunder Frank's terms, these were not 'undeveloped' spaces but rather, they were spaces that had been produced via the 'development of underdevelopment'. This latter concept is important because it implies Latin American development can be studied not as an accidental outcome of history, but as one of purposeful and continual agency. Power relations between core and periphery thus actively constructed this form of social space that was to be found in Latin American capitalism. Latin America was therefore a constituent, if peripheral part of the global capitalist mode of production.

It is important to stress that this dependency analysis was an explicitly political intervention that sought to change political praxis. The dependency view implied that it was precisely the links established with the global market since the time of capitalist colonisation, and continuing into the present via the extraction of wealth, that created subordination. This was the exact opposite position of orthodox Communist parties and of modernisation theory which argued that the reason for the backwards nature of the productive forces in Latin America came from a lack of integration and capitalist social relations (Dos Santos 1970). Therefore, according to the dependency position, the political solution to the problem of underdevelopment was to delink from the global capitalist system, as it was this very system that prevented development from occurring. Evidence showed that the greatest economic development in Latin America occurred when ties to the core countries were at their weakest (Gunder-Frank 2000: 10). Politically this inverted the orthodox Marxist formula of socialist revolution breaking out following the development of the productive forces. Here socialist revolution was necessary first, precisely in order to develop Latin America's productive forces for its own.

For Dos Santos (1970: 232), Latin America's relationship of dependency can be broken down into three distinct eras. These are:

1) Colonial dependence – this is linked to the export trade and dominance of European nations as well as the colonial state.
2) Financial-industrial dependence – consolidated from the end of the nineteenth century, this era was dominated by big capital from industrial centres who invested abroad in the production of raw materials including agricultural products for export.
3) Technical-industrial dependence – involving multinational corporations investing in industries linked to the expansion of the internal market in Latin America.

A major insight of dependency theory is its formulation of the geographical specificity of capitalism in Latin America. This has various elements to it. Key to this, however, is the structuring effect of the global division of labour. This creates a situation whereby Latin America is reduced to the role of a primary commodity producer. These products are exported primarily for the benefit of core capitalist economies. As this is where the dominant capitalist classes from Latin America realise their profits, they have little incentive to develop the domestic markets. Thus, an economy emerges characterised by, on the one hand, low productivity and primitive techniques and low output, oriented for the domestic market. On the other hand, there is a higher productivity sector with more advanced techniques but which is largely oriented to the export sector.

Another major innovative idea, from Ray Mauro Marini (2022), was that Latin American workers suffered from super-exploitation. In short, this means that the wages paid to workers was less than the value that they needed to replenish themselves. This was only possible due to the fact that the very model of capitalism developed in Latin America did not rely on the domestic market for demand for its products, but rather the demand from advanced industrial countries. Latin America economies were thus structured on foreign market requirements, not their own. An insight to be gleaned here is that whilst the immiseration of the worker in the centres of advanced capitalism that Marx and Engels predicted had not come to pass, the reason for this is because this process had been geographically relocated to peripheral countries.

ISI as a case study of dependent development

The experience of import-substitution industrialisation (ISI) provides a useful case study with which to examine some of the major ideas of dependency theory. ISI was the dominant model of development pursued by most Latin American countries after the Second World War. Its major goal, as its name suggests, was to replace the volume of goods currently imported from core countries by stimulating domestic industrialisation (see Figure 11.1). Production was to be re-oriented to a domestic market and domestic industries were to be protected.

ISI as a development project had been spurred on by the experience of the Second Word War. The war had severely impacted global trade. Capital was no longer exported to developing countries by European powers who now needed to retain capital to finance their war efforts (Inter-American Development Bank 2007: 74–75). After the war, European reconstruction was prioritised ahead of investment in Latin America owing to the geopolitics of the Cold War. By the 1960s, every republic had an ISI policy of sorts (Bulmer-Thomas 1994: 262). As Ricardo Ffrench-Davis (1994: 188) summarises, although the specific policy content differed, 'most Latin American countries from the 1950s to the 1980s have in common the basic features of relying upon the manufacturing sector as the main engine of growth. It is therefore possible to speak of a common Latin American experience of ISI.'

At first blush, the intellectual architects of ISI would appear to hold many of the same ideas as those of the dependency theorists. Its key intellectual figures were Raúl Prebisch (see Figure 11.2) and Hans Singer associated with the *Comision Economica Para America Latina y el Caribe* (CEPAL, Economic Commission for Latin America and the Caribbean) established in 1948. Prebisch was the first to popularise terms such as 'core' and 'periphery' and made the argument that income in the core grew faster

Figure 11.1 Ford Fiat factory, Argentina.
Source: Creative Commons.

Figure 11.2 Raúl Prebisch, intellectual architect of dependency theory.
Source: https://commons.wikimedia.org/wiki/Category:Raúl_Prébisch#/media/File:Raul_Prebisch_(1954).tif

than in the periphery, linked to the declining **terms of trade** of peripheral countries. This analysis, which became known as structuralism, identified the core problem of development in Latin America within the global division of labour, and stressed the need for industrialisation in Latin America to overcome poverty. The experience of ISI in many ways served to question the basic assumptions of structuralism, with dependency theory emerging as a more radical critique. Why was this?

One reason is that Latin America never really broke out of its relationship of dependency on core countries. For example, there remained a relationship of dependency on foreign capital to expand the accumulation process. For foreign capitalists investing in Latin America, the region provided new markets which were insulated from competition as well as providing cheap labour. These foreign firms also sought to repatriate profits rather than using them to help build up the domestic economy.

Second, the process of industrialisation never managed to become functionally independent. Whilst Latin America did become less dependent on consumer goods, this was only achieved by a new relationship of dependency on **capital goods** (Skidmore and Smith 1992: 56). According to dependency analyses, Latin America suffered from a range of problems, again linked to the power dynamics of capitalism within a global division of labour. This meant, for example, that owing to the pace of technical progress (linked to the imperatives of capitalism to innovate in order to produce profit) technology quickly became obsolete. The solution for the capitalists of the core countries was to export this obsolete technology before it lost all value. Latin America thus always occupied the lower stages of industrial production.

A related problem was that the whole process of industrialisation in Latin America was tied to both foreign patterns of investment as well as technology. This was labour saving technology that had been built up over a long period of time in core countries. In the context of places such as Latin America it meant that the use of such technology failed to generate sufficient employment opportunities and absorb mass migration from the countryside (Cardoso and Faletto 1979: 5). Instead, it created issues of urban marginalisation characterised by informal labour (see Chapters 19 and 20). Politically, the alliances between the domestic bourgeoisie, workers and peasants that had sustained ISI after the Second World War broke down as countries experienced unemployment and rising labour militancy. In this atmosphere, many Latin American countries moved away from a populist class alliance to more authoritarian forms of politics.

Whither dependency theory?

Despite the popularity of dependency theory during the 1960s and 1970s, its popularity began to wane during the 1980s and 1990s, and it is fair to say that it fell out of fashion as a radical theory of development. The reasons for this are multifaceted. One explanation lies in the empirical example of the newly industrialising countries of South East Asia.

Examining the tables below (see Tables 11.1 and 11.2) comparing the rates of growth and rates of per capita growth, one can see that a key argument of dependency theorists – namely that peripheral countries were largely unable to break out of their dependent position – was challenged by the experience of these countries. Whereas the countries of Latin America had undertaken state-led development largely in a climate of 'export pessimism', the South East Asian countries had used the power of the state to enhance higher value sectors of their economies in which they could be competitive

Table 11.1 GDP growth in Latin America and East Asia 1960–2005 (average annual percentage change).

Region	1960–65	1965–70	1970–75	1975–80	1980–85	1985–90	1990–95	1995–2000	2000–2005
Latin America	4.6	5.8	6.6	5.1	0.5	1.8	3.6	2.8	1.5
East Asia	5.0	7.5	6.8	7.6	7.1	8.2	8.8	4.9	6.2

Table 11.2 GDP per capita growth rates in Latin America and East Asia 1960–2005 (average annual percentage change).

Region	1960–65	1965–70	1970–75	1975–80	1980–85	1985–90	1990–95	1995–2000	2000–2005
Latin America	1.7	3.1	4.0	2.7	–1.6	–0.2	1.9	1.2	0.1
East Asia	3.0	4.7	4.4	5.9	5.4	6.4	7.5	3.8	5.3

Source: United Nation Conference on Trade and Development (2006).

(Amsden 1997). South East Asian countries thus appeared as dynamic players within the global political economy, and far from the static peripheral caricature that was associated with dependency theory.

Others sought to challenge dependency theory from a critical perspective on the grounds of its analysis departing from Marxism and also because of its political project. René Zavaleta Mercado (2018) argued that countries such as Bolivia did not contain one overarching mode of production, but rather multiple modes of production within what he termed a 'motley society.' Zavaleta Mercado furthermore critiqued dependency theory for what he regarded as an overemphasis on external causality. This led, he believed, to a lack of emphasis on geographical difference in the form of national histories which contained possibilities for thinking about self-determination and resistance.

Another objection was whether the conception of capitalism found in dependency theory accurately reflected Latin American society. This goes to the heart of why one's theory really matters for political praxis. Ernesto Laclau (1971) challenged the view that Latin American societies could simply be labelled as capitalist societies. He argued that dependency theory mistook the broader economic system in which the continent was embedded with specific modes of production that pertained within countries. He also claimed that Gunder Frank's definition of capitalism was far too loose, and conflated relations of exchange with relations of production and surplus extraction. For Laclau, defining the relationship of exploitation became a hallmark for understanding a mode of production. As extra-economic coercion remained in much of Latin America, Laclau concluded that various modes of production were articulated simultaneously, yet within a wider economic system in which they interacted and were reinforced. Meanwhile, others claimed that the conceptual vocabulary of dependency theory in terms of 'nation' and 'periphery' away from specific notions of 'class' and 'capitalist social relations' blurred our understanding of imperialism (Lazarus 2002).

Robert Brenner (1977: 91–92), in perhaps the most famous critique, argued that the dependency position, in identifying the main problem of development being in unequal exchange and not the logic of capitalism as an exploitative system of production per se, undermines the basis for interdependence among revolutionary movements of the world. Instead, it provides the logic of autarky (rather than socialism). Foster-Carter (1978: 50) thus dismissed dependency theory as 'the unhappy progeny of vague Marxist ideas coupled with Latin American bourgeois nationalism.'

In terms of political practice, Foster-Carter (1978: 49) notes that dependency theory encountered a difficulty in translating a general macro framework into a more micro analysis. In a similar vein, Henfrey (1981) notes that despite the focus on political strategy, dependency theory ended up providing abstract generalisations about the nature of capitalism in Latin America whilst having little to do with the substance of struggles taking place across the continent during the 1970s and 1980s.

Dependency theory redux

Despite the above criticisms, dependency theory has undergone a major revival in recent years, with many scholars re-appraising the contribution that it made. Slater (2004) contended that although dependency theory came to be seen as an outdated view, this ignores the importance of geopolitical memory (**see Chapter 5**). Taking this notion further, Reis and Oliveira (2021) argue that given the current trends to decolonise knowledge production, it is highly apt to revisit dependency theory and the Latin American authors that have been ignored in the Global North. As an example of this trend, Ray Mauro Marini's classic work *Dialectica de le Dependencia* (Dialectics of Dependency) – arguably the most sophisticated work in the canon of dependency theory – was translated into English for the first time in 2022.

In an important re-evaluation of dependency theory, Kvangraven (2021) has argued that, rather than focusing on individual theorists, we need to explore dependency theory more as a research programme. She submits that this can be linked to four key issues: 1) a global historical approach, 2) the polarising tendencies of capitalism, 3) the structures of production, and 4) the constraints that peripheral countries operate within in the context of the global political economy. Taken as a whole, she believes this makes dependency theory still highly relevant in today's world.

Looking at Latin America since the early 2000s, a new post-neoliberal model of development has often been pursued (**see Chapter 8**). This is sometimes more pejoratively labelled neo-extractivist development, and has involved states forming partnerships with transnational corporations with a view to increasing state revenue from the export of primary commodities (often in the case of left-leaning states who then channel such revenue into social programmes) (**see Chapter 17**). However, numerous critics of this model of development note that this has caused profound environmental conflicts, especially with Indigenous peoples in whose territories many natural resources are often found. Furthermore, critics charge that such a reliance once again on primary exports as the basis of wealth generation resigns Latin America to a subordinate role within the global economy, accepting the arguments of comparative advantage stressed by orthodox economic theory. In other words, despite some short-term success in raising state revenue, in terms of profound shifts in internal social relations or challenging the relationships of subservience between core and periphery, the current model of development does little to break with this (Hesketh 2023). Thus, the major intellectual and political concerns of dependency theory have been revitalised in recent years, often by intellectuals from Latin America concerned with long-term sustainable development (**see Chapter 10**).

Conclusion

Dependency theory has been one of the most powerful, home-grown theoretical innovations to come out of Latin America. Although not generated by geographers per se, it has clear and direct implications for how to understand the geographies of capitalist development in the region. It profoundly challenged the dominant ideas about development and exposed the highly Eurocentric ideas of such paradigms. Dependency theory inspired thinking about how alternative means of development and radical politics might be possible and what means might be necessary to achieve this. Ideas such as delinking from the global political economy and creating South to South means of trade and cooperation also inspired radical political proposals. Examples of this include the call for the New International Economic Order, essentially a manifesto for a new trading order taken up during the 1970s by developing countries, which included demands to provide fair prices for primary commodities, greater control over natural resources, debt relief and regulation of global corporations, all to address structural conditions of poverty in the global political economy.

Given the persistence of inequality within the world, the uneven distribution of productive resources and the conflicts taking places around extractive industries in a large volume of countries within the Global South, the core ideas of dependency theory are ripe for intellectual renewal.

Summary

- Dependency theory is a radical but plural school of political economy that provides a geographical analysis of Latin American development linked to a global division of labour.
- It provided a major challenge to orthodox development theory and practice in the 1960s and 1970s.
- Despite falling out of fashion in the 1980s and 1990s, recent years have seen a major intellectual re-appraisal of dependency theory. Many argue that its core ideas still retain relevance to explain uneven development and Latin America's subordinate role within the global division of labour.

Review questions

1. How did dependency theory challenge dominant ideas of development?
2. How is dependency theory an inherently geographical theory? Please provide examples.
3. How does dependency theory relate to Marxism?
4. Can dependency theory still adequately explain contemporary capitalist development in Latin America?

Further reading

Gunder-Frank, A. (1966/2000). The development of underdevelopment, in *Latin America: Underdevelopment or Revolution*. Monthly Review Press.
Perhaps the best known work of dependency theory. Gunder Frank challenged the key orthodoxy of the time including the idea that Latin America had a dual society
Kvangraven, I. H. (2021). Beyond the stereotype: Restating the relevance of the dependency research programme. *Development and Change*, *52*(1), 76–112.
A re-appraisal of dependency theory, this article makes the case for how we can consider dependency a research programme that still retains relevance in terms of its core ideas.
Laclau, E. (1971). Feudalism and capitalism in Latin America. *New Left Review*, *I*(67), 19–38.
One of the key works that challenged the theoretical ideas of dependency theory, especially in terms of Gunder Frank.

Marini, R. M. (2022). *The Dialectics of Dependency* (translated by A. Latimer). Monthly Review Press.

An English translation of one of the key distinctly Marxist works of dependency theory that introduces key ideas such as 'super-exploitation' of labour.

Taylor, J. G. (1979). *From Modernization to Modes of Production: A Critique of the Sociologies of Development and Underdevelopment*. Macmillan Press.

Comprehensive overview of the key debates around modernisation theory, dependency and capitalism in Latin America.

Keywords

Capital goods: these are assets and resources such as machinery, equipment, and tools used in the production of consumer goods and services.

Extra-economic coercion: a form of overt labour coercion (e.g. not one that is provided through the logic of the market)

Laitifundia: large privately owned forms of land, usually belonging to an individual or family and worked by peasants or serfs.

Mode of production: the social and technical means by which a society organises itself to produce surplus output.

Terms of trade: the relative price of exports to imports that a country has.

References

Amsden, A. (1997). The state and Taiwan's economic development. In G. Crane and A. Amawi (eds). *The Theoretical Evolution of the International Political Economy*. Oxford University Press.

Brenner, R. (1977). The origins of capitalist development: A critique of neo-Smithian Marxism. *New Left Review*, (104), 25–92.

Bulmer-Thomas, V. (1994). *The Economic History of Latin America since Independence*. Cambridge University Press.

Cardoso, F. H., and Faletto, E. (1979). *Dependency and development in Latin America*. Translated by M. Uruidi. University of California Press.

Castañeda, J. (1993). *Utopia Unarmed: The Latin American Left After the Cold War*. Vintage Books.

Dos Santos, T. (1970). The structure of dependence. *The American Economic Review*, 60(2), 231–236.

Ffrench-Davis, R. (1994). The Latin American economies 1950–1990. In L. Bethall (ed). *The Cambridge History of Latin America*. Vol. 6, pt. 1. Cambridge University Press.

Foster-Carter, A. (1978). The modes of production controversy. *New Left Review*, 107(1), 47–78.

Galeano, E. (2008). *Open Veins of Latin American: Five Centuries of the Pillage of a Continent*. Three Essays collective.

Gunder-Frank, A. (1966/2000). The development of underdevelopment. In *Latin America: Underdevelopment or Revolution*. Monthly Review Press.

Henfrey, C. (1981). Dependency, modes of production, and the class analysis of Latin America. *Latin American Perspectives*, 8(3–4), 17–54.

Hesketh, C. (2023). Indigenous resistance at the frontiers of accumulation: Challenging the coloniality of space in IR. *Review of International Studies*, 1–20, doi:10.1017/S0260210523000268

Inter-American Development Bank. (2007). *Economic and Social Progress Report*. Inter-American Development Bank.

Kay, C. (2010). *Latin American Theories of Development and Underdevelopment*. Routledge.

Kvangraven, I. H. (2021). Beyond the stereotype: Restating the relevance of the dependency research programme. *Development and Change*, 52(1), 76–112.

Laclau, E. (1971). Feudalism and capitalism in Latin America, *New Left Review*, I(67), 19–38.

Lazarus, N. (2002). The fetish of "the West" in postcolonial theory. In C. Bartolovich and N. Lazarus (eds). *Marxism, Modernity and Postcolonial Studies*. Cambridge University Press.

Marini, R. M. (2022). *The Dialectics of Dependency*. Translated by A. Latimer. Monthly Review Press.

Reis, N., and de Oliveira, F. A. (2021). Peripheral financialization and the transformation of dependency: A view from Latin America. *Review of International Political Economy*, 1–24.

Rostow, W. (1960). *The Stages of Economic Growth: A Non-Communist Manifesto*. Cambridge University Press

Ruccio, D. F. (2011). Radical theories of development: Frank, the modes of production school and Amin. In *Development and Globalization: A Marxian class analysis*. Routledge.

Skidmore, T., and Smith, P. (1992). *Modern Latin America*. 3rd edition. Oxford University Press.

Slater, D. (2004). *Geopolitics and the Post-Colonial: Rethinking North-South Relations*. Blackwell.

Taylor, J. G. (1979). *From Modernization to Modes of Production: A Critique of the Sociologies of Development and Underdevelopment*. Macmillan Press.

United Nations Conference on Trade and Development. (2006). *Trade and Development Report*. United Nations Publications.

Zavaleta Mercado, R. (2018). *Towards a History of the National Popular in Bolivia*. Translated by Anne Freeland. Seagull Books.

CHAPTER 12

Decolonising territory

Rogerio Haesbaert and Sam Halvorsen

Introduction

Territory is a core geographical concept that has taken on a prominent role in Latin American geographies in recent decades. In Anglophone geography, territory is typically understood as the spatial extent of the state's sovereign power. Elden (2013) traces the history of territory back to the Roman Empire in Europe when the concept of a centralised government exerting its control over a bounded geographical space emerged. This modern understanding of territory as state space is also a colonial one and was exported to what became known as Latin America in the context of European colonisation and, in particular, in the wake of decolonisation and the formation of national territories and their frontiers (see **Chapters 2 and 3**).

The imposition of territory was from the outset contested, most notably by Indigenous communities whose lands were now being carved up by national borders. However, by the late 20th century, the idea and practice of territory was also being recreated in the context of diverse struggles over ways of life and socio-political projects.

Today, territory is a core dimension to the practices, strategies and identities of diverse groups, communities and organizations across the region. Latin American experiences have challenged narrow Anglophone framings of territory as space dominated by the state and instead indicate a wider set of relations through which space is appropriated (i.e. occupied, produced or symbolically controlled) through the ideas and activities of people and nature. In this context, territory can be understood as any powerful attempt to rework space in pursuit of social or political objectives. Territory can be further understood as a simultaneous process of creation, destruction (or abandonment) and reconstruction of space, that is, of territorialisation-deterritorialisation-reterritorialisation (TDR).

Decolonising territory has become a core dimension to contemporary social and political life as a means of directly confronting, challenging and undoing the colonial legacies of how space is produced and organised in the so-called Global South, especially Abya Yala/Latin America. Decolonising territory is not, primarily, an academic

DOI: 10.4324/9781003430926-18

pursuit. It is an everyday activity of remaking space by dismantling colonial power relations, which intersect with class, gender and race.

The chapter begins by outlining the historical relationship between territory and coloniality, introducing the idea of territory as a concrete dynamic of TDR. It then examines four dimensions through which territory is being decolonised in the thought and practice of Latin American geographies: struggles for autonomy; re-existence to violence and ideas of body-territory; working relationally with territory as a more-than-human entity; and by opening up territory to a multiplicity of ideas and practices.

Coloniality and territory

Territory has a strong colonial content, to the extent that European territorial conquests occurred through processes of colonisation carried out by "central" or imperial nation-states that spread their political-territorial form throughout the world, with the intention of universalising it. The colonial conquest of Latin America was marked by dynamics of exploitation (of labour, especially via slavery), expropriation (extractivism of the land and its resources) and (patriarchal and racist) oppression, promoted by state power and the activities of large capitalist companies. The strengthening of many European modern-colonial states such as Portugal and Spain, in the 16th and 17th centuries, relied on these colonial territorial relations which, in Latin America, were strengthened by the riches of its raw materials (e.g. wood, gold, silver) and its capacity for livestock and cultivation (e.g. sugar cane, coffee, cocoa and other "commodities").

Two groups in particular were objects of this colonising process, with repercussions felt to this day: the Indigenous as original peoples and the deterritorialised slaves of the African continent (see **Chapters 13 and 15**). It is precisely these groups, expropriated from their territories of life, that today most clearly manifest the struggle for territory, often with women (and their "body-territories") at the forefront (see **Chapter 15**). Another group marked by the strong and ongoing coloniality of power are the subaltern classes of large urban peripheries, which make up the majority of Latin America's population, many being descendants from Afro and Indigenous people. Residents of urban peripheries (see **Chapter 19**) often develop strong bonds with their spaces of everyday reproduction, (re)claiming their processes of identification and resistance as territorial struggles.

Even after the process of formal colonisation was completed – in most countries, at the beginning of the 19th century (see **Chapter 2**) – the coloniality of power was maintained and sometimes even strengthened through new mechanisms of dispossession and exploitation, including financial dependence. In addition to such a coloniality of power, there exists what can be termed a coloniality of knowing (based on the a priori superiority of Eurocentric knowledge), the coloniality of nature (totally subordinated to exploitation by society), and the coloniality of being. The latter creates territories – or zones, as Franz Fanon (1967) said – of being and non-being, of those who have the right to existence and visibility and those to whom the possibility of being is denied.

Invisibility and dehumanisation are the primary expressions of the coloniality of being, as Maldonado-Torres (2017) tells us. Albán (2017) considers it more potent to speak of silencing, because somehow, through denial, these groups were seen negatively. The monocultural universality of capitalist colonial modernity has denied many cultures their manifestation and reproduction, provoking re-existence, understood by Albán (2017) as a praxis of decolonial liberation.

The coloniality of power is maintained through the exploitation of labour (with one of the greatest social inequalities in the world), resource extraction (with one of the highest concentrations of land ownership), the patriarchal system (with highest rates of femicide and violence against women, **see Chapters 14 and 23**) and religious fundamentalism (with the politicisation of Pentecostal religions in urban peripheries). All these dynamics are related to (post)colonial destruction and reappropriation of territories.

In the case of Indigenous peoples, the recovery of their original territories is a long process that intensified throughout the 20th century, but was always faced with strong resistance from national elites. This has been recognized by authors such as Pablo González Casanova (2007) as internal colonialism. In other words, in addition to the coloniality of power exercised by hegemonic groups external to Latin America, one can see the (re)production of elements of this coloniality inside Latin American countries. In the case of Indigenous peoples, whose existence depends on their cultural attachment with specific territories, based on its natural diversity, the destruction of these spaces corresponds to a "terricide" where the destruction of territory also means the disappearance of an entire ethnic group.

Terricide corresponds to the most violent and radical process of deterritorialisation (i.e. uprooting of peoples and the basis for their existence), as it destroys territories that correspond to certain ways of life. Territory disappears here in the two conditions highlighted by Gottman (1973): territory as a resource (source of water, land for cultivation, extraction of products from nature, etc.) and territory as shelter (living and working daily conditions).

While the hegemonic coloniality of power emphasises territory as a piece of land or a resource to be dominated and exploited (such as agricultural land, subsoil of fossil fuels, rivers for water, fishing or power, etc), the counter-power of the hegemonised groups values territory as shelter, a condition of life or, as many authors say in the Latin American context, of re-existence – where they not only resist the domination that is imposed on them but also seek to build other forms of existence capable of providing them with greater autonomy (**see Chapter 23**).

In the midst of this conflicting and contradictory history of the intense coloniality of power imposed by the modern-colonial capitalist, racist and patriarchal system, territory becomes, more than an analytical concept, a category of practice triggered by diverse subaltern groups/classes in their struggles for re-existence. In the academic field of Latin American geography, unlike much of Anglo-American geography, territory is understood through intense dialogues with these groups/classes that treat it, above all, as a tool for their struggles, based on a search for autonomy.

Autonomy and territory

Territory is central to people's capacity to govern and manage their lives. The occupation and control of space provides an opportunity for the bottom-up generation of identities, values and practices that do not necessarily adhere to dominant territorial logics that tend to be tied to the nation-state and the pursuit of producing and expanding capital. Porto-Gonçalves (2012) has identified autonomy as central to contemporary political struggles in Latin America but notes that such autonomy may be positioned both *with* and also *against* the state. This can be seen in two prominent struggles to decolonise territory in recent decades.

On New Year's Day, 1994, in Chiapas, South-East Mexico, the Zapatista Army of Liberation (known as the Zapatistas, after the agrarian revolutionary leader Emiliano Zapata) initiated an uprising that (re)claimed swathes of land from the Mexican state, declaring it as autonomous Zapatista territory. Here, territory is something more than land. More than a political or economic resource, it has a crucial symbolic dimension that signifies revolt and the search for alternatives. The date was chosen to mark the coming into force of the North American Free Trade Agreement (**see Chapter 6**) and their rebellion marked an explicit rejection of neoliberal globalisation. Yet it was also tied into a broader struggle, from a region with a high Indigenous population, to reclaim livelihoods, environments and dignity from the colonial logic of the Mexican state, long dominated by the authoritarian PRI party.

The Zapatistas initially attempted but then failed to reach an agreement with the Mexican government, and autonomy became central to their political strategy that rejected the notion of state sovereignty. Given the state's monopoly of violence, the Zapatistas felt it necessary to take up arms, even if this was more of a symbolic gesture given that they could never compete with Mexico's military power. They demarcated and governed territory through self-determined rules that included a commitment to deliberation and decentralised decision-making (see Figure 12.1). Despite the presence of a leader – Subcomandante Marcos (and, more recently, Comandante Insurgente Galeano) – the Zapatista army uses balaclavas and seeks to organise as a collective power without a sovereign centre or hierarchy.

Figure 12.1 The Rebel Territory of the Zapatistas, Chiapas, Mexico. Sign reads: "You are in Zapatista Rebel Territory – here the people rule, and the government obeys".

Source: Wikimedia.

The struggle to decolonise territory has included the development of alternative economic practices, based on fair-trade and solidarity agriculture (Naylor, 2019) as well as their autonomous system of education, health and so on. Autonomy from the state has never been complete, and the Zapatista community has interacted in a range of ways, including the receipt of state subsidies or the (involuntary) presence of roadblocks. Nevertheless, the continued existence of the autonomous, rebel territory of the Zapatistas three decades on from their first uprising remains a huge symbolic and material example of grassroots decolonial struggles for territorial autonomy.

The recent experiences of promoting the autonomy of Indigenous territories in Bolivia provides a contrasting example due to the central role of the state. In 2006, Evo Morales became the first Indigenous leader elected President, part of a broader tide of progressive governments in the region that were challenging neoliberal hegemony (see Chapter 8). Central to his agenda was the process of rewriting the constitution, which took place via decentralised constituent assemblies with a strong Indigenous component. This eventually led to the 2009 passing of a new constitution that re-assigned Bolivia as a pluri-national state in explicit recognition of Indigenous claims to territory that challenged modern, colonial sovereignty.

The establishment of a pluri-national state was of great historical significance, as it recognised Indigenous, decolonial claims to territory. However, in practice, claims for Indigenous autonomy have had to negotiate or at times resist the government's projects. Many Indigenous communities, such as the lowland Guaraní, reject any bureaucratic administrative attempts to control land and instead promote a view of autonomy that valorises relations between people and the environment (Postero and Fabricant, 2019).

In 2011, Indigenous communities from ancestral land in the Isiboro Sécure National Park and Indigenous Territory initiated a period of intense mobilisation against government plans to build a highway that would directly cut through their land (see Chapter 10). The justification for building it was to facilitate the transport of raw materials that would, in turn, fund the government's welfare state and allow for economic sovereignty beyond the neoliberal consensus. In this way, Indigenous struggles became entangled with environmental concerns that challenged the extractivist "commodity consensus" of progressive governments such as that of Morales (see Chapter 17).

Autonomy is core to any struggle to decolonise territory, yet this necessarily comes across tensions and contradictions with the nation-state that, in turn, may have ambitions to deconstruct certain colonial logics embedded in the state. This can be further understood when we shift our vantage point to the body and acknowledge the highly gendered experience of producing territory in and against the coloniality of power.

Body-territory and territory as body (of land/Earth)

One of the main contributions of Latin American thought on territory is that which relates it to the dimension or scale of the body itself. Contrary to a certain Eurocentric vision that emphasises state territory, through a territorial configuration by hegemonic groups, peripheral contexts such as Latin America allow for a (re)reading of territory from everyday scales where other forms of power are constructed. Domination, exploitation and oppression under slave regimes, for example, such as the one that prevailed for a long time in Latin America (Brazil was the last country to formally eradicate it, in 1888), took place from the control and even the torture of Black (or Indigenous) bodies.

Deterritorialisation begins with capitalist dynamics in relation to our body, deprived of land, appropriated and put "for sale" (literally, in the case of slavery) in the labour market (Federici, 2004). However, when we speak in defence of territory as a body, we move beyond the individual body and even the "social body" or the "political body" (of the nation). It responds to corporeality (or embodiment) in its broad sense, and the inseparability between spirit or consciousness and body-matter. Castro-Gómez and Grosfoguel (2007, 21) speak of a "body-politics of knowledge", since all knowledge is found "in-corporated, embodied in subjects crossed by social contradictions, linked to concrete struggles, rooted in specific points of observation".

Many of the social struggles that take place in Latin America are struggles that start from the body-territory and extend to the very body of the Earth, the consciousness of our territorial limits extending from the individual body to the entire ecumene (the habitable space) of the planet. An Indigenous group such as the Wayuu, near Lake Maracaibo, in Venezuela, analysed by Quintero-Weir (2019), clearly manifest the association between body, land and territory – to the point that a body-territory extends across several scales, from the individual body to the body of the Earth. From there, we must recognise that talking about body-territory on an individual scale is not enough: it is always inserted within broader socio-natural relationships. Following the Wayuu, territory is associated with the so-called natural conditions needed for our existence.

In the case of Latin America, under the dominant neo-extractivist economic pattern, there are countless territorial struggles involving the defence of resources, such as water, that are indispensable to the life of many social groups. The "body of the earth", therefore, becomes as fundamental as the "body of men", in a relationship of permanent inseparability. It is in this sense that the territory cannot be dissociated from the natural content that accompanies it, particularly in spaces marked by the brutal extraction of resources and dispossession of land. This visceral, existential link between body, land and territory leads to the proposition of a political ontology of territory.

Territory and political ontology

Activists and scholars increasingly emphasise that territory is not merely an object or a thing that can be decolonised; it is a way of understanding and living the world(s) that is constituted by relations between people and nature (see **Chapter 13**). The social and natural world are, for many, not conceived as separate from each other, and hence any understanding of territory must start from this perspective. For scholars such as Arturo Escobar or Mario Blaser this is the ontological realm of territory, a realm that is contested and hence highly political.

In 2017, Colombia's constitutional court granted rights to the Atrato River, a huge milestone in breaking down the anthropocentric understanding of territorial rights based on notions of private property centred on individuals (or corporations) (see Figure 12.2). This is significant in a country where rivers play a central role in the identities and political strategies of decolonising territory. Based on longstanding research along Colombia's Pacific coastline Oslender (2016) argues that local claims by Black communities over territory unfold through an "aquatic space", what Escobar (2016) terms their "relational worlds". Here, the territory is composed of multiple, more-than-human practices and ways of life that include the tidal rhythms of estuaries, the moon, mangroves, chants, rituals and spiritualities. These dimensions are central

Figure 12.2 The Atrato River in Colombia.
Source: Wikicommons.

to the political organisation of space and imply a shift towards decolonial understandings of territory.

The struggle to decolonise territory thus involves a recognition that the world contains many worlds, what may be referred to as a "pluriverse" (De la Cadena and Blaser, 2018). Agency is not solely located in human actors but is dispersed through the relational worlds and networks that traverse society and nature, part of what is best captured through the notion of Pacha Mama, Mother Earth, that is both a person and more-than-human. Such an approach is not an abstract, theoretical idea, but is grounded in a political perspective that the colonial production of territory is inextricably tied to the ecological crisis that define the current anthropocene or plantationocene (see Chapter 22).

In recent years, Latin America has seen an explosion of what Maristella Svampa terms eco-territorial movements (see Chapter 17) that are a response to the hegemony of neo-extractivism, an economic logic based on the mass extraction of primary materials for exportation. Resistance to extractivism is an embodied practice and often led by women who are on the frontline of extractivist industries. The body-territory is a key component to the life world of socio-natural relations. Decolonising territory, as a political project, must take this dimension seriously.

Andean states, most notably Bolivia and Ecuador, initiated a process of including this ontological dimension of territory in addition to Indigenous autonomy. The category of Buen Vivir (good living), also known as Sumak Kawsay in Quechua, is the most well-known attempt in this regard. As a guiding principle, or ontology, it strives to capture the knowledge, health and vitality of people and the planet, proposing a harmonious relationship that could govern modes of political organisation and well-being. It also points to the multiple ideas and practices of territory that co-exist.

Multi-transterritoriality of territory

Decolonising territory involves a struggle against binary conceptions, such as the dualisms of society-nature, spirit-matter and theory-practice. It provides a critique of abstract universalist views, often defended by Eurocentric theories and capitalist practices that destroy the economic, political and natural-cultural diversity of peoples. In this sense, decolonising also means defending the plurality of forms of life and territorial organisation, the pluriverse instead of the simple uni-verse.

Territory is no longer conceived in a single, universal and hegemonic way, based on the state's imagined nation and dominated land. Instead of a singular process of territorialisation as synonymous with political-economic domination within a large capitalist world-system, multiple territorialisations are recognised through the various forms of appropriation of space, a space – or a territory, if we emphasise the power relations that constitute it – which cannot be understood without its great natural diversity.

Decolonising territory requires us to acknowledge the multiple modalities/cultures through which it is conceived and constructed. Each Indigenous people, Afro-descendant group, or traditional people (in the Brazilian case) has its own form of territorialisation, of domination and appropriation of space, exercised through a power that is not only political and economic but also symbolic and affective, in relation to the biodiversity of nature that is part of it.

For the capitalist, who pursues a private and mercantile relationship with territory, multiterritoriality is merely functional: they circulate through multiple territories (residences, hotels, companies) to reproduce the same economic, political and cultural way of life. For many subaltern classes, multiterritoriality is often forced upon them, as they migrate in search of work, and involves adapting to culturally diverse spaces. Thus, many migratory diasporas are multi- or trans-territorial – transiting through different territories (regional, national, continental) leading to diverse economic (at the level of solidarity, for example) and cultural exchanges (see **Chapter 9**).

Subaltern groups, often residing in urban peripheries, may be forced to move between territories of legality and illegality through economic circuits such as drug trafficking. At the same time, in a more positive sense, the diversity of these groups gives rise to cultural dialogues that favour the phenomenon identified by Cuban Fernando Ortiz as "transculturation": a mixture of identities of which Latin America is a veritable laboratory. The so-called miscegenation (mestizaje) process is not only a colonial hegemonic political strategy, in search of a "whitening" (and cultural homogenisation) of the territories, but also a counter-hegemonic one, in the sense of building mutually enriching cultural mixes, without, however, losing elements of original cultures, as defended by Bolivian Indigenous intellectual and activist Silvia Rivera Cusicanqui.

Transterritoriality is both a category of analysis to understand the intersection of these multiple spaces and also a category of practice, of concrete struggle, as among Guaraní Indigenous peoples on the borders between Brazil, Argentina, Paraguay and Bolivia. They claim, on the part of the governments of these countries, the formal

Figure 12.3 Central Square of Cherán with San Francisco de Asis Church.
Source: Rogerio Haesbaert.

recognition as transterritorial peoples, thus allowing them to maintain the historic circulation through the territory currently divided between those four countries.

This can be seen in the Purépecha Indigenous people from the autonomous municipality of Cherán (see Figure 12.3), now a community governed by customs and traditions inside the Mexican state. Their rebellion of 2011 achieved relative autonomy only at the municipal level, similar to the Zapatistas, yet their achievement was multi- or transterritorial. The revolt against the violence of narco-businessmen (narcoempresarios) who exploited the region's forests and against the political-economic oppression of the state and companies succeeded by appealing to legal loopholes in the Mexican Constitution, measures from the UN global level (especially the 169 convention on Indigenous peoples), and also the resources from Purépecha migrants living abroad (half the population lives in the US). Today, in Mexico, other communities can also choose to adopt the customs and traditions ("usos y costumbres") approach to governance.

Conclusion

Decolonising territory is a political struggle that, from a Latin American perspective, requires us to take on board a multiplicity of perspectives tied to the dismantling of colonial power relations in and beyond the region. Central to such a project are the diverse peoples – subaltern, Indigenous, Afro-descendent – who appropriate territory as a political tool for greater autonomy as well as an ontology based on alternative ways of knowing and living.

Decolonising territory involves a recognition that territory is always multiple, opening up to different scales and territories, starting from the body itself (especially in the case of Indigenous women and/or feminists defending their body as their first territory) and extending to the earthly ecumene, which is the limit of our exercise of power, always inseparable from the natural forces that also define us. In this sense, experience is always multi- or transterritorial, as we transit, compulsorily or voluntarily, through multiple territories.

Territory, as a process of deterritorialisation and reterritorialisation, is inherent to our own existence, part of our ontology, and is not something external to us. Instead of simply being dominated and exploited, as in the hegemonic colonising view implies, territory must first and foremost be appropriated (symbolically) and cared for. This provides the starting point for any struggle to decolonise territory.

Summary

- Territory is part of a contradictory and plural process of spatialised power relations (territorialisation, deterritorialisation, and reterritorialisation) that involves not only hegemonic domination but also multiple forms of resistance from subalternised groups.
- Decolonising territory involves overcoming the Eurocentric view that overemphasises territory from the perspective of hegemonic groups/classes, especially through the role of the State and large corporations.
- It also involves giving voice to those who use territory as a tool for fighting for greater autonomy, such as many Indigenous and Afro-descendant peoples in Latin America who have been – and continue to be – expropriated from their lands.
- Finally, decolonising territory is not just an analytical and intellectual issue, but a practical-political one, including construction of new ways of delimiting, producing and signifying our so-called territories of life.

Review questions

1. What distinctions can be made between a Eurocentric view of territory and a decolonial perspective based in Latin America/Abya Yala? And what is the social and political importance of such distinctions?
2. In what way is territory relational? Between who or what are the relations and what implications does this have in practice?
3. Using examples, discuss how decolonial conceptions of territory have implications for political action and social transformation among subalternised groups/classes.

Further reading

Guillen, A.L.Z. (2021). Maroon Socioterritorial Movements. Annals of the American *Association of Geographers*, 112(4), 1123–1138.
Explores how maroon communities – descendants of fugitives from slavery – have produced territory in urban and rural spaces

Haesbaert, R. (2013). A Global Sense of Place and Multi-Territoriality: Notes for Dialogue from a 'Peripheral' Point of View. In: Featherstone, D., & Painter, J. (eds.). *Spatial Politics: Essays for Doreen Massey*. Wiley-Blackwell.
English-language overview of recent debates in Brazilian geography on the multiple dimensions of territoriality and its relation with conceptions of place in Anglophone literature.

Halvorsen, S. (2019). Decolonizing Territory: Dialogues with Latin American Knowledges and Grassroots Politics. *Progress in Human Geography*, 43(5), 790–814.
Overview of recent Latin American literatures and attempt to build dialogue with approaches to territory that are more dominant in Anglophone scholarship (such as that of Stuart Elden).

Maldonado-Torres, N. (2017). The Decolonial Turn. In: Poblete, J. (ed.). *New Approaches to Latin American Studies: Culture and Power*. Routledge.
Summary of the debate on the so-called decolonial turn in the context of Latin American social sciences.

Zaragocin, S., & Caretta, M.A. (2021). Cuerpo-Territorio: A Decolonial Feminist Geographical Method for the Study of Embodiment. *Annals of the American Association of Geographers*, 111(5), 1503–1518.
Summary of recent feminist work that starts from the body as a scale of territorial thought and action.

Keywords

Body-territory: our first territory, defended mainly by (Indigenous) women, and which extends relationally to other scales in which these bodies transit and fight, in short, expand and become effective.

Multiterritoriality: a relational and multi-scalar strategy that engages within and beyond the state, providing an important dimension to struggles to decolonise territory, as experienced by many migrants and Indigenous groups. When the movement of transit between different territories/territorialities is emphasised, we can also use the term transterritoriality.

Re-existence: social resistance as a struggle not only to maintain an already existing form of existence but to create a new, more autonomous and liberating form of life.

Territorialisation, deterritorialisation and reterritorialisation (TDR): the simultaneous process of construction, destruction (or abandonment) and reconstruction of territories through different (political, economic and symbolic) spatialised power relations.

Territoriality: the attempt to control people, information and things by controlling geographical space. Also used to refer to the necessary condition for the concrete existence of territory, for example, the identity of a certain group that claims a corresponding territory.

Territory: space built by the exercise of power in its multiple dimensions (legal-political, economic, symbolic-affective), scales (starting with the body) and relationships (class, ethnic, gender and with forces of nature), always involving relations of domination and resistance.

Transterritoriality: see multiterritoriality

References

Albán, A. (2017). *Prácticas educativas de re-existencia basadas en lugar: más allá del arte, el mundo de lo posible.* El Siglo.

Castro-Gómez, S., & Grosfoguel, R. (2007). Prólogo. Giro decolonial, teoría crítica y pensamiento heterárquico. In: Castro-Gómez, S. and Grosfoguel, R. (eds.) *El giro decolonial: reflexiones para una diversidad epistémica más allá del capitalismo global.* Siglo del Hombre Editores.

De la Cadena, M., & Blaser, M. (eds.). (2018). *A World of Many Worlds [FMR1] [SH2].* Duke University Press.

Elden, S. (2013). *The Birth of Territory.* Chicago University Press.

Escobar, A. (2016). Thinking-Feeling with the Earth: Territorial Struggles and the Ontological Dimensions of the Epistemologies of the South. *Revista de Antropología Iberoamericana*, 11(1), 11–32.

Fanon, F. (1967). *Black Skin, White Masks.* New York: Grove Press.

Federici, S. (2004). *Caliban and the Witch: Women, the Body and Primitive Accumulation.* Autonomedia.

González Casanova, P. (2007). Colonialismo interno (uma redefinição). In: Borón, A. et al. (eds.). *Teoria Marxista hoje: problemas e perspectivas.* CLACSO.

Gottman, J. (1973). *The Significance of Territory.* The University Press of Virginia.

Naylor, L. (2019). *Fair Trade Rebels: Coffee Production and Struggles for Autonomy in Chiapas.* University of Minnesota Press.

Oslender, U. (2016). *The Geographies of Social Movements: Afro-Colombian Mobilization and the Aquatic Space.* Duke University Press.

Porto-Gonçalves, C.W. (2012). *A Reinvenção dos Territórios: a experiência latino-americana e caribenha.* Universidad Autonomas de México.

Postero, N., & Fabricant, N. (2019). Indigenous Sovereignty and the New Developmentalism in Plurinational Bolivia. *Anthropological Theory*, 19(1), 95–119.

Quintero-Weir, J.A. (2019). *Fazer comunidade: notas sobre território e territorialidade a partir do sentipensar indígena na bacia do Lago de Maracaibo, Venezuela.* Deriva.

CHAPTER 13

Relational Indigenous spatialities

Astrid Ulloa

Introduction

In Latin America, Indigenous peoples with academic training in geography have increasingly demanded the recognition of their territories and Indigenous territorialities, and questioned the denomination of Indigenous geographies, due to the colonial trajectories of the discipline. Consequently, there is a call from Indigenous peoples, native peoples, or nations, to decolonize and/or re-signify core geographical categories such as territory, territoriality, place, human/non-human, man/woman, and public/private, among others. Indigenous peoples make this call based on their ontologies and epistemologies, highlighting the relationality between humans and non-humans as living beings (Chindoy, 2021; Villamil, 2020; Ulloa, 2021). Likewise, the emergence of Indigenous women's movements has positioned their daily actions and the defence of territories as other ways of doing politics.

These demands have emerged for several reasons. Firstly, they respond to the rethinking of geographical and territorial debates from Indigenous perspectives, which have generated diverse conceptual and methodological approaches. Likewise, they appeared due to the emergence of Indigenous feminisms, which have confronted academic traditions based on nature/culture and male/female binarism, among others, and have positioned new debates around territorial defences against extractivism and the body-territory relationship. Finally, critical, cultural, feminist, and decolonial geographies have opened debates on other territorial ontologies and feminist spatialities (**see Chapters 12 and 14**). These debates are related to the national political processes of recognition of Indigenous territories and autonomies.

The above processes have influenced the disciplinary developments of geography in each country in specific ways. For example, in Colombia, territories, territorialities, and territorial representations have been approached by anthropology and only recently have some people approached them from geography. The opposite is the case in Chile, where geography has approached Indigenous territories in a more central way. Therefore, addressing decolonial Indigenous geographies goes beyond disciplinary

DOI: 10.4324/9781003430926-19

Figure 13.1 Extractive process of coal mining and the transformation of Wayuu people's territory, La Guajira, Colombia.

Source: Astrid Ulloa, 2019.

rethinking, to position demands from the contexts of the recognition of the rights of Indigenous peoples in accordance with ILO Convention No. 169 on Indigenous and Tribal Peoples (1989), and the United Nations Declaration on the Rights of Indigenous Peoples (2007).

In Latin America, Indigenous peoples demand the recognition of territorial rights, the inclusion of the diversity of territorial relations, and the rethinking of hegemonic notions of space. Likewise, they propose the ontological and epistemological rethinking of the territorial, considering their situated and historical contexts following cultural differences, and the structural and intersectional inequalities that have affected them. Therefore, contemporary critical and radical geographies imply rethinking the debates on national territory and cartographic representations as homogeneous, to include in their analyses the power relations and the imposition of notions of territory associated with national territorial constructions and the political, economic, and extractivist demands of the States, which have ignored Indigenous rights and demands (see Figure 13.1).

The emergence of decolonial Indigenous geographies is related to Indigenous geographies as a sub-discipline, Indigenous proposals around relationality, the emergence of Indigenous feminisms, and the current decolonial approaches proposed by Indigenous academics, which transform not only geography but also the social and human sciences. I will focus in a general way on Latin America, specifically with some examples from Colombia and Chile.

Indigenous geographies

Institutionalized geography has begun to look at Indigenous territories and has focused on the problems and proposals that emerge in national contexts for the recognition of

Indigenous territories and territorialities. These processes in countries that have ratified the conventions that cover Indigenous rights have generated recognition/misrecognition by the states of rights, forms of collective property of the territory, and ancestral authorities, as well as territorialities in rural and urban contexts. These dynamics have given geography new perspectives for thinking about national territory in terms of the diversity of territorial notions, autonomies, and self-determination. It has also analyzed the implications of institutional actions of public policies with respect to Indigenous peoples.

Researchers in geography have begun to analyze Indigenous mobility and its articulation with urban spaces (Bernal, 2012; Sepulveda and Zuniga, 2015), which has implied rethinking territorial planning processes and collective rights in urban contexts. Another axis that has emerged is focused on the analysis of Indigenous territorial reconfigurations and historical changes, due to the ways in which other territorialities are superimposed, generating dispossessions, and fragmenting the territories of Indigenous peoples since the conquest and the colony (Caniguan, 2020). All these new analyses have implied changes in cartographic representations from the ethnolinguistic, territorial, and cultural location of Indigenous peoples and the processes of asymmetries in the production of knowledge around territorial representations (Hirt, 2012) to proposals for counter-mappings or radical cartographies from Indigenous notions (Offen, 2009; Sletto, 2014; Sletto et al., 2020) and as political tools for territorial defence and decolonization (Melín et al., 2019; Romero-Toledo and Sambolin, 2019).

However, critical geographies (Zaragocin et al., 2018) and Indigenous perspectives (Mavisoy Muchavisoy, 2018) demand the decolonization of territorial categories and/or their resignification, and therefore of geography itself, given that they have served the processes of appropriation and territorial occupation of Indigenous peoples. Therefore, they question the name of Indigenous geographies and instead point towards the emergence of critical and decolonial spatial proposals that focus on Indigenous ontologies and epistemologies, and the proposals of Indigenous academics not only from geography but also from the social sciences, which I call relational Indigenous spatialities.

Relational Indigenous spatialities

The questioning of external views of Indigenous territories and territorialities gives way to perspectives that focus on Indigenous ontologies and epistemologies from their territorial concepts and proposals of Indigenous geographers and Indigenous academics, who do not find in geography a perspective that accounts for their territorial dimensions and spatial practices. Likewise, Indigenous peoples demand the recognition of their worldviews, and ways of knowing, which implies the recognition of their autonomy and self-determination. Positioning their livings, experiences, actions, and territorial, social, political, and economic practices, which are based on the principles that are born from the law of origin or major law of each people. For example, Pewenche territorial practices occur through mobility and the interconnection with footprints and trails used by the communities (Huiliñir-Curío, 2015; Huiliñir-Curio and MacAdoo, 2014).

To position their relational ontological and epistemological perspectives, Indigenous peoples use Spanish words for body-territory, territory, nature, living beings, or nonhumans, but they do not always correspond to the debates that have been taking place

Figure 13.2 Mapuche people's demands of autonomy (Ancestral territory in recovery, Lof Tri-weche, reads the sign).

Source: Astrid Ulloa.

in various disciplines or in social movements. This ontological and epistemological difference is due to the incommensurability of certain concepts, and the problems of comparison/appropriation from the academic disciplines, in terms of notions of time, space, scales, materiality, spirituality, and agency of the non-human, among others. As Gladys Tzul Tzul (2015) states, "Our interpretations make use of certain theoretical tools produced by some university, but that have their ways of knowing own elaboration and that are taking shape according to the territorial and temporal space from where they are produced" (Tzul Tzul, 2015, p. 12).

Indigenous universes and worlds are defined according to the ontology of each Indigenous nation, such as *Wallmapu* (Mapuche ancestral territory, **see** Figure 13.2), *Tabanok* (place of departure and return of the Camëntšá people), among others. These respond to a relationality that states that all human and non-human beings are alive and in permanent interrelation and interdependence, which allows the circulation of life. Under their relational ontology, human-non-human territorial relations imply ways of being, remaining, doing, and feeling in and with the territory. They have embodied relations among all beings.

The relational ontological conceptions imply an intertwining between humans and non-humans and the dynamic interrelation of life, through rhythms, colours, textures, smells, and flavours, crossed by feelings, emotions, and affections. Indigenous territoriality is relationally given such that "they insert human beings in a network of relationships with animals, ancestors, and divinities, where exchange and reciprocity play a fundamental role" (Benciolini, 2017, 7). Indigenous peoples position their cultural and territorial policies of life, with a political ontology (Escobar, 2015) centred on autonomy and self-determination over their territories, and in permanent interaction with other living beings (Ulloa, 2021).

Intertwined human-non-human temporalities

Territories are living beings with memories where practices and relations of living beings are inscribed and where diverse symbolic, political, economic, and social relations, among others, are articulated. They are relationships traversed by temporal and spatial notions of movement, where memory is becoming (in which past, present and future merge permanently), and places where spiritual and material dimensions of multiple worlds converge, making up their universes. Non-humans also relate to specific places, which give them meaning and identities, and exercise territoriality.

In Camëntšá thought, for example, relationships with plants allow us to understand the interaction that transcends human corporeality and is in permanent fusion and continuity with them, becoming part of them. Living beings have their territories (worlds), their knowledge, and their territorialities, emotions, and affections, which imply negotiations or conflicts among them (Chindoy, 2021).

Similarly, in the Wayúu thought and their political proposals for the positioning of their ways of life, they propose a collaborative relationship with different beings to continue with life. In the Wayúu ontology, according to Guerra-Curvelo (2019), all living beings have the capacity for agency and maintain relationships of affinity and kinship. The worldview and practices of the Wayúu are embedded in their territory. Guerra-Curvelo (2019) states, "The winds are beings that can be masculine or feminine, have alliances, loving or antagonistic relationships with each other, and transit through ancestral places and paths" (Guerra-Curvelo, 2019, 89).

Vertical, spiral, or discontinuous territorialities

Peoples maintain a relationality with their territories that implies recognizing local geopolitics or Indigenous alter-geopolitics (see Chapter 5), which refers "to the daily and symbolic spatial relations and practices that are inscribed in the territories of Indigenous peoples, which allow them to consolidate defence strategies in the face of the intervention of economic and political actors that confront them" (Ulloa, 2012, 15). The dynamics of autonomy and self-determination of Indigenous peoples in their territories include spiritual and sacred dimensions, based on relational ontological concepts and on the intertwining of the human and non-human worlds. This entanglement implies that territories can be discontinuous and can establish relationships that come and go according to the times and territorialities of all beings; these relationships, which condense time and space, are called spiral times and territorialities by several Indigenous peoples.

Spiral territorialities involve diverse times and spatial practices of living beings. There are also spiral territorialities when Indigenous peoples come and go between their ancestral places, and the new territorial dynamics, imply diverse territorialities, which can be discontinuous, vertical, interrelated, or embedded. From these perspectives, new approaches to the territorial practices of Indigenous peoples in urban contexts have been proposed, making visible the Misak spatial practices in the spiral movements between their ancestral territory and Bogotá (Acosta, 2016), and the territorial reconfigurations of the Emberá in contexts of territorial dispossession in Puerto Boyacá that implies mobility practices in different moments and returning to their ancestral territory for specific symbolic and cultural practices (Díaz Santamaría, 2020).

Living territories of conviviality

All living beings have their domains in specific places (land, sea, air), and they exercise territoriality, but at the same time, they create links with other beings (human, non-human, ancestors) establishing a network connection of mutual affectation and reciprocity in specific places. They are places that articulate memories, knowledge, and temporalities of shared relationships and experiences, which in turn are related to other places. Indigenous peoples have a relationship with the non-human, including territory as a living being, according to conceptions and categories related to the feminine and the masculine. Among Indigenous peoples, these relationships with territories and non-humans are based on the practices of both women and men and on their ontological and epistemological notions. Such gendered relations establish the ways of use, access, control, decision-making, and rights around territorial processes or practices related, for example, to agriculture or climate variability.

Although the debate is very complex, what is certain is that, in the face of global changes, Indigenous women, through their daily practices, are making these changes more evident and have mobilized to position the care of the body, territory, and life. The Nasa people re-signify the territory in the face of extractive processes, based on political-organizational proposals and other ways of relating to the non-human, consolidating their autonomy, self-determination, and worldview, which implies a demand for historical territorial justice, based on their autonomy and self-determination (Caro, 2021).

Indigenous women and territorial defences

It is important to highlight the emergence of Indigenous feminism and Indigenous women's networks on territorial defences. The demands for political spaces of and for Indigenous women have been shaped through national and international meetings (Sciortino, 2011), the political formation of female leadership (Gómez and Sciortino, 2018; Ulloa, 2020), participation in national and transactional scenarios and in Indigenous women's networks (Galeano and Werner, 2015), and political proposals articulated to territorial defence (Vallejo and Duhalde Ruiz, 2019; Ulloa, 2020). The political and daily actions of Indigenous women have positioned the body-territory relationship and new spatial representations.

Indigenous women argue that there are differences and inequalities between men and women in their cultural practices and through external impositions since colonial times, yet they argue that within some Indigenous contexts there is also complementarity between men and women under different notions of being a woman and a man. For these reasons, Indigenous feminisms (communitarian, communitarian-territorial, communitarian-territorial-antipatriarchal) have rethought the debates on gender and feminisms, generating criticism and distancing themselves from hegemonic feminisms, and positioning what I call territorial feminisms (Ulloa, 2016). These territorial feminisms seek, based on women's territorial experiences and collective or individual dynamics, the defence of their territories and the recognition of their differences as women, in political contexts within their organizations or in dialogue with external and non-Indigenous actors. These territorial feminisms seek equality and justice within their peoples and the recognition of their autonomy and body-territory relations based on their conceptualizations, and in dialogue with networks and alliances of other Indigenous women

(Cabnal, 2010, 2015; Guzmán, 2021). In this way, Indigenous women rethink not only feminism but also demands around body-territory relationships.

Multiple bodies-territories

Indigenous women's political proposals of the defence of body-territory respond to their ontological and epistemological notions, which position networks of life and interrelated worlds based on their relational territorial practices. The body-territory interrelationship raises the defence of worlds that are part of their universes and makes visible the relationships between living beings with the capacity of agency, which allow the continuity of life.

Body notions and practices transcend human corporeality and establish permanent relationships with other beings, which respond to Indigenous perspectives of interrelation between body-territory-non-humans in the framework of their cultural categories, which relate dimensions and meanings in specific places with the sense of belonging, collective identity, and emotions. Broadly speaking, they argue that human becoming is always in relation to the non-human. Given those living beings are embedded in multiple intertwinements, body-territory invokes interactions through reciprocities, emotions, affections, and even confrontations between living beings. Thus, humans-non-humans become "bodies as living and historical territories that allude to a cosmogonic and political interpretation" (Cruz Hernández, 2016, 44).

Interwoven aesthetic representations

One of the strategies being used by Indigenous women in territorial defence, which has gained strength in recent times, is the use of art and audiovisual media to make their denouncements through aesthetic representations and media practices (Belotti, 2022; Orobitg, 2020; Magallanes Blanco and Ramos Rodríguez, 2016). Representations that position other ways of doing politics become tools to disseminate, denounce, resist, value, and transmit their thoughts and political positions in defence of their bodies-territories. That is the case of the political strategies of Maya women in Mexico and Guatemala through theatre and dance in creative collective spaces as a response to the racism, exclusion, discrimination, and violence they experience daily. These ludic-political strategies allow them to engage in collective healing processes and to position their knowledge and autonomous processes as part of political proposals for the defence of their rights (Méndez, 2020). Similarly, in response to this violence and dispossession in the territories of the Camëntšá people, some women express themselves using audiovisual media and their own aesthetic representations to defend their bodies, thoughts, spirits, and places (Chindoy and Ulloa, 2023) (see Figure 13.3).

They are decolonial aesthetics or an Indigenous aesthetic of resistance that allows the positioning of resistance against colonial, patriarchal, and capitalist processes. Perspectives that rethink the idea of politics and the political and focus on everyday practices and other ways of doing politics, such as the defence of life and body-territories (Ulloa, 2021, 2023), reconfigure the representations on their territories and bodies, and position their political agency, and their feminine and spiritual strength as expressions of an aesthetic insurgency since they position their ways of representing and their daily and symbolic practices in an intertwined way. Already from their knowledge and

Figure 13.3 Camëntšá women and political participation.

Source: Yanitza Chindoy, 2017.

relations with the non-human, they propose decolonial aesthetics, which present other ways of seeing their bodies-territories (Chindoy and Ulloa, 2023).

Conclusions: diverse ways of rethinking space

In contexts of territorial violence and environmental injustices, the territorial demands of Indigenous and women peoples manage to make territorial injustices visible and demand a territorial justice that considers their temporal and spatial notions and the historical injustices processes that have affected their worlds. They demand respect for cultural and spiritual practices in accordance with their spatial dynamics and with the relationality and interdependence between living beings. In other words, Indigenous peoples demand relational territorial-environmental justice. Indigenous relational spatialities make visible the body-territory relationship from political and aesthetic dynamics, which implies rethinking the political, collective, and spatial practices. These spatial relationships are associated with ways of being, being, and feeling in and with the territory, allowing the recognition of their rights to the continuity of life in their territories (Chindoy, 2021).

The conceptual contributions from Indigenous relational territorial perspectives rethink the notions of territory and territoriality from multiple dimensions and spatial representations of human-non-human from the embodied relationship between body-territory. In these processes, Indigenous peoples and women position their actions as Indigenous alter-geopolitics (Ulloa, 2011, 2012), based on their relational autonomy. In this way, they are changing the geographical thought (Ulloa, 2024) and their categories and representations in Latin American geography.

Summary

- Indigenous peoples demand their territorial rights based on their relational ontology and their interactions with all living beings.
- Indigenous feminisms reconfigured the categories of gender and feminism proposing body-territory defences.
- The political actions of Indigenous peoples propose not only to decolonize binary and time/spatial categories, but also territoriality by position alter-geopolitics based on their autonomy and self-determination from their knowledge and political proposals.
- Aesthetic representations of Indigenous women are putting forth a new way of defending their territory against violence due to extractive processes.

Review questions

1. What has been the importance of Indigenous territorial perspectives for not only Latin American but also global geography?
2. How have Indigenous feminisms contributed to the geographical categories of body, corporeality, and territory?
3. What does the notion of territory as a living being imply for geography?

Further reading

Quintero-Weir, J., P. Mansilla-Quiñones and A. Moreira Muñoz (2023). The Exile of Juyá: Decolonial Geonarratives of Water. *GeoHumanities*. https://doi.org/10.1080/2373566X.2022.2155561.
Takes up Wayuu's knowledge from a decolonial perspective and its relation to territory in contexts of climate change, to position other ways of thinking and relating to the non-human.
Romero-Toledo, H. and K. Jenkins (2022). Contested waters, extractivisms, and territories: Indigenous people in Chile and the neoliberal crisis. In: *Indigenous water and drought management in a changing world*, edited by Miguel Siou, Elsevier, 189–208.
Presents resistance and demands for the recognition of the ancestral territories of Indigenous peoples in Chile, in the face of colonization and coloniality processes related to extractivism.

Sletto, B., J. Bryan, A. Wagner and C. Hale (Ed.). (2020). *Radical cartographies: Participatory mapmaking from Latin America*. University of Texas Press.
Positions other ways of cartographic representation from the voices and representations of Indigenous peoples, Afro-descendants, and local communities.
Ulloa, A. (2020). The rights of the Wayúu people and water in the context of mining in La Guajira, Colombia: Demands of relational water justice. *Human Geography*, 13(1), 6–15. https://doi.org/10.1177/1942778620910894.
Addresses the rights of water as a living being and the body-territory-water interrelationship from the perspectives of Wayuu women, in contexts of coal mining.

Keywords

Indigenous aesthetics of *resistance*: Indigenous representations that express their knowledge and relationships with the non-human, which become tools to disseminate, denounce, resist, value, and transmit their thoughts and political positions in defence of their bodies-territories.

Indigenous alter-geopolitics: Indigenous collective strategies that confront the logic of transnational and national economic and political appropriation, by demanding recognition of their autonomy over their territories, including the air and the subsoil, throughout the vertical and horizontal political control and protection of their territories.

Indigenous relational spatialities: Indigenous peoples' conceptions of territory, territoriality, and spatial practices according to their relational ontology and under cultural notions of time-space in specific places where develop human-non-humans' interrelations.

Indigenous territorial feminisms: the political dynamics of Indigenous women and their defence of life and territories, centred on the defence of the care of the body-territory and the non-humans that allow the continuity of life in their worlds.

References

Acosta, G. (2016). *Entre territorios admitidos y territorios blindados: Reconfiguraciones espaciales del pueblo Misak entre el valle de Pubenza y el borde urbano de Bogotá, DC*. Bogotá, Colombia: Universidad Nacional de Colombia. https://repositorio.unal.edu.co/handle/unal/58213

Belotti, F. (2022). *Indigenous Media Activism in Argentina*. Routledge Focus.

Benciolini, M. (2017). Territorialidades relacionales: Conflictos ambientales y cosmopolíticas en el occidente y norte de México. *Frontera Norte*, 29(58), 5–23.

Bernal, M. (2012, enero-junio del). Territorialidad nasa en Bogotá: apropiación, percepción y sentido de lugar. *Cuadernos de Geografía. Revista Colombiana de Geografía*, 21(1), 83–98. ISSN: 0121–215X, Bogotá, Colombia.

Cabnal, L. (2010). *Feminismos diversos. El feminismo comunitario*. Acsur Las Segovias.

Cabnal, L. (2015, de septiembre 11). Lorena Cabnal, feminista comunitaria. *SUDS*. https://suds.cat/es/experiencias/lorena-cabnal-feminista-comunitaria/, consultado el 1 de febrero de 2020.

Caniguan, N. (2020). Construcciones sociopolíticas del territorio. Movimientos indígenas y políticas públicas, la configuración de los espacios locales. *CUHSO*, 30(2), 19–40. https://doi.org/10.7770/cuhso-v30N2-art2140

Caro, C. (2021). *"La piquiña de la minería." Prácticas territoriales y transformaciones socio-espaciales en los resguardos indígenas nasa del Cerro Munchique Santander de Quilichao, Colombia 2009–2019.* Bogotá, Colombia: Universidad Nacional de Colombia. https://repositorio.unal.edu.co/handle/unal/82163

Chindoy, Y. (2021). *Tejidos, resistencias y resignificación de la mujer Camëntšá y sus territorios. Cabëngbe Waman Luar, Nuestro sagrado lugar de origen. Sibundoy – alto Putumayo Colombia. Trabajo de grado. Departamento de Geografía.* Universidad Nacional de Colombia.

Chindoy, Y. and A. Ulloa (2023). Defensa cuerpo-territorio de las mujeres Camëntšá, Colombia: Acciones estético-políticas y propuestas conviviales. *Revista: Iberoamericana*, XXIII(84), 31–56. https://doi.org/10.18441/ibam.23.2023.84.31–56

Cruz Hernández, D. T. (2016). Una mirada muy otra a los territorios-cuerpos femeninos. *Solar*, 12(1), 35–46.

Díaz Santamaría, J. M. (2020). *Producción de comunidades y territorios ambivalentes, entre el reconocimiento y la repulsión de las familias embera asentadas en Puerto Boyacá, Colombia.* Bogotá, Colombia: Tesis Maestría en geografía, Universidad Nacional de Colombia. https://repositorio.unal.edu.co/handle/unal/79151

Escobar, A. (2015). Territorios de diferencia: la ontología política de los "derechos al territorio". *Cuadernos de Antropología Social*, (41), 25–38.

Galeano, K. and M. Werner (2015). Mujeres indígenas y aborígenes del Abya Yala. Agendas solidarias y diversas. *Ciencia Política*, 10, 227–252. https://doi.org/10.15446/cp.v10n19.52379

Gómez, Mariana y Silvana Sciortino, comp. (2018). *Mujeres indígenas y formas de hacer política: un intercambio de experiencias situadas en Brasil y Argentina.* Tren en Movimiento.

Guerra-Curvelo, W. (2019). *Ontología Wayuu: categorización, identificación y relaciones de los seres en la sociedad indígena de la península de La Guajira, Colombia.* Tesis doctoral en antropología. Bogotá: Universidad de los Andes. https://repositorio.uniandes.edu.co/handle/1992/41315

Guzmán, A. (2019). *Descolonizar la memoria. Descolonizar los feminismos.* Tarpuna Muya-Feminismo comunitario antipatrialcal-Quillasuyo Marca, Tarpuna Muya, Bolivia.

Hirt, I. (2012). Mapping dreams/dreaming maps: Bridging indigenous and western geographical knowledge. *Cartographica*, 47(2), 105–120. https://doi.org/10.3138/carto.47.2.105

Huiliñir-Curío, V. (2015). Los senderos pehuenches en Alto Biobío (Chile): articulación espacial, movilidad y territorialidad. *Revista de Geografía Norte Grande*, (62), 47–66.

Huiliñir-Curio, V. and A. MacAdoo (2014). Las disputas del espacio y los senderos pehuenche en Alto Bíobío. *Revista Geográfica del Sur*, 5(7), 95–112.

Magallanes Blanco, C. and J. M. Ramos Rodríguez (2016). *Miradas propias: pueblos indígenas, comunicación y medios en la sociedad global.* Ediciones Ciespal, Universidad Iberoamericana de Puebla y Abya-Yala.

Mavisoy Muchavisoy, W. J. (2018). El conocimiento indígena para descolonizar el territorio. La experiencia *Kamëntšá* (Colombia). *Nómadas*, 48, 101–115.

Melín, M., P. Mansilla and M. Royo (2019). *Cartografía cultural del Wallmapu. Elementos para descolonizar el mapa en territorio mapuche.* Lom Ediciones.

Méndez, G. (2020). Espacios colectivos-políticos-creativos y la producción de pensamiento de las mujeres indígenas en México y Guatemala. In: *Mujeres indígenas haciendo, investigando y reescribiendo lo político en América Latina*, editado por Astrid Ulloa. Universidad Nacional de Colombia, 319–344.

Offen, K. (2009). O mapeas o te mapean: Mapeo indígena y negro en América Latina. *Tabula Rasa*, (10), 163–189. Bogotá – Colombia.

Orobitg, G. (Coord.). (2020). *Medios indígenas teorías y experiencias de la comunicación indígena en América*. Iberoamericana – Vervuert.

Romero-Toledo, H. and A. Sambolin (2019). Indigeneidad y territorio: los aymaras y quechuas en el norte de Chile. *Scripta Nova*, XXIII(611), 1–32.

Sciortino, S. (2011). Saberes y prácticas situadas: la experiencia de las mujeres mapuches en los encuentros de mujeres. In: *Saberes situados/Teorías transhumantes*, edited by María Luisa Femenías and Paila Soza Rossi. Fundación de Humanidades y Ciencias de la Educación de la Universidad de la Plata, 115–139.

Sepulveda, B. and P. Zuniga (2015). Geografías indígenas urbanas: el caso mapuche en La Pintana, Santiago de Chile. *Rev. geogr. Norte Grande*, (62), 127–149.

Sletto, B. (2014). Cartographies of remembrance and becoming in the Sierra de Perijá, Venezuela. *Transactions-Royal Geographical Society*, 39, 360–372. https://doi.org/10.1111/tran.12038

Tzul Tzul, G. (2015).¿Cómo construyen crítica las comunidades indígenas? Un acercamiento a las formas de la exclusión epistémica. *Lasaforum Winter 2015*, XLVI(1),12–13.

Tzul Tzul, G. (2019). La forma comunal de la resistencia. *Revista de la Universidad de México*, 847, 105–111.

Ulloa, A. (2012, de noviembre de 1). Los territorios indígenas en Colombia: de escenarios de apropiación transnacional a territorialidades alternativas. *Scripta Nova. Revista Electrónica de Geografía y Ciencias Sociales*, XVI(418), 65. [En línea]. Barcelona: Universidad de Barcelona. <www.ub.es/geocrit/sn/sn-418/sn-418-65.htm>

Ulloa, A. (2016, octubre de). Feminismos territoriales en América Latina: defensas de la vida frente a los extractivismos. *Revista Nómadas*, (45), 123–139.

Ulloa, A. (2020). Mujeres indígenas: participando y haciendo política. En: *Mujeres indígenas haciendo, investigando y reescribiendo lo político en América Latina*, editado por Astrid Ulloa. Bogotá: Universidad Nacional de Colombia, 11–25.

Ulloa, A. (2021). Repolitizar la vida, defender los cuerpos-territorios y colectivizar las acciones desde los feminismos indígenas. *Ecología Política*, 61, 38–48. https://doi.org/10.53368/EP61FCep03

Ulloa, A. (2024). Destabilising geographies in Colombia: Trajectories and perspectives. *Transactions of the Institute of British Geographers*, 49, 1–8. https://doi.org/10.1111/tran.12588

Ulloa, A. (2023). Cuerpos-territorios en movimiento: Mujeres indígenas y espacialidades relacionales. En: *Corpos e Geografia: expressões de espaços encarnados*, organizado por Joseli Maria Silva, Marcio Jose Ornat e Alides Baptista Chimin Júnior. Brasil, TodaPalavra Editora, 325–342.

Vallejo, I. and C. Duhalde Ruiz. (2019). Las mujeres indígenas amazónicas: actoras emergentes en las relaciones Estado-Organizaciones Indígenas Amazónicas durante el Gobierno de Alianza País en el Ecuador. *Polis*, 18(52), 30–44.

Villamil, J. (2020). *La territorialidad del pueblo kamëntšá de Sibundoy (Putumayo, Colombia). Una dimensión cultural para la construcción política*. Tesis de la Maestría de Investigación en Estudios de la Cultura, mención en Políticas Culturales. Universidad Andina Simón Bolívar.

Zaragocin Carvajal, S., M. Moreano Venegas and S. Álvarez Velasco (2018, mayo). Hacia una reapropiación de la geografía crítica en América Latina. *Íconos. Revista de Ciencias Sociales*, (61), 11–32. Quito.

CHAPTER 14

Feminist geographies

Sofia Zaragocin

Introduction

Latin American feminist geographies focus on the social construction of space and territories from Latin American feminist theories and praxis. Contemporary Latin American feminist theories are vast and plural, making the field of Latin American feminist geographies highly diverse. Key topics include the geographies of gender-based violence, feminist activist geographies and feminist debates on territory that will be further explored in this chapter. One of the central features defining Latin American feminist geography is that it is critical of the dominance of Anglophone feminist geography, defined as feminist geography that is written in English and addressed to a Global North audience. In recent years there has been a push towards developing feminist geographies from the Global South and Latin America in particular.

The creation of GEOFEMSUR, a group of feminist geographers in different countries of Latin America focused on developing a Latin America feminist geography, is an example of this. The first virtual gathering in 2021, organized by Astrid Ulloa, Diana Ojeda and the Francia Marquez collective in Colombia, focused on decoloniality and antiracist geographies. GEOFEMSUR is part of a larger apparatus of feminist geography networks and alliances established in the region. In 2023, GEOFEMSUR met online, hosted by the Institute for Advanced Studies in Inequalities at the Universidad San Francisco de Quito, Ecuador. Prior meetings included the Latin American Seminar on Geography, Gender and Sexuality, which has had regional meetings in Brazil, Mexico, Argentina and Chile since 2011, while universities in Mexico have hosted five Gender and Space International annual conferences since 2018.

Major topics in these regional and international conferences happening in Latin America include: geographies and cartographies of gender-based violence, feminist political ecology and geography, systems of care, intersectionality, domestic space and new ways to inhabit, sexualities and space, safe spaces, extractive industry and the defence of territory, spatial dynamics of migration processes, masculinities, ethical environmentalism, new urban and rural spatialities, decoloniality and anti-coloniality,

DOI: 10.4324/9781003430926-20

anti-racism and critiques to cisgender and heteronormative spaces. These established regional meetings act as spaces of care and for the production of collective knowledge construction across the Americas, linking feminist geography praxis and reflection from the Global South. Lastly, the creation of the Latin American Journal of Geography and Gender (Revista Latino-America de Geografia e Género) in 2009 demonstrates a sustained effort by feminist geographers in the region to foster feminist geography from Latin American perspectives.

Within Latin America there are countries with more established feminist geography knowledge production and praxis. In the region, there are at least three ways that feminist geography is produced and disseminated. These three outlets of feminist geography knowledge and praxis pertain to the academy, activist groups and feminist geography collectives and, more recently, by public policy sector. Within the academy, geography departments in Brazil, Mexico, Argentina, Chile and Colombia are making important contributions on feminist and gender geographies based on the work of leading feminist geographers in these countries (see the work of Joseli Silva, Paula Soto, Monica Coimbra, Yasna Contreras and Astrid Ulloa, respectively). Nevertheless, as feminist geographers in these countries highlight, despite several decades of knowledge production in gender studies and feminist theories, questions on the legitimacy of the field by masculinist critical geography persist (Hanson, 2020).

Feminist geography networks within Latin American academies are an important way for feminist geographers to support one another and create mentor-mentee relationships. Another area where geographical knowledge is produced in Latin America is through strong connections to feminist activist groups, which benefit from social cartographic methods to depict long-standing struggles such as gender-based violence. Feminist geographic collectives are present in many Latin American countries, usually outside of academic settings and are dedicated to developing methods and methodologies to facilitate research activism on gender-based violence and anti-extractivist activity amongst other topics. Lastly, feminist geography cartography and mapping methodologies are being used in public policy, particularly for the prevention and eradication of gender-based violence.

For example, femicide maps are being used to monitor and heavily influence national women's public policy at the local and national governmental levels in many different countries in Latin America. These maps have proven to be crucial because they are depicting a particular type of violence that is defined as the killing of a person because of their gender identity. Latin America has been vital in positioning specific debates such as feminicide through the work of Marcela Lagarde on the central role of state negligence in the systematic collective death of people because of their gender identity. Placing the death of women and trans people onto national cartographic representations is a way to hold the state accountable, but also a novel way to communicate gender-based violence. Members of the Critical Geography Collective of Ecuador have characterized this spatial analysis as geographies of femicide (Zaragocin et al., 2019). Another area within public policy is the recent use of feminist geography and cartography analysis to further develop an understanding of embodied disaster-risk studies as they intersect with migration dynamics (Contreras Gatica and Seguel Calderón, 2022). The use of feminist geography methods and methodologies are used to assure an embodied response and analysis that will prove crucial to the future problems presented by climate change.

Geographies of gender-based violence

An important pillar of Latin American feminist geographers is understanding the spatial elements of gender-based violence and developing innovative and critical cartographic methods. The main issues regarding gender-based violence in Latin America consider the criminalization of abortion, femicide and the prevention of gender-based violence (GBV) in all sectors. As Joseli Silva et al. (2020) mention, the urban landscape of feminist demands are visible throughout all Latin America via transnational activism against femicide (#NiunaMenos) and campaigns to decriminalize abortion (#Niñas-NoMadres) (see Chapter 23). Within the University, large-scale protests denouncing institutional negligence on cases of sexual harassment and other forms of gender-based violence are happening in many Latin American universities (Silva et al., 2020).

As femicide became legally recognized in most Latin American countries, official data also became available that was then used for mapping this most extreme type of GBV as well as other types of GBV and in some cases the continuum of GBV. Official state data on femicide was also used in conjunction with knowledge stemming from NGOs and newspaper articles, a methodology used in the femicide mappings (Lan and Rocher, 2020; Zaragocin et al., 2019). In the case of Argentina, Diana Lan and Heder Rocha demonstrate how official data is used for the mapping of femicide in Argentina as they also draw on the work by the collective *Geógrafas hacienda Lugar* (Woman Geographers Creating Place) to show countermapping of fear (Lan and Rocher, 2020). In most Latin American countries, official quantitative data is used in conjunction with qualitative analysis from feminist geography perspectives to understand the complexity of cartographies of GBV (Zaragocin et al., 2019). The emphasis on method and methodologies points to ways of *doing* feminist geography. Feminist geography collectives and the alliances between feminist geographers and feminist social movements are playing a central part in defining contemporary feminist geography praxis.

Cartographies of GBV are an important contribution from Latin America to the field of feminist geographies elsewhere. Struggles against rampant femicide, or the killing of women or gender diverse peoples because of their gender, and feminicide, which Marcela La Garde defined as hate crimes against women forged by social and state tolerance to systemic gender-based violence (2016), are at the top of the feminist agendas in the Americas. Mapping femicide by artists, activist and collectives is an important feminist praxis in Latin America (see https://mlf.mundosur.org). Collectives not only focus on femicide but all types of gender-based violence, developing powerful cartographic representations of gender-based violence that in turn are impacting public policy in important ways.

Feminist movements and geography

In Latin America the relationship between feminist activism and geography is strong. Contrary to other parts of the globe where feminist geography lies solely within the academy, in Latin America, feminist academic geographers usually partake in activism(s). In some cases, like that of Ecuador, feminist geographers have developed the first undergraduate program on human geography from outside of the academy and from critical geography activist dynamics. Collectives self-identified as feminist geography groups are in many cases positioned outside the university setting. Feminist geography collectives are happening in Chile, Argentina, Brazil, Colombia, and critical geography collectives that work on feminist geography are occurring in Mexico, Ecuador, Uruguay and at times at a transnational scale between countries. The work of these collectives tends to focus on methodologies and methods such as social cartography from feminist perspectives. Usually the topics addressed are gender-based violence, anti-extractivist activities and the intersections between both. More recently, Latinx and Latin American feminist geographies across the Americas are in dialogue to foster and promote discomfort feminism at a hemispheric scale in bilingual publications (Zaragocin et al., 2019).

Latinx geographies represent the spaces of the Latin American diaspora that have embodied and generational ties to Latin America. Decolonial antiracist feminist geographers are advocating for sustained efforts to bring together these distinct yet overlapping geographies (Zaragocin et al., 2019). The relationship between Latinx and Latin American feminist geographies is recent, but given historical migration flows across the Americas, will most definitely be present in feminist geography knowledge production and praxis in the future. An upcoming edited collection on feminist geography activism in English highlights the work of the Critical Geography Collective of Ecuador and Geobrujas, a community of women geographers as examples of feminist geography activisms happening in Latin America. That collective feminist geographical praxis and knowledge is being showcased in English and from critical Anglophone feminist geography is telling of the important connections being developed across the Americas.

Important large-scale mapping exercises are underway highlighting *where* feminist geographers are in Latin America. See the work of Selene Yang Rappaccioli and others from Geochicas for a detailed account on where feminist geographers and feminist collectives are located in Latin America (Geógrafas feministas de América Latina, n.d.). This work is key for making visible the existing feminist geography activism and academics, that in some instances overlap.

Feminist debates on territories

Territory and its relationship to the body is the preferred spatial category for feminist movements in the region (see **Chapter 12**). Body-territory, the body as territory and territory as a social body (Cruz and Bayón, 2020), has defined feminist geography perspectives across the Americas, demonstrating the inseparable co-constitution between bodies and territories. Body as the first territory is enunciated as part of urban feminist activisms, for example, in the struggle against the criminalization of abortion. Meanwhile, body-territory stems from Indigenous and communitarian feminist frameworks that position the ontological unity between territories and bodies, with a focus on healing (Cabnal, 2018).

Methodologically, body-territory does the opposite of traditional social cartography exercises, where bodies are drawn on specific territories to highlight human activity on a particular place. Instead, the body-territory method draws territories on bodies as way to prioritize the scale of the body and highlight emotions and affect that bring bodies and territories together. Feminist geographers in Latin American are using body-territory methodologies and conceptualizations in many ways, but one of the main areas where the concept and method has been used has been with regards to struggles against extractivist industry activity.

In feminist geography, body-territory as a method has been argued as exemplary of a decolonial feminist geographical method because it centres Indigenous feminist conceptual proposals from Latin America that prioritize the scale of the body in relation to territory (Zaragocin and Caretta, 2021). Recent publications on body-territory also point to understanding embodied feminist responses to risk-disaster studies (Satizabál and Melo Zurita, 2022) and complex dynamics bringing together gender-based violence and migration experiences (Lopes Heimer, 2021). Lastly, body-territory has further embodied understandings of aquatic space through proposals of water-body-territory (agua-cuerpo-territorio) as well as ideas concerning the space-time of death through the concept death-body-territory (muerte-cuerpo-territorio), where the death of territories and bodies are intrinsically linked (Zaragocin, 2018, 2021).

The centrality of body-territory within feminist geographies in Latin America is demonstrative of decolonial undertakings in the field. Recognition and sustained engagement with concepts and praxis stemming from Indigenous and communitarian feminisms such as body-territory and others demonstrate that feminist geography is looking to pluralize the ontological and epistemological frameworks from which it constructs idea of space. Feminist geography is also in very close dialogue with Latin America feminist politics based on diverse feminisms, feminist spatialities and movements, prioritizing Indigenous women's movements and worldviews to understand the relationship between the environment, nature and gender dynamics (Ulloa, 2020, see Chapter 13). Indigenous and communitarian feminisms in Latin America have called for a re-centring of the community and displacement of the individual for desired gender relations (Cabnal, 2018). At the intersections between Latin American feminist political ecology and Indigenous and communitarian feminisms is the possibility of a decolonial feminist geography framework that focuses on anti-racism thought and praxis.

An example of anti-racist feminist geographies underway in Latin America is the RECLAMA (Harnessing Afro-Ecuadorian Women's Heritage) project carried out in Esmeraldas, Ecuador. This project is a decolonial, anti-racist and feminist research project based on geography and critical history using critical social cartography and oral history methods to strengthen the re-existence of Afro-descendant-Black women (see Chapter 15). The aim of this project was to visibilize the knowledge of Afro-descendent Black women to document the rich heritage of Afro-descendant Black women living in Esmeraldas Province on the north coast of Ecuador. Through social cartography, oral histories and art, community and peer researchers explored memories, experiences and knowledge with Afro-Ecuadorian women (see: https://proyectoreclama.wixsite.com/reclama). Methodological pluralisms, whereby feminist geographies in Latin America are moving beyond traditional social cartography to engage with different qualitative research methods, allow for the furthering of decolonial perspectives in geographical knowledge construction (see Figure 14.1).

Both from Latin American feminist geography and feminist political ecology perspectives, an incipient conversation with Black feminisms from Brazil and Colombia are impacting the field in positive and transformative ways. The work of Sharlene Mollet, for example, is advocating for political ecologies of race as hemispheric, intersectional and relational across the Americas (Mollet, 2020) that place Black feminist ideas of Latin America at the centre. The RECLAMA project amongst others are moving

Figure 14.1 Map of the African diaspora in the Esmeralda Province, Ecuador.

Source: https://proyectoreclama.wixsite.com/reclama RECLAMA project. Authors: Dayana Paulette Castillo Chichande, Maria Preciado y Nayely González.

Latin American feminist geography and feminist political ecology to engage not just with Indigenous feminisms but also different strands of Black feminisms. See the bilingual website on the project: https://proyectoreclama.wixsite.com/reclam.

Conclusion

Latin American feminist geography is moving in the direction of decolonial and anti-racist feminist geography that deeply question androcentric understandings of geography in the region (Hanson, 2020). Despite still dealing with dominant androcentric and masculinist views of geography, feminist geography in Latin America is gaining traction within Latin American critical geography and human geography worldwide. In this chapter I have highlighted three of the ways that feminist geography in Latin America is positioning itself on a global scale through geographies of gender-based violence, feminist geographies of activism and feminist debates on territory in the region. Important methodological contributions by feminist geographers on mapping femicide, understanding body-territory as a decolonial feminist method and centring Indigenous and Black feminisms when considering the social construction of space through the body are just some examples of what Latin American feminist geography looks like today. Sustained conversations and praxis are happening with Latinx populations in the US and Latin American diaspora in the Global North that undoubtedly will show Latin American feminist geographies paving the way for feminist geographies at the global scale.

Summary

- Feminist geography occurs within the university, collective and activist circles as well as in public policy.
- One of the defining features of Latin American feminist geographies is the strong link between feminist activism and feminist geography in the academy.
- Mapping femicide, feminicide and the continuum of gender-based violence is a priority for feminist geography scholars and practitioners.
- The interlocking nature of body-territory is central to feminist debates on territory throughout all Latin America.
- The future of Latin American feminist geography will centre Indigenous, Black and Latinx diaspora spatial knowledge and praxis across the Americas.

Review questions

1. What is the relationship between feminist geography activism and the academy in Latin America?
2. Can we imagine geographies of femicide in other parts of the world?
3. What are the ways that body-territory has been used as a decolonial feminist geographical method?
4. What does anti-racist feminist geography look like?

Further reading

Colombara M (2019) Gender Geography in Argentina: A Brief Overview. *Gender, Place & Culture* 26(7–9): 935–944.

A review of the main debates on feminist geography in Argentina.

de Geógrafas GC & Mason-Deese L (2021) Bodies, Borders, and Resistance: Women Conjuring Geography through Experiences from the Other Side of the Wall. *Journal of Latin American Geography* 20(2): 168–178.

An important collective reflection on border regimes between Mexico and the United States from embodied feminist geography perspectives from the Global South.

Silva JM & Ornat MJ (2020) Feminist Geographies in Latin America: Epistemological Challenges and the Decoloniality of Knowledge. Translated by Liz Mason-Deese. *Journal of Latin American Geography* 19(1): 269–277.

Brazilian feminist scholars Joseli Silva and Marcio Ornat give an important overview of contemporary feminist geography in Latin America, in particular questions to the coloniality and whiteness of the discipline within Latin America.

Soto Villagrán P (2019) Geographies of Gender and Feminism in Mexico: A Field in Construction. *Gender, Place & Culture* 26(7–9): 1170–1181.

A review of the main debates on feminist geography in Mexico

Ulloa A (2019) Gender Feminist Geography in Colombia. *Gender, Place & Culture* 26(7–9): 1012–1031.

A review of the main debates on feminist geography in Colombia.

Zaragocin S (2019) Feminist Geography in Ecuador. *Gender, Place & Culture* 26: 7–9.

A review of the main debates on feminist geography in Ecuador

Zaragocin S, Ramirez M, Garcia M & González Y (2022) Bilingual Intervention – "LatinX and Latin American Geographies: A Dialogue"/"Dialogo entre las geografías LatinX y lati-noamericanas". *Antipode Online*. https://antipodeonline.org/2022/08/08/latinx-and-latin-american-geographies/

A bilingual dialogue between US based LatinX feminist geographers and feminist geographers in Latin America that are interested in link both fields of geography.

Keywords

Body-territory: the ontological unity between bodies and territories.

Decolonial feminist geography: a feminist geographical framework that prioritizes antiracist geographies and looks to do away with the coloniality of gender in space.

Feminist geography collectives: groups of geographers working on feminist theory and gender perspectives in geography at the theoretical, methodological or political level(s) within or outside the university.

Latinx geographies: represent the spaces of the Latin American diaspora that have embodied and generational ties to Latin America.

References

Cabnal L (2018) TZK' AT, Red de Sanadoras Ancestrales Del Feminismo Comunitario Desde Iximulew-Guatemala (Network of Ancestral Healers of Communitarian Feminism from Iximulew-Guatemala). *Ecología Política* 54: 98–102.

Contreras Gatica Y & Seguel Calderón B (2022) Territorio informal. Una nueva lectura del acceso a la vivienda y al suelo en Chile. *Revista de Geografía Norte Grande* 81: 113–136.

Cruz, Hernández Delmy Y & Bayón, Jiménez M (Coords) (2020) *Cuerpos, Territorios y Feminismos. Compliación latinoamericana de teorías, metodologías y prácticas políticas*. Abya Yala.

Geógrafas feministas de América Latina, n.d. https://umap.openstreetmap.fr/es/map/geografas-feministas-de-america-latina_828853#3/3.78/-68.03

Hanson A-M (2020) Feminist Futures in Latin American Geography. *Journal of Latin American Geography* 19(1): 215–224.

La Garde M (2016) Del femicidio al Feminicidio. In: *Desde el jardín de Freud*. Bogotá: Universidad Nacional de Colombia.

Lan D & Rocher H (2020) Metodologías feministas para el mapeo de geografías oprimidas en Argentina. *Geopauta* 4(4): 46–67.

Lopes Heimer R (2021) *Travelling Cuerpo-Territorios:* A Decolonial Feminist Geographical Methodology to Conduct Research *with* Migrant Women. *Third World Thematics: A TWQ Journal* 6–4(6): 290–319.

Mollet S (2020) Hemispheric, Relational, and Intersectional Political Ecologies of Race: Centring Land-Body Entanglements in the Americas. *Antipode* 53(3): 810–830.

Reclama Project. https://proyectoreclama.wixsite.com/reclama

Satizabál P & Melo Zurita M (2022) Bodies-Holding-Bodies: The Trembling of Women´s cuerpo-territorio-tierra and the Feminist Responses to the Earthquakes in Mexico City. *Third World Thematics: A TWQ Journal* 6(4–6): 267–289.

Silva JM, Ornat MJ & Mason-Deese L (2020) Feminist Geographies in Latin America: Epistemological Challenges and the Decoloniality of Knowledge. *Journal of Latin American Geography* 19(1): 269–277.

Ulloa A (2020) Feminist Political Ecologies in Latin America Context. In: Naples NA (Ed), *Companian to Feminist Studies.* John Wiley & Sons Ltd.

Zaragocin S (2018) Espacios Acuáticos desde una Descolonialidad Hemisférica Feminista. *Mulier Sapiens* 10: 6–19.

Zaragocin S (2021) La geopolítica del Útero: hacia una geopolítica feminista decolonial en espacioa de muerte lenta. In: Cruz Hernández DT & Bayón M (Coords), *Cuerpos, territorios y feminismos. Compilación latinoamericana de teorías, metodologías y prácticas políticas.* Abya Yala.

Zaragocin S & Caretta M (2021) *Cuerpo-Territorio*: A Decolonial Feminist Geographical Method for the Study of Embodiment. *Annals of the American Association of Geographers* 111(5): 1503–1518.

Zaragocin S, Silveira M & Arrazola I (2019) Construyendo una geografía del feminicidio en el Ecuador. Geografías del femicidio en el Ecuador. In Navas M & Garza M (Eds), *Appropiaciones de la ciudad. Género y producción urbana. La reivindicación del derecho a la ciudad como práctica espacial.* Barcelona: Pol-len edicions, pp. 47–75.

CHAPTER 15

Afro-descendant geographies

Ana Laura Zavala Guillen and
Nadia Mosquera Muriel

Introduction

The latest historical research estimates that, from 1501 to 1866, around 12,500,000 Africans were enslaved and forcibly displaced to Latin America, a territory colonised and exploited by the European empires (Candioti, 2021) (**see Chapter 2**). However, records in colonial archives, such as the General Archive of Indies in Seville, demonstrate that this figure might be higher when considering the slavery traffic carried out without royal permissions condemned by the empires. After enslavement, African men, women, and children found themselves in new geographies. They faced the challenge of recovering freedom and reformulating their belonging to strange lands. Thus, making space in contested territories has been a political survival task that started in colonial times and continues until today. For example, these were those who dared to escape slavery to create territorialised societies (McFarlane, 2008). Although these societies functioned as shelters for fugitives from slavery and their descendants born in these territories, they continued to be exposed to a precarious life of war with the risk of re-enslavement and a new separation from their families, comrades, and communities (Zavala Guillen, 2021).

In modern times, direct violence against Afro-descendant people and their territories has been exercised by new hegemonies: capital, its extractive industries, and illegal armies, with the support and acquiescence of the state. Capital and the state exploit their natural resources and destroy their environments and landscapes while forcibly displacing men, women, and children from their lands (Oslender, 2008). Therefore, the risk of uprooting and the experience of it has been a constant in the geographies of the Afro-descendants in Latin America. Dispossession has left Afro-descendants unable to exercise basic human rights in extreme deprivation and racial violence generation after generation.

As a response, countering dispossession demanded the development of political creativity to envisage and implement strategies of resistance. Space-making has been, therefore, a form of Black collective mobilisation intrinsically linked with the possibility of

DOI: 10.4324/9781003430926-21

existing with joy and freedom in their territories. Furthermore, Black spaces have intertwined with Indigenous, *campesino*, and State-based territories in cooperation, tension, and violent disputes to create territories that distance market-driven forces (Escobar, 2008). These spaces have crossed urban and rural areas, showing that Black resistance is transterritorial and is everywhere in Latin America despite the cultural violence that has pursued its erasure. Cultural violence has been, in practice, a consistent whitening of the national political projects that made invisible Black territories in public policies and maps. For example, official cartography has portrayed Latin America as *blanca* [white] and *Mestiza* [a blend of Afro-descendant, European, and Indigenous elements] with no room for Afro-descendant spaces (Dunnavant, 2020; Zavala Guillen, 2023).

This cultural Black erasure has also occurred in academia, denying Afro-descendant people as subjects able to produce knowledge and epistemologies to understand their own social, political, and geographical processes, and overlooking their social products. For example, their political contribution to social movements has been largely dismissed compared to Indigenous and *campesino* groups. Therefore, this chapter delves into three Afro-descendant concepts and practices – **vivir sabroso** [living richly], *cimarronaje* [marooning], and **quilombismo** [body as a Black space of resistance] – used as lenses to understand the struggles for the territory(ies) of the Black social movement in Latin America through the cases of Brazil and Colombia.

Vivir sabroso, living richly

Vivir sabroso, or living richly, is a concept that emerged from the daily life of the Afro-descendant communities living in the Medio Atrato (comprising Chocó and Antioquia departments) in the northern Pacific region in Colombia. In this chapter, we use the concept *vivir sabroso* as a geographical epistemology that weaves spatial and spiritual conceptions of territory. It encompasses a range of "place-making" practices shaped by politics, kinship, religion, and culture. *Vivir sabroso* is also a response by Afro-Colombian communities who seek alternative perspectives on development rooted in local epistemologies, challenging dominant narratives of market-driven economic growth. The concept of *vivir sabroso* is similar to the community-driven, anti-capitalist concept of *Sumak Kawsay* or *"buen vivir"* (good living) developed by Andean (Quechua) populations from Bolivia and Ecuador, who also emphasise alternatives to development anchored in respect to nature (**see Chapter 10**). Both frameworks incorporate ancestral knowledge and community well-being at their core. However, *vivir sabroso* interweaves Black epistemologies, which encapsulate the unique ways of knowing and experiencing the world that are specific to the Afro-diaspora, which is the basis of contemporary political mobilisation for these communities. In what follows, we briefly clarify the central tenets of *vivir sabroso* or living richly, whose names we use interchangeably.

Delving into the *vivir sabroso* worldview

The concept of *vivir sabroso* draws upon the intrinsic connection between kinship and territoriality, forming a philosophical foundation deeply embedded in Afro-Colombian communities. Anthropologist Natalia Quiceno Toro ethnographically examined in her work "Vivir sabroso: Luchas y Movimientos Afro-Atrateños en Bojayá Chocó, Colombia" (2016) how this philosophy enabled Afro-Colombian communities to develop a place-based understanding that epitomises the creation of community through various facets such as territory, cultural identity, kinship, spirituality, and well-being in resistance to anti-Black structural violence. Afro-Atrateños nurture their relationships through a diverse range of kinship practices. Among Afro-Atrateña communities, inheritance extends beyond mere land ownership and encompasses intricate networks of care within specific territories. A prime example of this is the practice of *compadrazgo*, a special bond formed between a child's godparents and their parents.

Additionally, marriages and friendships strengthen these interwoven ties (Quiceno Toro, 2016: 58). Also, Afro-Atrateños draw on the profound influence of mobility for constructing their territories, which is nourished by their interactions with waterscapes. Navigating the riverine ecosystems of the Pacific facilitates the establishment of meaningful relationships with residents from different locations, regardless of their geographical proximity. In this manner, Afro-Colombians overcame geographical barriers by seamlessly knitting together Black territories under the shared understanding of Blackness as *"familia"* (family) and *"mi gente"* (my people). Thus, *vivir sabroso* is a geographical concept. Living richly implies a mobile life that follows the logic of rivers and bridges, different spaces, the rural and the urban, families in distant places, and the Indigenous and the Afro-descendants' attachments to lands.

Living richly also recognises that ancestral knowledge holds both spiritual and therapeutic importance. It incorporates the spiritual notions of Kalunga, which Afro-Colombians in the Pacific region adapted to the local context. Kalunga is a spiritual concept associated with the ocean and ancestral connections in Afro-descendant cultures. This knowledge encompasses botanical wisdom involving the understanding of plants (**see Chapter 18**), as well as Afro-Catholic devotions to saints, practised through the transformative power of prayer. Within this context, each plant and root possess a specific purpose or "service" when needed. Plants' temperature also plays a role in spiritual practices, distinguishing between cold and hot plants. Techniques such as *serenar* involve leaving the plant, leaves, or roots outdoors to soak overnight, activating their properties. Baths, in particular, serve as valuable sources of protection. Botanical knowledge, whether for restoration or harm purposes, has always served as a form of resistance for Black populations (Sergio Mosquera, as cited by Quiceno Toro, 2016: 103). Negative spiritual influences can be dissolved through these practices, justice can be pursued, and protection and well-being can be cultivated (Quiceno Toro, 2016: 108). These ancestral ways of engaging with spirituality and healing comprise how Afro-Atrateños embody the concept of living richly.

The opposite of this way of *vivir sabroso* is *vivir enmontado* (living under siege). This refers to the deadly effects of necropolitical governance in Black communities. Since the mid-twentieth century, multinational corporations, drug cartels, illegal miners, guerrillas and paramilitaries, and the state's resource extraction activities have threatened Black communities and their territories. Criminal groups, multinationals, and state groups have seized lands, polluted water sources, threatened livelihoods, displaced, and brought harm, terror, and death to Black communities and their territories in the Pacific. These invasions, alongside the militarisation of black territories, have led to *vivir enmontado*, which reflects the deeply racialised restriction of freedom of movement that prevent Black communities from transiting their multiple territories (Quiceno Toro, 2016).

Scholars have observed that *"vivir sabroso"* is the galvanising response to *"vivir enmontado"*, which has ideologically guided the political struggles of Afro-Colombian populations from the Pacific region. Quiceno Toro (2016) observed that *vivir sabroso* emerged as an ideology that drove the political struggles of Afro-Colombian social movements. Notably, the state-imposed development frameworks have compelled Black communities to form alliances and establish community organisations with a shared goal: the defence of life or "*defensa de la vida*" (Quiceno Toro, 2016: 74). There-fore, the usurpation of Black territories coupled with the employment of necropolitical tactics has exacerbated the displacement suffered by Black communities and has given rise to deeply entrenched territorial disputes.

We want to stress that "*vivir sabroso*" is also a worldview that traces its roots back to marooning practices and the historical experiences of Afro-Colombian communities in the Pacific region. Still, it is hard to disentangle it from its long genealogy linked to the *Ubuntu* philosophy in Africa. Contemporary Afro-Latin American struggles, and mainly Afro-Latin American feminist political mobilisation from the mid-2010s, have begun to explicitly draw on the concept of *Ubuntu* or "I am because we are". Notably, Black feminists in the region who have stepped into formal politics have drawn on *Ubuntu*'s sense of community to dismantle the colonial legacies that have shaped their countries' political landscape (white and masculinist by default).

Vivir sabroso in Colombian national politics

The upsurge in interest in the *vivir sabroso* philosophy responds to political changes in Colombia. In 2021, Colombia's Vice President Francia Márquez integrated during her presidential campaign the slogans of "*Soy porque Somos*" (I am because we are) into the philosophy of *vivir sabroso* within her political agenda. Francia Márquez, an award-winning Afro-Colombian lawyer and grassroots environmental leader, formally entered Colombia's politics in July 2021 when she began her campaign for president. The political movement that supported her candidacy often referenced the need to change Colombian socio-economic structures through the framework of *vivir sabroso*. Rejecting the insensitive misrecognition of *vivir sabroso*, which was often mocked as laziness by the mainstream media and political pundits of the Pacto Histórico political party (in power since 2022), Márquez defined "*vivir sabroso*" as "living without fear, with dignity, and with protected human rights". Marquez's radical political agenda, embodied by her lived experience as a miner and domestic worker in rural territo-ries of the Colombian Pacific (see Figure 15.1), embraced a project that called for an anti-racist, anti-patriarchal, and anti-commodification of nature approach, challenging colonial legacies in Colombia. While the concept of *vivir sabroso* has received more attention in the last decade, this historical philosophy developed by Black communities is best understood as a foundational component that underpins the historical devel-opment of ethno-territorial claims by Afro-descendant organisations in Colombia's Pacific region. This worldview is a quintessential example of the epistemological and geographic production of Black communities in Colombia's Pacific lands.

Marooning

Marooning, or *marronage*, started as a resistance strategy as soon as enslaved Afri-cans were violently brought as a labour force after the large extermination of Indig-enous populations in Latin America. Indigenous geographies, decimated by violence,

Figure 15.1 Francia Márquez's symbolic inaugural ceremony in Cauca as Vice President of Colombia.

Source: Author.

hunger, and disease, became territories where African fugitives found precarious spaces to exist away from slavery in cities, mines, and large farms. These communities of fugitives from slavery – or Maroons – received different names in Latin America, such as *quilombo* in Brazil and *palenque* in Colombia. Despite their political diversity, their common characteristic was that their mere existence subverted the conception that enslavement was the only fate possible for Africans. While many of these communities perished due to the systematic colonial warfare against them, others survived by negotiating and achieving peace treaties with the colonisers. These peace treaties allowed them to regain freedom and territory as freed Maroon towns. Some of these towns are still part of the postcolonial states in Latin America, such as San Basilio del Palenque in Colombia (Cassiani Herrera, 2014) (see Figure 15.2).

As an ethic of living in resistance through escape from oppression, *cimarronaje* [marooning] has inspired and informed the modern struggles of the Black social movements in Latin America. For example, the knowledge of those African men and women who produced space and territory capitalised on nature to defend their lives in a permanent state of war against the colonial authorities passed through the oral history of Afro-descendant communities. This knowledge can still be found in the present strategies of those who survived forced displacement in the context of armed conflicts

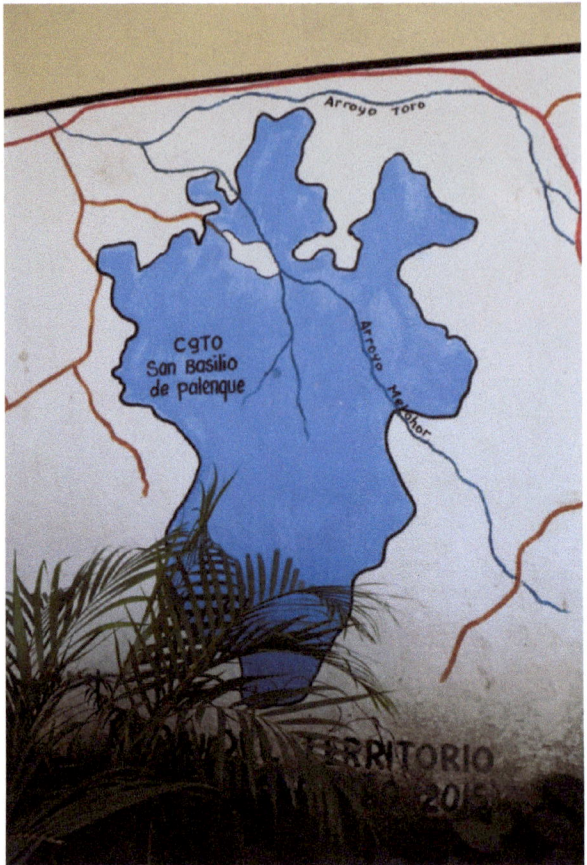

Figure 15.2 Map of San Basilio del Palenque.
Source: Author.

through establishing new settlements away from war. Furthermore, ideologically, *marronage* became a Black activist possibility of flight from hegemonies, such as capitalism, heteronormativity, patriarchy, and the mainstream European knowledge system. The following sections will delve into examples of present *marronage* in relation to the struggles for land and Black feminist activism in Colombia.

Space-making as resistance: the Case of San Basilio de Palenque, Colombia

Understanding the history of endurance of San Basilio de Palenque, a community of descendants of fugitives from slavery in the Colombian Caribbean, requires critically grasping their intrinsic connection with their territory. Territory made *marronage* possible. A landscape of hills, dense vegetation, and wetlands, with dried and wet seasons, allowed African fugitives to be difficult to reach, and they built fortified spaces with sticks and traps to avoid being retaken into slavery.

Marronage also produced a specific territory, the *palenque*, an anticolonial space where precarious freedom was feasible for the runaway enslaved people. Therefore, defending the *palenque* and defending freedom were synonymous. The colonial authorities, aware of this, undertook military campaigns to erase the community of rebels, kill their leaders, and re-enslave the survivors. Maroon leaders quickly understood that the physical defence of the *palenque* was not enough to survive. Alliances with the Catholic church's members and negotiations with the Spaniards co-existed with a permanent state of war against their oppressors. This political ability to weave different strategies of resistance allowed the Maroons to secure a peace treaty with the colonial authorities. They became one of the first colonial towns of freed fugitives from slavery in 1714 in Latin America.

The Maroon ancestral knowledge to resist through the production of territory has persisted in the spatial practices of their descendants who, in modern times, struggled against land encroachment at the hands of the State, white and Mestizo-elite landowners, businesses, and armed groups. Consequently, Maroon descendants of San Basilio de Palenque have built new settlements or *palenques* outside their ancestral territory in rural areas and cities in Colombia and Panama to regain space after uprooting. In these settlements, Maroon consciousness awakened, making members of San Basilio de Palenque participate in the largest Black social movement in Colombia, particularly in mobilising land rights and pursuing the collective title of their territories. In the case of San Basilio de Palenque, the community obtained a promise of collective titling of their ancestral territory from the Colombian State in 2012. Although sometimes forgotten, the history of San Basilio de Palenque demonstrates how Black struggles for territories have long inspired, along with Indigenous struggles, the new social movements in Latin America (**see Chapter 21**).

Recovery land as Maroon-descendant women's daily practice of territorial care after displacement

In 2001, the Maroon-descendant community of La Bonga, which was part of the town of San Basilio de Palenque, was displaced by paramilitary forces, the *Autodefensas Unidas de Colombia* [The United Self-Defence Forces of Colombia] and forced to leave behind their lands, traditional crops, and small cattle. Some of them found shelter in the built-up area of San Basilio del Palenque, and the rest in a nearby former airstrip where they started progressively to construct accommodations. The displaced Maroon descendants decided to remain in physical proximity to their La Bonga to ensure a quick return to their community when the armed violence stopped. Nevertheless, almost two decades after the forced displacement, the return did not happen as a group. Fear of experiencing new armed violence, traumatic memories of the war, and attachment to the new settlements were some of the reasons for this. However, in the last few years, María de los Santos started a symbolic and material reappropriation of the space of La Bonga to demonstrate to their own that recovering this territory is still possible, and with that, a life associated with ancestral agriculture and small cattle. In this sense, María de Los Santos started to return to La Bonga to crop, staying there during the night and returning after a few days to the settlement in the former airstrip where their children are. María brings the products of her cropping to feed her own, showing that another life is possible where hunger and dependency are not the rules. Slowly others started to follow María de los Santos to La Bonga to repair what was left of the community buildings, such as a school, and prepare the land for new crops. The daily spatial resistance of María de los Santos in La Bonga recalls the tradition of women who have performed acts of care, resistance, and reparation after mass human rights violations in war and dictatorships in Latin America.

Quilombismo

From a Latin American decolonial perspective, the intrinsic relationship between territory and body *(cuerpo-territorio)*, particularly the female body or bodies, has been largely explored in relation to Indigenous women and their struggles against extractive industries destroying their environments and ancestral places, physically and symbolically (**see Chapters 13 and 14**). The mutilation of their territories is also experienced in the body as an intimate space in connection with the Earth. To geographers, the body has been central in understanding the material effects of power dynamics and the spatial politics of violence against racialised bodies and their territories (Alves, 2021; McKittrick, 2006).

In Latin America, the boundaries between the body and space are understood as fluid. According to Zaragocín and Caretta (2021), the concept of *"cuerpo territorio"* (body-territory) centres on the relational nature of the body and territory as a scale. It recognises the interconnectedness of epistemologies developed through the relationship between the body, lived experience, embodiment (including emotions and bodily sensations), and the land (**see also Chapter 14**). This concept finds its roots in Indigenous feminism, as developed by the Guatemalan Indigenous scholar Lorna Cabnal and further developed as a decolonial feminist methodology for geographies of Latin America by Zaragocín and Caretta (2021). However, an essential connection between understanding the body and social struggles amongst Afro-Latin American social movements is the idea of *quilombismo*.

Quilombismo in Latin America encompasses a dual meaning: first, it refers to the spaces of refuge, autonomy, and freedom established by Africans who were forcibly brought to the Americas to work under brutal colonial conditions on haciendas and plantations (in this chapter, it is also known as marooning). Second, *quilombismo* is a concept embraced by contemporary Afro-Latin American social movements as part of their epistemic and political projects, prioritising modes of self-determination. When Afro-descended fugitives escaped chattel slavery, they founded the Quilombo dos Palmares in Pernambuco Brazil, which was the largest Black anti-colonial bastion of resistance in Latin America and lasted nearly a hundred years between the sixteenth and seventeenth centuries (Gomes, 2005). In the present day, these initiatives of Black self-determination seek to *aquilombar-se* or create safe spaces that resist oppressive systems such as racism, patriarchy, heteronormativity, capitalist domination, and exploitative modes of development and modernisation (Nascimento, 2023).

Quilombismo represents a political movement in which figures such as the Black feminist theorist Beatriz Nascimento has reflected throughout her work. She proposed *quilombismo* as an anti-colonial, anti-racist alternative, defining *quilombos* as spaces that embody a social and political alternative system based on intrinsic values, rooted in some political-social institutions in Angola. Nascimento envisioned *quilombos* as sites for reclaiming and constructing an Afro-diasporic identity and fostering Black empowerment within the context of Black consciousness movements in Brazil. As Smith et al. (2021) noted, one of her most innovative analytical interventions was the territorialisation of the Black body. Nascimento theorised *quilombos* by foregrounding the Black body in cartographic terms, meaning viewing it as an extension of the land able to subvert legacies of colonial violence in Black bodies (Smith et al., 2021). Thus, to Nascimento, the Black body is a scale. Here, the body is central in establishing and maintaining *quilombos*, which represent spaces of resistance, resilience, and cultural autonomy for Afro-descendant communities. Using the Black body/*quilombo* as a geographic analytic, Nascimento conceptualises it as a vessel for expressing and embodying ideas of movement, providing a place of safety and protection, shaping individual and collective identities, and serving as a means of achieving liberation rooted in Afro-descended worldviews (Smith et al., 2021: 286). Moreover, *quilombos* are embodied, transatlantic, and spiritual spaces (Smith, 2016). Amongst the latter, Nascimento centred spirituality through *Candomblé*, an Afro-Brazilian religion widely practised in northeast Brazil that syncretises Yoruba, Bantu, and Fon spiritualities with Catholicism and Indigenous beliefs. Particularly useful to the growing field of Black geographies in Latin America is Nascimento's engagement of the body as a scale in the performance of spatial politics of resistance. Particularly, Nascimento was interested in developing a revisionist analysis of *quilombos* as ongoing spaces for forging Black subjectivity and transnational connections through a Black feminist perspective that decentres male-centric analysis of Black fugitivity (Smith, 2016). *Quilombismo* through the body foregrounds a unique perspective on Afro-descended practices of resistance in the Americas, which is crucial for understanding contemporary struggles of Afro-Latin American social movements.

Conclusion

Black space-making was almost simultaneous with the forced arrival of Africans into the Latin-American region, for example, through escape and the building of territorialised societies of fugitives away from slavery. The manner of producing and reproducing territories from colonial times to the present days has been informed first by African knowledge and practices that were reformulated in the new geographies. Some Afro-descendant concepts, such as *vivir sabroso*, marooning, and *quilombismo*, which are not exhaustive of the Black people of Latin America, have interwoven spatial resistance in Latin America. Indeed, contemporary Afro-Latin American movements continue to struggle against patriarchy, anti-Blackness, and socio-economic exclusion drawing upon their place-based rooted epistemic projects and worldviews. By examining Colombian and Brazilian cases, the chapter argued that a deeper understanding of Black mobilisation in Latin America entails adopting an anti-colonial perspective that is attentive to the multifaceted epistemology that Afro-Latin American populations have historically produced. The chapter has also proposed reorienting our approach to understand space, embodiment, structures of feeling, and Afro-diasporic spirituality as essential in recognising the significance of epistemic political projects generated within specific Afro-diasporic struggles in the region.

Summary

■ After enslavement, Africans contested colonised geographies by producing territories; this act of regaining freedom through territory sustains political resistance grounded in Afro-epistemologies, such as *vivir sabroso*, marooning, and *quilombismo*.

■ Traditionally, geographic scholarship on Latin America's social movements overlooked Black space-making as a form of resistance; this elision responded to dominant racial ideologies that portrayed the region as white-Mestiza.

■ Historical and contemporary structural forms of violence have forcibly displaced and hindered Latin America's Black communities to live fully and peacefully in their ancestral lands. These populations were inspired by marooning practices to create territories free of oppressions and hegemonies, from colonisation and slavery to present-day challenges of capitalism and patriarchy.

■ *Vivir sabroso* is a geographical Afro-Colombian epistemology that combines spatial and spiritual understandings of territory which encompasses political, kinship, religious, and cultural practices that challenge dominant narratives of market-driven economic growth while seeking alternative perspectives on development grounded in their epistemologies.

■ *Quilombismo* is an Afro-Brazilian framework of resistance rooted in practices of refuge against colonial violence by centring the body as scale embodying mobility, refuge, identity, spirituality, and liberation rooted in place-based Afro-descended worldviews in Brazil.

Review questions

1. How have Afro-descendant ideas and practices impacted the production of space and territory in Latin America?
2. What are the similarities and differences between the Black struggles for land and territory in comparison to the Indigenous and *campesino* ones?
3. What are the ideologies that have whitened the Latin American spaces and practices? How have Black social movements resisted this?

Further reading

Asher, K. (2007). 'Ser y tener: Black Women's Activism, Development, and Ethnicity in the Pacific Lowlands of Colombia', *Feminist Studies*, 33(1), pp. 11–37. https://doi.org/10.2307/20459116.
Historical overview of the Women Black Social Movement in Colombia during the 1990s and its racial struggles linked with development and ethnoracial policies.

Bledsoe, A. (2017). 'Marronage as a Past and Present Geography in the Americas', *Southeastern Geographer*, 57(1), pp. 30–50.

Historical overview of present and past Maroon geographies in Brazil and the US and their struggles for freedom, land, and human rights.

Perry, K.K. (2013). *Black Women Against the Land Grab: The Fight for Racial Justice in Brazil*. University of Minnesota Press.

A comprehensive overview that delves into the processes of black women's mobilizations for housing and land rights in Brazil, highlighting the strategies galvanised by black women when confronting projects of dispossession.

Schwaller, R.C. (2022). *African Maroons in Sixteenth-Century Panama: A History in Documents*. University of Oklahoma Press.

A documentary compilation of archival records related to Maroon struggles and settlements in Panama during the sixteenth century.

Keywords

Afro-descendant people: Black people; those whose ancestors were African enslaved people in Latin America. Within the Black social movements, these terms have been politically used, appropriated, and mobilised as alternatives to describe a racial identity in opposition to *Mestizaje* and whiteness and its cultural, political, social, and economic oppression.

Marooning: the process of escaping from slavery to create territorialised societies away from those who claimed ownership over the fugitives. It aimed to recover freedom – although precarious – through space-making.

Mestizaje: a racial mixture between Afro-descendant, Indigenous, and European people that sustains the myth of racial democracies in Latin America and the subsequent hegemony of white and white-*Mestizo* elites in the region.

Quilombismo – body as a quilombo: anti-colonial space of refuge and self-determination political project that understands Black body(ies) as a continuation of the land, a particular one, a *quilombo* – a place of Maroon resistance against coloniality.

Vivir sabroso: Afro-descendant idea and practice that promotes a mobile life across territories in urban and rural spaces, families, and the Afro and Indigenous worlds.

References

Alves, J.A. (2021). 'Fatal Blow: Urbicidal Geographies, Pax Colonial and Black Sovereignty in the Colombian City', *Environment and Planning D: Society and Space*, 39(6), pp. 1055–1072.

Candioti, M. (2021). *Una historia de la emancipación negra. Esclavitud y abolición en la Argentina*. Siglo ventiuno editores.

Cassiani Herrera, A. (2014). *Palenque Magno: Resistencias y Luchas Libertarias del Palenque de la Matuna a San Basilio Magno 1599–1714*. Incultur.

Dunnavant, J.P. (2020). 'Have Confidence in the Sea: Maritime Maroons and Fugitive Geographies', *Antipode*, 53(3), pp. 1–22. https://doi.org/10.1111/anti.12695.

Escobar, A. (2008). *Territories of Difference: Place, Movements, Life, Redes*. Duke University Press.

Gomes, F.D.S. (2005). *Palmares: escravidão e liberdade no Atlântico Sul*. Contexto.

McFarlane, A. (2008). 'Cimarrones and Palenques: Runaways and Resistance in Colonial Colombia', *Slavery & Abolition*, 6(3), pp. 131–151. https://doi.org/10.1080/01440398508574897.

McKittrick, K. (2006). *Demonic Grounds: Black Women and Cartographies of Struggle*. University of Minnesota Press.

Nascimento, B. (2023). *The Dialectic is in the Sea. The Black Radical Thought of Beatriz Nascimento*. Princeton University Press.

Oslender, U. (2008). 'Another History of Violence: The Production of "Geographies of Terror" in Colombia's Pacific Coast Region', *Latin American Perspectives*, 35(5), pp. 77–102. https://doi.org/10.1177/0094582x08321961.

Quiceno Toro, N. (2016). *Vivir sabroso. Luchas y movimientos afroatrateños, en Bojayá, Chocó, Colombia*. Editorial Universidad del Rosario

Smith, C. (2016). 'Towards a Black Feminist Model of Black Atlantic Liberation: Remembering Beatriz Nascimento', *Meridians: Feminism, Race and Transnationalism*, 14(2), pp. 71–87.

Smith, C., Davies, A., & Gomes, B. (2021). '"In Front of the World": Translating Beatriz Nascimento', *Antipode*, 53(1), pp. 279–316.

Zaragocín, S., & Caretta, M.A. (2021). 'Cuerpo-Territorio: A Decolonial Feminist Geographical Method for the Study of Embodiment', *Annals of the American Association of Geographers*, 111(5), pp. 1503–1518.

Zavala Guillen, A.L. (2021). 'Afro-Latin American Geographies of in-betweenness: Colonial Marronage in Colombia', *Journal of Historical Geography*, 72, pp. 13–22. https://doi.org/10.1016/j.jhg.2021.02.002.

Zavala Guillen, A.L. (2023). 'Feeling/Thinking the Archive: Participatory Mapping Marronage', *Area*, 55, pp. 416–425. https://doi.org/10.1111/area.12869.

PART III

Uneven processes

Ecologies

CHAPTER 16

Political ecology of cities and urbanization

Marcelo Lopes de Souza

Introduction

Cities and urbanization cannot be properly understood without understanding society's relation with the environment and wider ecological processes (what can be termed 'socio-ecological metabolism') – be it in the (sub)continent traditionally and mistakenly called 'Latin' America or elsewhere. Just as supposedly 'natural' spaces (such as national parks, some of them located in urban areas) are permeated by social relations – processes of territorialization and deterritorialization, acid rain and air pollution caused by fossil fuels, and so on – the social production of urban space does not occur in the absence of geobiophysical processes, dynamics, forms and cycles. In the cities of Latin America, located in peripheral or semi-peripheral capitalist countries (see Chapter 11), the socio-ecological materiality of space reflects the inequalities and injustices that impregnate and shape the societies in whose context these cities exist.

From landslides and floods to environmental contamination to the lack of basic sanitation, the metabolism of these cities is closely linked to the reproduction of human suffering, including what has been called *environmental suffering* (discomfort, diseases and mutilations caused by pollution, disasters, and lack of infrastructure). All this has repeatedly been a source of conflict. However, those problems are not the only sources of conflict, nor of suffering or injustice. The very way in which access to resources and 'natural amenities' (for instance, proximity to beaches and green areas, access to sources of drinking water, milder mesoclimate, pollution-free places, scenic beauty, etc.) and environmental protection objectives have been instrumentalized often reveals an authoritarian component that aggravates situations of residential segregation and socio-spatial stigmatization. In recent decades, Latin American urbanization has shown increasingly elitist and socially excluding traits (see Chapter 7), in the wake of the growing self-segregation of elites and middle classes and the socio-political-spatial fragmentation of the urban fabric (see Chapter 19). The way that 'nature' is perceived, transformed, and manipulated within this framework is and will increasingly be an essential part both of capital's strategies and of social struggles.

DOI: 10.4324/9781003430926-24

Over recent decades, human and physical geography have gradually grown apart. It is interesting to note how, for some geographers, 'ecology' was synonymous with political alienation, as if the consideration of naturogenic factors (that is, those not created or fully controlled by humans) would be as such synonymous with not taking seriously society in its real complexity. These critics, even though they are right on several points, forget that 'ecological' knowledge and concerns have been disputed and appropriated by critical strands of thought (including anarchists and Marxists) for generations. Indeed, there is no such a thing as a single 'ecology' (or a single 'ecologism,' or a single 'environmentalism'), but several. Ecological knowledge is undoubtedly an arena of ideological struggle between emancipatory intellectual and political currents, on the one side, and retrograde intellectual and political currents, on the other. The inevitable overlaps, confusions and contradictions must not blind us to the fact that 'ecology' is *plural*. Fortunately, a renewed interest in 'ecological' knowledge – not only about extra-human nature, but also about nature-with-society, or about the imbrications between human and non-human agents and processes – has been promising.

Amidst the plurality of 'ecologies,' a current – or rather a set of currents – stands out, which is in fact an interdisciplinary field, and also (at least under certain circumstances) a praxis: *political ecology*. Regrettably, political ecology has been slow to give due attention to cities and urbanization (Murray Bookchin's neoanarchist version of political ecology, called 'social ecology,' was the great exception until the 1980s and 1990s [see Bookchin, 1974, 1992]). The overwhelming majority of researchers, both in geography and anthropology, have focused, for decades, on studies on the problems, challenges, and socio-ecological conflicts of rural communities in peripheral and semi-peripheral countries. It was not until the beginning of the 21st century that, driven especially by the work of geographers, an *urban political ecology* gained international visibility (see e.g. Swyngedouw and Heynen, 2003; Heinen et al., 2006).

In its usual, neo-Marxist version, however, urban political ecology finds it difficult to carry out a more radical critique of the urbanization process itself, since it typically reproduces an 'urbanophilic' theoretical body, very frequently animated by a somewhat simplistic and Eurocentric assessment of the superiority of urban life and the political and cultural importance of social classes associated with cities and industrialization. From the *Communist Manifesto* to Henri Lefebvre's reflections on the rise of the 'urban society,' the 'urbanization of society,' the 'urban revolution' and the 'right to the city' (Lefebvre, 1983, 1991) to the thesis about a so-called 'planetary urbanization' (Brenner, 2014), Marxist thought has commonly been biased in this sense – and 'urban political ecology' often pays tribute to this legacy. To overcome Eurocentrism and 'urbanocentrism,' it is necessary to expand analytical horizons and adopt a culturally more generous look at the 'ethnogeodiversity' of the world. This is what is intended in this chapter, in the wake of an analysis of a *political ecology of urbanization* applied to the task of elucidating the ecological and social problems and contradictions of 'Latin' America – or *Abya Yala*, as Indigenous peoples and decolonial/anti-colonial academics increasingly prefer and insist.

Environmental suffering: 'sacrifice zones' everywhere

Latin American cities concentrate wealth and poverty. Sophisticated consumption patterns in shopping malls frequented by the middle classes and the bourgeoisie contrast to lack of infrastructure and even basic sanitation in poor neighborhoods and informal settlements. In these cities, bubbles of comfort and security (albeit shaken, from time to time, by the violence that floods and saturates the daily life that surrounds them), represented by gated communities (known as '*condomínios exclusivos*' in Brazil and '*barrios cerrados*' in Argentina, to mention just two examples), and the living spaces of the poor working class (formal workers and, in large numbers, also informal workers), subjected to constant environmental suffering, coexist close to each other – and sometimes literally side by side.

'Environmental suffering' (in Spanish, *sufrimiento ambiental*) is a concept introduced by Argentine researchers (see, for example, Auyero and Swistun, 2007, 2009). It concerns the suffering that affects victims of environmental injustice (**see Chapter 18**). Poor workers (and particularly women who do not work outside the home, young children, the elderly and people with physical disabilities) are usually intensely exposed to various types of pollution, as well as to environmental risks linked to landslides and floods. In the case of pollution, this is because segregated spaces, especially on the urban outskirts, are seen as an attractive location for polluting factories, because of cheap land; in the case of landslides and floods, the pattern of residential segregation 'pushes' many urban poor to locations such as the slopes of hills and the banks of rivers and canals, where residents will be quite vulnerable in the face of the consequences of heavy rains and extreme weather events. The way space is produced in cities in peripheral and semi-peripheral countries creates, in many parts of cities and metropolises, 'sacrifice zones.'

The lives of the people who live in a sacrifice zone are structurally devalued. In other words, it is a space whose inhabitants are tacitly seen as inferior, 'less human' or 'less important,' and, therefore, individually disposable, however much they correspond to a useful and necessary social group as a source of cheap and exploitable labor by the urban economy (Souza, 2021). The noun 'zone' points to a relevant aspect: either by turning a blind eye, or by officially legitimizing and regulating sacrifice zones in the context of territorial planning – through zoning – the state apparatus is an unavoidable instance of power in the analysis of how such spaces are created. The capitalist firms that operate and make profits – from industrial plants to garbage incineration plants – could not implant and operate anything and obtain any profits without the state's consent. This consent is legalized and (pseudo)legitimized by achieving, through environmental legislation and permissive and lenient regulatory institutions, an official character.

Let us now consider a concrete case, systematically studied by me and my research team (see e.g. Souza, 2021): the sacrifice zone of Santa Cruz, a place at the periphery of Rio de Janeiro (**see** Figure 16.1). With an area of 125 km² and bordered to the west by the waters of Sepetiba Bay, Santa Cruz is a huge district. It shows the features of a typical urban periphery landscape (**see Chapter 19**), with its irregular subdivisions (*loteamentos irregulares*), slums (*favelas*), and working-class housing complexes (*conjuntos habitacionais*). In the mid-1970s, the Industrial Zone of Santa Cruz was officially established, and the cornerstone of the steel mill ThyssenKrupp Companhia Siderúrgica do Atlântico (TKCSA) was laid in 2006, with the first blast furnace starting operations in 2010.

TKCSA's operations were marked, from the beginning, by controversies and conflicts. A target of criticism and protest by activists from social movement organizations,

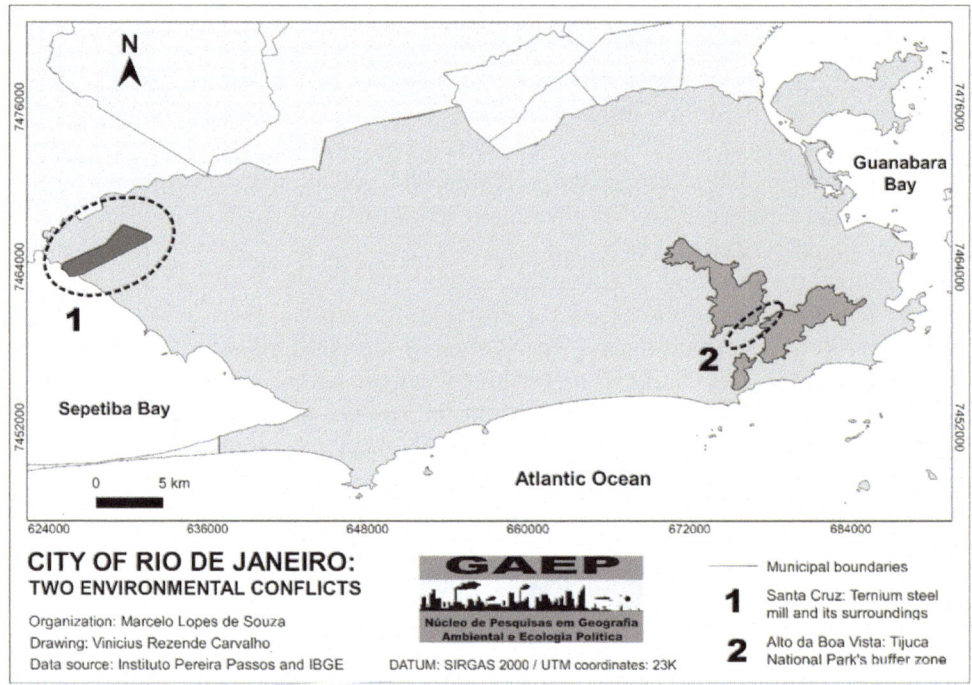

Figure 16.1 Ternium steel mill (dark grey) and its surroundings in the working-class district of Santa Cruz and Tijuca National Park (medium grey) and part of its buffer zone in the socially heterogeneous district of Alto da Boa Vista: two settings of environmental conflicts.

NGOs, and state inspection and control bodies (state and federal Public Prosecutor's Office, Public Defender's Office, and the State Institute for the Environment [INEA]), the steel mill operated for a few years without having a proper license. For non-compliance with environmental regulations, the company was fined R$1.8 million in 2010; the following year, it was punished with a new fine of R$2.8 million; finally, in 2012, the fine was no less than R$10.5 million. The company was also obliged to reimburse fishermen (affected by the resulting silting up and pollution of the São Francisco canal, the local name of the Guandu River, near its mouth) and make investments in flood control, and it also had to invest in a health clinic. None of this, however, even remotely solved the problems: emissions of pollutants (particulate material, gases and effluents) which, in a way that is now well documented, began to affect the health of residents living in the vicinity of the steel plant (due to air contamination) and dramatically impacted the subsistence of artisanal fishermen (see Figure 16.2).

After numerous frictions and controversies, ThyssenKrupp sold the steel plant in 2017, which passed to the control of Ternium (a transnational firm based in Luxembourg), and the industrial plant was renamed Ternium Brasil. Public relations work has improved since then, with the company stressing (and exaggerating) the generation of direct and indirect jobs, as well as committing itself to 'sustainability' while denying accusations of violating environmental standards. Moreover, in an effort to polish its corporate image, Ternium has launched several 'social initiatives for the community,' from internships aimed at young residents of the neighborhood to support for cultural

Figure 16.2 A resident shows her hands dirty with particulate matter emitted by the steel plant (the second photo shows a part of the plant).
Source: Thiago Roniere Tavares and Marcelo Lopes de Souza, 2019.

projects. For a population that is poor and precariously assisted by the state, the co-option potential of this new strategy is certainly great. Nonetheless, pollution problems continue to plague residents, causing them environmental suffering. What changed after 2017 was mainly the company's strategy: instead of ThyssenKrupp's stubbornness and lack of flexibility, a new approach was implemented, based on an attempt to divide the residents and weak their resistance. The local population has been therefore pushed to make a sordid bargain: exchanging health for crumbs.

Ecofascism and the city: ecologizing the 'phobopolis'?
Despite the laudable efforts of some currents of critical thought and several emancipatory social movements, political conservatism (or even reactionaryism) on the part of 'ecologically concerned' people has historically been at least as significant as progressivism. From Ernst Haeckel himself, the German scientist who coined the term 'ecology' (*Ökologie*) in the mid-19th century, to the conservationist and bioethical concerns during the Third Reich to the contemporary conservative and neo-Malthusian environmentalism of a large portion of 'deep ecology' followers and activists, there is a broad right-wing lineage committed to the defense of ecological values and principles (environmental protection, agro-ecological forestry, animal rights, etc.) at the expense of values and principles concerning human rights – in other words, committed to ecofascism. There is a trend in urban Brazil that is increasingly closely related to the ecofascist view of the poor and 'nature,' and the case of Rio de Janeiro will be used to illustrate it in this section.

While *favelados* (i.e. *favela* inhabitants) were primarily seen as 'parasitic' in the 1960s–1980s, in the last 30 years they have come to be increasingly seen as 'thugs' and 'outlaws' or, at the very least, as 'conniving with gangsters' (Souza, 2000, 2008, 2009). Prejudices increased and social and socio-spatial stigmatization intensified. This 'evolution' took place in the context of a growing fragmentation of the socio-political-spatial fabric of the metropolis (Souza, 2000, 2008), in which context we can see an increasing contrast between segregated spaces (usually *favelas* territorialized by retail drug dealers and later by paramilitary groups or 'militias') and the circuit of self-segregation (gated communities, shopping malls, etc.). The circuit of self-segregation became the breeding ground, *par excellence*, of what Boaventura de Sousa Santos (2002) called 'social fascism' (*fascismo societal*) in the late 1990s. Such 'everyday life fascism,' in given circumstances, overflows from the undergrounds of social life and reaches the level of institutional and party politics, as has been seen in Brazil mainly since 2019. The generalized fear that flourishes in such a context of inequality, prejudice and violence (and that feeds it back) configures the type of city that I called a 'phobopolis.'

'Phobopolis' (Port. *fobópole*) is a new word; it means 'city of fear.' I introduced this concept in order to emphasize the degree of intensity in terms of violence and fear prevailing in some cities today, especially (but by no means only) in (semi-)peripheral countries (Souza, 2008, 2009). The fear of suffering physical aggression, of being the victim of a violent crime, is surely nothing new in history; it has always been present and is present today in many cities. However, in some more than in others, and in some much, much more than in others. A 'phobopolis' is, bluntly put, a city dominated by the fear of violent crime. More and more cities are, these days, becoming 'phobopolises.' Brazilian metropolises can be seen as privileged 'laboratories' in this regard, starting with the two biggest ones, São Paulo and Rio de Janeiro. Both are cities whose inhabitants experience a situation which could be described as one of generalized fear of violence and the widespread perception of growing risk in terms of public safety, concerns that have assumed an increasingly prominent role in daily life conversations, in the mainstream press, etc. This is related to decisions and measures made or carried out by the state apparatus and by civil society, which have severe implications for the further deterioration of urban life – from the decision to move to a gated community or to go shopping only in presumably 'secure' shopping malls to the decision to avoid new investments in certain cities and metropolises to the decision to migrate (inside a city or to another city) in order to escape notorious so-called 'risk places' to the governmental decision to send in the army itself to fight drug traffickers in *favelas* and in the streets, as it has already occurred several times in Rio de Janeiro.

However, beyond the accusation of being 'outlaws,' there has been yet another kind of stigmatization that is of central relevance here: *favelados* have often been seen as 'destroyers of the environment.' In matters of conservative and elitist governmentalization of nature (in other words, in terms of ecofascism), Rio de Janeiro can also be mentioned as a useful illustration. In this city, environmental protection has been put at the service of agents directly or indirectly linked to a potential gentrification. Thus, the spectrum of 'green evictions' comes into play, a phenomenon that has also been observed in other cities around the world.

A clear example concerns the pro-environment and at the same time tacitly anti-poor alliance that has evolved in Rio de Janeiro since the 2000s, relating to the fate of Tijuca National Park's buffer zone (**see Figure 16.1**). Located in the heart of the city, the slopes of the Tijuca massif spread across several districts, and include a variety of neighborhoods that range from some *bairros* of the affluent, elitist South Zone to many *favelas*. With an area of 39.5 km², Tijuca National Park was established in 1961; it corresponds to the largest replanted urban forest in the world, and it is Brazil's most visited national park. There are 13, most small or very small, *favelas* located in the park's buffer zone, especially in the district Alto da Boa Vista. These *favelas* (all of them having been there for generations) began to be seen as targets for removal by the state since the 2000s. This threat, however, is associated with injustice and distortion of reality.

The corporate media has promoted an asymmetrical treatment of social classes by the state apparatus in Rio de Janeiro. Although the Public Prosecutor's Office for environmental and cultural heritage issues of the state of Rio de Janeiro has been the main institutional agent of the current attempt to promote the total or partial eviction of the *favelas* located in the park's buffer zone, it can be said that its role has been not only highlighted but perhaps even stimulated by the corporate media.

According to the Public Prosecutor's Office almost 20 years ago, those *favelas* were expanding rapidly, threatening the park and its biodiversity. However, a comparison of such alarmist forecast with census data and even the data offered by the planning office of Rio de Janeiro's Municipality itself made clear that the spatial growth of almost all *favelas* have not been significant (see Souza, 2016, 2019). After almost two decades since that forecast was proclaimed in 2006, the expected growth is far from being a reality. Hence, the alleged 'threat to biodiversity' represented by those *favelas* seems to be rather an excuse than a credible fact.

Symptomatically, while the Public Prosecutor's Office and the corporate media continue their anti-*favela* crusade, residential occupation by the middle class is left undisturbed, even in those situations where it occurs close to a *favela* targeted for removal (see Figure 16.3). The Municipality of Rio de Janeiro has made it clear on more than one occasion how desirable it would be to re-attract financially well-endowed citizens to that area. In fact, the *Prefeitura* (City Hall) had already tried to 'contain' the expansion of *favelas* through highly controversial so-called '*ecolimites*' (walls or more usually fences surrounding *favelas*) since the beginning of the 2000s, allegedly for the purpose of protecting the remnants of the Atlantic Rainforest. The state government of Rio de Janeiro followed the same steps, but its attempt to build a wall around Rocinha, Rio de Janeiro's biggest *favela*, earned strong criticism not only from Brazilian society but also from abroad (for instance, from the United Nations agencies), and the project was eventually abandoned – but 'ecofences' still exist, especially in the form of steel cables surrounding several *favelas*.

The aforementioned example shows (among several others in Rio de Janeiro alone) that ecofascism goes well with the 'everyday life fascism' typical of a 'phobopolis.' However, in the 'phobopolis' – be it in Rio de Janeiro or elsewhere in Latin America – the role of ecological concerns tends to be episodic and secondary for the elites, most of the time. It is important not to exaggerate the relevance of 'ecological' arguments, motivations, and discourses in the midst of the growing importance of criminal-informal capitalism and criminal gangs as well as paramilitary groups in cities and metropolises of the Global South. In the cities of Latin America, gangsters of petty criminal-informal capitalism (retail drug trafficking and 'militias'), often in alliance with corporate gangsters of big capital and with agents of the capitalist state, seem to have little or no inclination towards concerns about 'environmental protection' and the

Figure 16.3 The photo on the left shows a residence in the small *favela* Vale Encantado, while the photo on the right shows middle-class residences and a club. Vale Encantado and the middle-class residences are located very close to each other, but only the *favela* is threatened with eviction. The larger photo shows a panorama seen from the *favela*.

Source: Marcelo Lopes de Souza, 2015.

like. Their gangsterism has less to do with neo-Malthusianism and ecofascism than with extreme right-wing populism. This circumstance does not eliminate the possibility that, in the future, Latin American 'phobopolises' will witness the flourishing of ecofascist speeches and measures ('green evictions,' 'ecofences,' and the like) in many situations, whenever this proves to be convenient and useful for the interests of real estate capital (as it is the case of gentrification) or another fraction of capital.

Conclusion

Many tensions and conflicts that we can observe in Latin American cities relate to socio-ecological metabolism: the way in which pollution affects people's health; tragic environmental disasters; the formation of heat islands that negatively affect the quality of life, especially for the poorest; and disputes over land use and vital resources such as water.

In the context of global climate change, several of these problems (landslides, floods, etc.) tend to increase in the wake of the increasingly frequent succession of extreme events; at the same time, some new conflicts are already appearing on the horizon, related to the trend of sea level rise, which is already affecting and will further affect the densely populated coastal areas of the continent. The socio-spatial consequences of all these phenomena will be unequal, due to residential segregation (and the corresponding unequal distribution of social vulnerability) and the certainly unequal distribution of resources for works to mitigate the effects of global warming.

Several environmental conflicts aggravate or 'update' pre-existing problems (such as socio-spatial stigmatization), and all of them are part of the context of an interplay between crass social inequality (islands of affluence in an ocean of poverty), segregation and violence – particularly typical of big cities and metropolises.

Just as the role of residential segregation and the dispute over public resources for infrastructure will become clearer amid the trend of worsening effects of global warming, the ecofascist ideology also tends to become more present in the context of disputes over green areas, natural amenities, and water. It is no exaggeration to predict that the Latin American cities of the future will be more and more violent arenas of resistance and, consequently, of growing (socio)environmental conflicts.

Be it in Latin America or on other continents, urbanization raises a crucial question, which political ecology should not underestimate. Sure, it does not seem at all reasonable to deny the socially and spatially hegemonic presence of the capitalist mode of production and the historical trend of sprawl and dominance of capitalist urbanization. Nor would it be a case of cultivating 'urbanophobic' values, performing a simplistic inversion in comparison with 'urbanocentrism': instead of overestimating the 'civilizing' role of urbanization, a romanticization of rural life and pre-capitalist modes of production and societies. However, whether in *Abya Yala* or anywhere else, why on earth would it be necessary to think in a *binary* way, even when complex social and spatial hybridisms constantly suggest to us the convenience of doing exactly the opposite? Only an ideology that makes us believe that the political horizon given by current reality determines the entire range of possibilities (and even tells us what we can desire), can make us blind to the idea of an emancipatory project that encompasses a theory of socio-spatial change that transcends the terms of the debate as it has been usually placed – 'against' or 'for' the city/urbanization or the countryside – in favor of a vision committed to a *radical reconfiguration of social relations and geographic space*. For this reason, it is necessary to surpass a mere 'right to the city.' If we want to go beyond the impositions of the Western cultural matrix and the ideology of 'development' and modernization (**see Chapters 10 and 11**) – processes that lie in the root of the serious social and ecological problems we find in peripheral and semi-peripheral countries – it is urgent to call for a more generous, pluralistic right to the *planet* (Souza, 2015, 2019).

Summary

- The production of urban space involves both social and ecological processes; these cannot be separated.
- Cities, as socio-ecological spaces, contain inequalities that reflect broader society.
- Urban inequalities are exacerbated by socio-ecological processes such as floods or landslides, which in turn generate new conflicts.
- In Latin America (or *Abya Yala*), cities have witnessed a growing authoritarianism, creating forms of 'eco-fascism,' which intersect with disputes over urban nature.

Review questions

1. Why have geographers, when studying cities and urban problems, often not adequately taken into account the relationships between the geobiophysical dimension of reality and the social production of space?
2. Which aspects of the 'socio-ecological metabolism' in Latin American cities may be seen as problematic? What are the factors that contribute to this?
3. What is a 'sacrifice zone,' and how important is this phenomenon is Latin American cities?
4. How can environmental protection be instrumentalized to favor the interests of the dominant classes?

Further reading

Auyero, J., & Swistun, D. (2009). *Flammable: Environmental Suffering in an Argentine Shantytown*. Oxford University Press.

This book, in addition to being an excellent analysis of a situation of environmental injustice in a slum (*villa*) in Greater Buenos Aires, and in addition to providing an example of how to deal with the relevant concept of 'environmental suffering,' discusses the problem of social asymmetries in a complex and multifaceted way.

Souza, M. L. de. (2016). Urban eco-geopolitics: Rio de Janeiro's paradigmatic case and its global context. *City*, 20(6), 765–785.

In this paper, the concept of 'urban eco-geopolitics' is introduced. Many situations, not only in Latin America, but also in the Global South in general (and at the same time in the Global North) can be better understood with the help of this approach. 'Eco-geopolitics' refers to the governmentalization of 'nature' and the 'environment,' using the 'environmental protection' discourse (and often even that of 'environmental security') as a tool for socio-spatial control.

Souza, M. L. de. (2021). 'Sacrifice zone': The environment-territory-place of disposable lives. *Community Development Journal*, 56(2), 220–243.

This paper not only synthetically presents the concept of 'sacrifice zone,' but also presents it from a theoretical point of view that emphasizes the specificities of the Global South. Sacrifice zones are, moreover, treated multidimensionally in their threefold condition as environments, territories, and places. The main empirical example discussed there is precisely that of Santa Cruz.

Keywords

Ecofascism: as early as in the 1970s, Michel Bosquet (pseudonym of André Gorz) had coined the term '*éco-fascisme*' (Bosquet, 1978), and it seems that, at about the same time, or perhaps a little later, Murray Bookchin had also begun to use the expression, which would come to be used equally by his collaborators (see Biehl and Staudenmaier, 1995). In a nutshell, ecofascism consists of promoting or trying to promote, by authoritarian means and in disregard of human rights, measures that supposedly guarantee an 'ecological balance,' the 'protection of the natural environment' and the protection of 'wilderness.' These measures include birth control and rigid control of the spatial mobility of the population, mainly the poorest. There has been no lack of examples of ecofascist practices and interpretations in Latin America, despite the fact that on an ideological level the ecofascist ideology has been much less articulated than, say, in Third Reich Germany (and, to some extent, still in today's Germany) or in the United States.

Governmentalization of nature: 'governmentality' is a concept developed by philosopher Michel Foucault, basically meaning the 'art of government,' or the methods, strategies and techniques through which people are governed (not only by the state, but also at the level of firms, families, etc.). To 'governmentalize' nature corresponds to transforming it (or the discourse about it) into a means of controlling or influencing human behavior and social groups.

Socio-ecological metabolism: inspired by German agrochemist Justus von Liebig (1803–1873), who popularized the term 'metabolism' (*Stoffwechsel*), Karl Marx (1818–1883) used the expression 'social metabolism' (*gesellschaftlicher Stoffwechsel*) to refer to the embeddedness of society in its concrete material context, amidst of society's (self-)production on the basis of an incessant transformation of nature. However, *socio-ecological* metabolism has the advantage of suggesting more clearly, already at first glance, the conceptual content of the term.

References

Auyero, J., & Swistun, D. (2007). Expuestos y confundidos: Un relato etnográfico sobre sufrimiento ambiental. *Íconos: Revista de Ciencias Sociales*, 28, 137–152.

Biehl, J., & Staudenmaier, P. (1995). *Ecofascism: Lessons from the German Experience*. AK Press.

Bookchin, M. (1974). *The Limits of the City*. Harper Colophon Books.

Bookchin, M. (1992). *Urbanization without Cities: The Rise and Decline of Citizenship*. Black Rose Books.

Bosquet, M. [André Gorz] (1978). *Écologie et politique*. Seuil.

Brenner, N. (ed.) (2014). *Implosions/Explosions: Towards a Study of Planetary Urbanization*. Jovis.

Heinen, N. et al. (eds.) (2006). *In the Nature of Cities: Urban Political Ecology and the Politics of Urban Metabolism*. Routledge.

Lefebvre, H. (1983 [1970]). *La revolución urbana*. Alianza Editorial, 4th edition.

Lefebvre, H. (1991 [1968]). *O direito à cidade*. Moraes.

Santos, B. de S. (2002 [1998]). *Reinventar a democracia*. Gradiva, 2nd edition.

Souza, M. L. de. (2000). *O desafio metropolitano: Um estudo sobre a problemática sócio-espacial nas metrópoles brasileiras*. Bertrand Brasil.

Souza, M. L. de. (2008). *Fobópole: O medo generalizado e a militarização da questão urbana*. Bertrand Brasil.

Souza, M. L. de. (2009). Social movements in the face of criminal power: The socio-political fragmentation of space and "micro-level warlords" as challenges for emancipative urban struggles. *City*, 13(1), 26–52.

Souza, M. L. de. (2015). From the 'right to the city' to the right to the *planet*: Reinterpreting our contemporary challenges for socio-spatial development. *City*, 19(4), 408–443.

Souza, M. L. de. (2019). *Ambientes e territórios: Uma introdução à Ecologia Política*. Bertrand Brasil.

Swyngedouw, E., & Heynen, N. (2003). Urban political ecology, justice and the politics of scale. *Antipode*, 35(5), 898–918.

CHAPTER 17

Socioecological crises and transitions

Maristella Svampa (trans. Sofia Negri)

Introduction

We live in an age in which uncertainty and the absence of parameters with which to compare our world with other historical periods imply an abysmal leap, a hitherto unknown danger threshold regarding the possible responses of Nature. We have long since left the Holocene, an age characterized by climatic stability that lasted approximately between ten and twelve thousand years and allowed the expansion and dominance of humans on Earth. Today we are in the age of the Anthropocene: a new era in which we have become a transformative force with a global and geological scope whose destructive impacts endanger the reproduction of life on the planet. This chapter introduces key debates on the socioecological crisis and transitions out of it, written from the perspective of Latin America situated within the Global South.

Rather than being trapped by a fatalistic vision, this chapter focuses on the socioecological crisis, its scope and challenges, across five sections. It opens with a discussion of the COVID-19 pandemic and how it recast the ecological crisis. The chapter then provides further characterisation of the Anthropocene as an underlying concept for thinking through not only crisis but socioecological transition from the Global South. The third section provides a situated understanding of both the global asymmetries and the local and territorial dimensions of the Anthropocene. Fourth, the chapter emphasises the role of eco-territorial movements in the debate over the socio-ecological transition. Finally, it provides an analysis of some of the debates related to the transition – its social, ecological and economic scope – in a (post-)pandemic world.

Socio-ecological crisis and COVID-19

The COVID-19 pandemic centred problems that were previously on the periphery, minimised or invisible. On the one hand, it heightened social, economic, ethnic and regional inequalities, making them more unbearable than ever. After several decades of neoliberalism, it made bare the rollback of basic welfare services, not only in relation to

DOI: 10.4324/9781003430926-25

health but also to education (e.g. the digital divide), access to housing and habitat degradation. The spread of the virus also showed the failure of a neoliberal globalisation model, consolidated in the last 30 years under the World Trade Organization (WTO) (see Chapter 6).

On the other hand, the pandemic demonstrated the close link between the socioecological crisis, models of "bad development" and human health. Until March 2020, the term zoonosis was not part of our language, and perhaps for some it is still a somewhat technical or distant concept, yet it is key to understanding the pandemic. Deforestation has destroyed ecosystems, expelling wild animals from their natural environments, and released zoonotic viruses that had been isolated for millennia, putting them in contact with other animals and humans in urban environments, thus enabling them to jump between species.

Several years on from the COVID-19 outbreak, and with the consequences of Russia's invasion of Ukraine, what is glimpsed under the name of the "new normal" reveals a worsening and exacerbation of the existing conditions – social, economic, ecological and geopolitical – indicating the entry into the era of collapse, not only ecological but also systemic. The crisis includes economic recession, increasing concentration of wealth, an explosion of inequalities, the climate crisis and threat of extinction, to which war and its consequences in terms of global energy and food crisis have recently been added. We are thus living through a polycrisis, whose multiple dimensions make up a dangerous matrix, since, as Adam Tooze (2022) affirms, they present strong macroscopic elements of uncertainty, which tend to intertwine and reinforce each other.

The energy crisis aggravated by the consequences of the Russia-Ukraine war has produced a huge setback in the ecosocial transition agenda, particularly in relation to the energy shift. With the specter of the 2023 winter lurking behind its back, Europe made the transition from energy transition to energy security. Not only did it enable extreme energies (natural gas, the product of fracking and offshore exploitation), but it also authorised the reopening of coal plants (Germany, signed by a minister from the Green Party) and bet even more on nuclear power plants as energy source (especially France). The international scenario is therefore much more complex, since it confronts us not only with a setback in the environmental agenda but also with growing militarisation and what can be characterized as a "new Cold War" that increasingly divides the world system into blocs led by the United States and China, faced with possibilities (and realities) of war that could lead to a nuclear holocaust (Lander, 2022, np).[2]

At the same time, from the Global South, we warn that historical asymmetries and colonial legacies not only persist but have deepened. Within the framework of the polycrisis, the pressure from the capitalist centres to extract natural resources in the periphery has been exacerbated, which entails an amplification of the ecological debt. This is not only based on the expansion of the already known extractive paradigm in Latin America and elsewhere. What is new today is the multiplication of corporate megaprojects and new extractive pressures on the territories of the South, now in the name of the "Green Transition": cobalt and lithium (see Figure 17.1) for the production of high energy density batteries, balsa wood for turbines for wind power, transition minerals, rare earths and green hydrogen. All of this is export-oriented, that is, to ensure the decarbonisation process of the central countries, without taking into account what this means in terms of a new phase of dispossession – recharged – in the

Figure 17.1 A lithium mine in Chile.
Source: Creative Commons.

Global South. History repeats itself once again: the Global South is seen as a sacrifice zone, as an inexhaustible basket of resources for the countries of the North, not only for those lacking the so-called critical strategic resources (such as Europe), but also for those that do have them (China, the United States, Russia). In the face of this new green extractivism or energy colonialism we cannot simply jump on the bandwagon of any transition proposal.

Anthropocene times

The Anthropocene can be understood through multiple factors such as climatic instability as a result of global warming; the mass extinction of species and the consequent loss of biodiversity; changes in biogeochemical cycles, essential to maintain the balance of ecosystems; the increase in world population and urban concentration; the expansion of an unsustainable consumption model; and a toxic global food regime controlled by large corporations (see Svampa, 2019; Svampa and Viale, 2020).

One of the most worrying factors is global warming, a product of the increase in carbon dioxide emissions and other greenhouse gasses. Compared to 1750, the atmosphere currently contains 150% more methane gas and 45% more carbon dioxide, the product of human emissions. The consequence of this is that since the middle of the twentieth century the temperature has increased by 0.8°C, and the *Intergovernmental Panel for Climate Change* (IPCC) forecasts a temperature increase of between 1.2 and 6°C towards the end of the twenty-first century. Experts consider that an increase of 2°C establishes a danger threshold. As stated by the UN Environment Program, even with the fulfilment of the commitments acquired in the Paris Agreement (2015), global temperatures could increase up to 3.4°C, which would make human adaptation extremely difficult due to the new extreme weather conditions patterns (United Nations Environment Programme Report, 2020).

Currently, CO2 emissions produced from the burning of fossil fuels represent 65% of total greenhouse gasses (GHG) emissions. This does not have a uniform expression, since there are four main emitters (China, the United States, the 27 members of the European Union and India) that have contributed 55% of the total GHG emissions (United Nations Environment Programme Report, 2020). On the one hand, agriculture, forestry, and other land uses cause nearly a quarter of human-caused greenhouse gas emissions, with deforestation and forest degradation representing 11%, according to FAO data (2019). This is especially so in the so-called developing countries in the Global South and is due not only to the increase in classic extractive activities, but also to the notorious shift towards a large-scale food model, focused on high productivity and the maximization of economic benefit, built by large agrifood companies. The expansion of the agrarian frontier entails a degradation of all ecosystems: an expansion of monocultures, that produces the annihilation of biodiversity; overfishing; contamination by fertilizers and pesticides; deforestation; and land grabbing, among others.

On the other hand, during COVID-19, the activation of the emergency brake was relative. In Latin America, despite the increasing importance of socioenvironmental conflict, several extractive activities were declared as essential (such as mining), and clearing and deforestation advanced, and with it also forest fires, as can be seen in El Pantanal in Brazil, the largest continental wetland on the planet, or the fires that have occurred in Argentina.

The Anthropocene turn also has profound philosophical, ethical and political repercussions, leading us to reconsider the link between society and nature, between humans and non-humans. As we understand it here, it questions the cultural paradigm of modernity, based on an instrumental vision of nature, functional to the logic of expansion of capital. Geography, in dialogue with anthropology and critical philosophy from recent decades, reminds us of the existence of other modalities of constructing the link with nature, associated both with the relational vision of native peoples and with popular, territorial feminisms and eco-feminisms (see **Chapters 13 and 14**).

The Anthropocene signals the imminence of a point of no return and warns us that ecosystem collapse has already begun. The climate alerts are of such magnitude that it is difficult to carry out a survey that is not immediately surpassed by new tragedies. For example, the forest fires in the Amazon and Australia between 2019/2020 showed new phenomena classified as "fire storms": fires that release so much energy that they modify the meteorology of their surroundings. In July 2021, the world's newspapers reported that the Amazon (the southeastern region) went from being a plant system that captures carbon dioxide, thus preventing global warming, to now becoming an emitter of it (see Figure 17.2). The causes of this worrisome reversal are linked to deforestation, drought, fires and the advance of the agrarian frontier, among other factors. The fact itself is not unique nor is it isolated. As climatologists warn, there are inflection points (tipping points) for each subsystem, "a value from which a successive cascade of destabilizing events begins in each of the different subsystems that the reorganization of the whole entails" (Puig Vilar, 2021, np).

Figure 17.2 Deforestation in the Amazon, Brazil.
Source: www.grida.no/resources/3127

Thus, more than three years after the COVID-19 pandemic was declared, what is glimpsed under the name of "new normality" reveals a worsening and exacerbation of existing social and ecological conditions. In everyday life, we are facing the proliferation of catastrophic and dystopian images of society, many of them devoid of political language (or openly anti-political), alluding to collapse and chaos. The collapse also implies a reduction in complexity as well as a loss of democratic values, which generates social behaviours linked to fear and lack of solidarity, a breeding ground for the expansion of ultra-right expressions. Its transition involves, however, different levels (ecological, economic, social, political) as well as different degrees (it does not have to be total) and geopolitical, regional, social and ethnic differences (not all collapse in the same way) (Diamond, 2006; Fernández Durán and González Reyes, 2018; Servigne and Stevens, 2015; Taibo, 2017).

However, becoming aware of the collapse does not mean abandoning oneself with open arms to dystopia, falling into action paralysis or cognitive closure, thus renouncing the imagination of another possible future. Rather, it constitutes an urgent call for the need to redouble the commitment to resilience and the sustainability of life. The collapse opens a civilising dispute and should be seen as an opportunity to rethink what Anthropocene we want to go through as humanity. It is possible to reactivate the political imagination from the principle of hope, not from a naive attitude, but from the individual and collective understanding of what our limits are as humanity and the need to move towards an ecosocial transition that bets on the sustainability of a dignified life.

Critical lexicon of the Anthropocene from Latin America and the Global South

It is useful to think about the Anthropocene also in terms of *commodification* and *frontiers*: of the disembedding of the economic in relation to the social and the uncontrolled expansion of the frontiers of capital. As such, the Anthropocene is also the Capitalocene (Moore, 2017), understood as an internal phase of capitalist globalisation (Altvater, 2014), which expresses both the advance of the commodification of all the factors of production, as well as the material and ecological limits of the planet. In Latin America, and the Global South, this is reflected in the exacerbation of multiple models of (bad) development where extraordinary profitability, destruction of territories and dispossession of populations are combined. In that sense, there is a critical lexicon of the Anthropocene/Capitalocene which includes certain key concepts such as Social Metabolism, Anthropocene Geopolitics, Neo-extractivism, among others.

Social Metabolism refers to the way in which human societies organise the increasing exchanges between energy and materials in the environment (Martinez Alier and Walter, 2015). In general terms, capitalism has been deepening an unsustainable metabolic profile through the acceleration of social metabolism. Globalisation consolidated an unsustainable consumption model that, in order to be maintained in the richest countries, requires a greater quantity of raw materials and energy. Currently as humanity we are consuming 1.75 times what the planet can provide in a sustainable way (Overshoot Day, 2023).

The study of social metabolism reveals an Anthropocene Geopolitics through which the central industrialised countries are importers of nature; a role that is now also being assumed by the large emerging economies (such as China). These countries import commodities (or primary and secondary products) thus externalising the impacts in the name of caring for the environment in their own territories. For its part, Latin America, and the Global South, bears the burden of the costs of appropriation and extraction of commodities, turning its territories into sacrifice zones.

In Latin America, since around 2000, we have witnessed a strong return of the developmentalist imaginary. Along with the increase in the prices of raw materials, neo-extractivism updated social imaginaries linked to the (historical) abundance of natural resources in our region. Thus, within the framework of a new phase of expansion of the frontiers of capital, due to economic opportunities (the rise in the prices of raw materials and the growing demand coming mainly from China), the region entered the era of the Commodities Consensus.

In this context, neo-extractivism designates something more than those activities traditionally considered extractive, since it ranges from open-pit mega-mining, to the expansion of the oil and energy frontier (through fracking), the construction of large hydroelectric dams and other infrastructure works – waterways and ports, among others – to the expansion of different forms of monocultures or monoproduction, through to the generalisation of the agribusiness model (soybeans, palm leaves, etc.), overfishing or forest monocultures.

Finally, reflections on the Anthropocene also involve an assessment of dialogues of knowledge. New currents of critical and decolonising thought have emerged that seek to build a modest and cooperative vision, aiming at expanding dialogues of knowledge not only between different disciplinary approaches, but also with local knowledge and the vision of the affected communities (see **Chapter 12**). Thus, in Latin America it is possible to find a heterogeneous space in which geography is part of a dialogue with political ecology, environmental history, ecological economics, environmental sciences, critical sociology of social movements, climate change scholars, the post-development, ecofeminism and agroecology. Part of these new critical approaches are attempting to build an alternative paradigm based on a different relationship between society and nature, between space and social relations, between collective subjects and democracy, while proposing another epistemology. It is an inter- and trans-multidisciplinary field under construction, with great potential that is undoubtedly built in debate with other dominant perspectives.

Socioenvironmental movements

Since environmental problems entered the global agenda around the 1970s, there has been a proliferation of what Joan Martínez Alier (2005) terms "popular environmentalism": socio-environmental mobilisations led by the most vulnerable populations above all, affected by the expansion of the mining and oil frontier, as well as by polluting industries. In Latin America, over the last twenty years, the exponential expansion of popular environmentalisms is closely associated with the rise of neo-extractivism in the twenty-first century (Svampa, 2019). This has led to an eco-territorial shift in the region, illustrated by the innovative crossing of different narratives, the Indigenous community with the environmentalist regional one, to which we can add, towards the end of the century, the (eco)feminist narrative. In this way, common frameworks for collective action were configured that function not only as alternative interpretation schemes, but rather as producers of a collective subjectivity that install new themes, valuation, languages and slogans.

A key field of action, at a regional and global level, are Climate Justice movements, that include grassroots organisations (local and cultural socio-environmental movements, environmental NGOs, Indigenous Peoples' organisations, among others), networks of organisations and social movements that are born through coordination for carrying out specific and simultaneous protest actions in different parts of the world; youth protests in the form of "climate strikes"; and spontaneous mobilisations or actions of civil disobedience that demand changes in climate policies and/or denounce the inaction of the respective governments in the face of certain environmental crimes.

It is no coincidence, then, that notions such as Socioecological Transition, Climate Justice and Just Transition occupy an increasingly relevant role within popular environmentalisms. In Latin America, the transition proposed by social movements entails the challenge of thinking about alternatives to the dominant neo-extractivism both in the energy field, due to its dependence on fossil fuels, and in the productive field, due to the expansion of agribusiness. For this, it is considered necessary to overcome those hegemonic visions that continue to see development from an instrumental and productivist perspective as if natural goods were inexhaustible, and assume a relational vision that – as the native peoples and territorial feminisms emphasize – place interdependence, eco-dependence and caring for life at the centre (see **Chapter 13**).

Debating the socioecological transition

In general terms, the transition designates a process with a certain extension in time that includes stages and can refer to a change of social system (such as the transition from feudalism to capitalism), or political regime (such as the transition from a dictatorship to democracy in Latin America). The transition must be understood in terms of comprehensive and profound change, which encompasses all spheres of social life (Brand, 2011, 146). The transition implies a democratic and democratising transformation of the energetic, productive and urban spheres towards models that articulate social justice with environmental justice, towards economic and productive practices based on reciprocity, complementarity and care and towards a new pact with the nature of sustainability of a good life.

To the extent that COVID-19 put what had been on the periphery at the centre, it also enabled debates on the urgency of the ecosocial transition, in which the energy transition must be included. Since early 2020 there have been multiple proposals for ecosocial transitions, of which this section only mentions a few. In the USA, through the left wing of the Democratic Party, a Green New Deal has been promoted, aiming to decarbonise the economy and create green jobs, for which it proposes a democratic state. During 2020, the proposal was translated into a "Green Stimulus Plan" whose objective is to recover the economy using public resources for the energy transition.

At the international scale, the Progressive International was established under the motto "Internationalism or extinction". Their first virtual summit was held in September 2020 and proposed an "international ecological agreement" that, with a budget of 8 trillion dollars a year, could carry out the transition from fossil energies to renewable energies, reducing meat consumption and investing in organic food. However, the Progressive International brings together a very heterogeneous conglomerate of intellectual and political figures: from well-known environmentalists who promote the ecosocial transition to leaders of Latin American extractivist progressivism (e.g. Rafael Correa or Álvaro García Linera), recognised for the persecution of environmental sectors in their countries. As such, it is unclear how this vision would successfully articulate between social justice and environmental justice.

In Latin America, there have been many ecosocial transition proposals. For example, the "Ecosocial and Intercultural Pact of the South" has been promoted by different activists, intellectuals and social organisations from countries such as Argentina, Brazil, Bolivia, Ecuador, Colombia, Peru, Venezuela and Chile, linked to the eco-territorial struggles of the continent. The Ecosocial Pact was launched in June 2020 based on the key axes of: the paradigm of care; the articulation between social justice and ecological justice (basic income, comprehensive tax reform and suspension of foreign debt); the comprehensive socio-ecological transition (energy, food and productive), and the defence of democracy and autonomy (in terms of ethnic and gender justice). It is a collective platform that invites us to build social imaginaries, agree on a shared course of transformation and a basis for platforms of struggle in the most diverse spheres of our societies (see Ecosocialpact, 2020).

The ecosocial pact is in dialogue with other proposals such as the Green New Deal or degrowth, but is designed from the South and denies attempts to articulate proposals from the North that do not centre the question of ecological debt. The Ecosocial Pact affirms that the problems of Latin America are different from those of the North, that there are strong historical and geopolitical asymmetries and that in the heat of the socio-ecological crisis and with the increase in social metabolism, the ecological debt of the North has increased exponentially in relation to the South. Along these lines, it also warns us about the impossibility of simply jumping on the bandwagon of any transition if it promotes a corporate and concentrated model and not a democratic and popular one that ensures a just transition for the South. Lastly, far from being an abstract proposal, this is linked to concrete struggles, to processes of re-existence and the concepts-horizons forged in recent decades in the Global South and in Latin America in particular (**see Chapter 23**).

Conclusion

The pandemic has shown the need to transform the relationship between society and nature, to overcome the dualistic and anthropocentric paradigm that conceives humanity as external to nature, a conception that is at the origin of the bad development models that we suffer today. It is no coincidence that our gaze pays more and more attention to other relational paradigms or narratives that place interdependence and reciprocity at the centre. In this line, one of the great contributions of ecofeminisms, popular feminisms of the South and feminist economics, together with native peoples, is the recognition of other valuation languages, other possible links between society and nature that place care and sustaining life at the centre.

The care paradigm, as the basis of an ecosocial transition, aims to be conceived from a multidimensional perspective, including the articulation with the different spheres of social life: care and health, care and education, care and work, care and access to housing, care and community management, among others. It is the keystone of the ecosocial transition and reveals the power of the different feminisms being mobilised today. The care paradigm provides a radical questioning of patriarchy and a denunciation of capitalism as a war machine against life and is instead committed to the sustainability of a dignified life.

In conclusion, the experience of the accelerated collapse during the COVID-19 pandemic, and now in the context of the polycrisis, places us in a civilisational dispute that challenges us to think about ways out and through the socioecological transition. It is shocking that in the face of the worsening of the climate crisis and after having gone through a pandemic with undoubtedly zoonotic roots, the political and economic elites in a large part of Latin America continue to deny the importance of socio-environmental problems. Despite the fact that there is a growing socioenvironmental awareness from below, the environmental issue continues to be a blind spot for many governments in the region, an issue that goes beyond political-ideological differences.

Summary

- Our planet has abandoned the Holocene, which lasted for some ten to twelve thousand years, and moved into the age of the Anthropocene.
- The pandemic has heightened the ecological crisis alongside the related crisis of globalisation and development.
- However, numerous transitions out of the current socioecological crisis have been proposed in Latin America and beyond.
- The socio-ecological transition is being mobilised through social movements and grassroots actors, yet most governments in Latin America are failing to take this agenda seriously.

Review questions

1. What are some of the characteristics of the Anthropocene?
2. What is Latin America's role within global ecological crises?
3. How can, and is, Latin America generating alternatives from which to lead a socio-ecological transition?

Further reading

Moore, J. (Ed.). (2016). *Anthropocene or Capitalocene? Nature, History, and the Crisis of Capitalism*. PM Press.

An overview of key debates in relation to the anthropocene and capitalocene.

Svampa, M. (2019). *Neo-Extractivism in Latin America Socio-Environmental Conflicts, the Territorial Turn, and New Political Narratives*. Cambridge University Press.

A detailed overview of neo-extractivism in Latin American and its contestations through a range of conflicts and mobilizations.

Keywords

Anthropocene: a new era in which humans have become a transformative force with a global and geological scope whose destructive impacts endanger the reproduction of life on the planet. It is a concept in dispute, as there is no unified vision, although we can distinguish the more critical perspectives (which include elements of class, ethnicity and the North/South divide and the critique of development models) from the more generalist ones.

Capitalocene: an internal phase of capitalist globalisation (Altvater, 2014), which expresses both the advance of the commodification of all the factors of production, as well as the material and ecological limits of the planet.

Ecological debt: the debt accumulated by the countries of the industrial North towards the countries and peoples of the South as a result of the plundering of resources, environmental damage and the free occupation of environmental space as a repository of waste, such as greenhouse gases. In accounting terms, the climate debt is only one line in the much larger balance sheet of a wider ecological debt.

Neo-extractivism: designates something more than those activities traditionally considered extractive, since it ranges from open-pit mega-mining, to the expansion of the oil and energy frontier (through fracking), the construction of large hydroelectric dams and other infrastructure works –waterways, ports, bioceanic corridors, among others – to the expansion of different forms of monocultures or monoproduction, through to the generalisation of the agribusiness model (soybeans, palm leaves, etc.), overfishing or forest monocultures.

Social Metabolism: refers to the way in which human societies organise the increasing exchanges between energy and materials in the environment. Social metabolism allows us to account for the degree of pressure exerted by societies on natural resources and even the inequalities with which this pressure is expressed.

References

Altvater, E. (2014). El capital y el Capitaloceno, *Mundo Siglo XXI*, IX(33), pp. 5–15.

Brand, U. (2011). El papel del estado y de las políticas públicas en los procesos de transformación, en AAVV (Eds.), *Mas allá del Desarrollo*. Grupo Permanente de Trabajo sobre Alternativas al Desarrollo, América Libre-Abya Yala.

Diamond, J. (2006). *Colapso: Por qué unas sociedades sobreviven y otras desaparecen*, Spanish edition. Debate Editorial.

Ecosocialpact. (2020). *Pacto Ecosocial del Sur.* https://pactoecosocialdelsur.com/ and https://pactoecosocialyeconomico.blogspot.com/2020/05/ (Accessed: 3 May 2023).

FAO. (2019). *El trabajo de la FAO sobre el cambio climático.* www.fao.org/3/ca7126es/ca7126es.pdf (Accessed: 3 May 2023).

Fernández Durán, R. and González Reyes, L. (2018). *En la espiral de la energía*, 2a edición. Ed. Libros en Acción.

Lander, E. (2022, September 29). Keynote Lecture: Just Transitions in the Context of a New Cold War: Towards a Neocolonial Green New Deal? *Keynote Lecture.*

Martínez Alier, J. (2005). *El ecologismo de los pobres: conflictos ambientales y lenguajes de valoración.* Icaria.

Martinez Alier, J. and Walter, M. (2015). Metabolismo social y conflictos extractivos, in Castro, F., Hogenboom, B. and Baud, M. (Eds.), *Gobernanza ambiental en América Latina.* CLACSO, pp. 73–104.

Moore, M. (2017). Anthropocenes & the Capitalocene Alternative, *Azimuth*, 5, pp. 71–80. Rome.

Overshoot Day. (2023). www.overshootday.org/newsroom/press-release-june-2023-english/ (Accessed: 13 March 2024).

Puig Vilar, F. (2021, March 8). *Peor de lo esperado: puntos críticos superados, y Gaia en peligro. (2): El punto crítico global, Usted no se lo cree.* https://ustednoselocree.com/2021/03/08/peor-de-lo-esperado-tipping-points-superados-y-gaia-en-peligro-2-el-tp-global/ (Accessed: 3 May 2023).

Servigne, P. and Stevens, R. (2015). *Comment tout peut s'effondrer: Petit manuel de collapsologie à l'usage des générations présentes.* SEUIL.

Svampa, M. (2019). *Neo-Extractivism in Latin America Socio-Environmental Conflicts, the Territorial Turn, and New Political Narratives.* Cambridge University Press.

Svampa, M. and Viale, E. (2020). *El colapso ecológico ya llegó: Una brújula para salir del (mal) desarrollo.* Siglo XXI Editores.

Taibo, C. (2017). *Colapso. Capitalismo terminal, transición ecosocial, ecofascismo.* Libros de Anarres.

Tooze, A. (2022). Welcome to the World of the Polycrisis. *Financial Times.* www.ft.com/content/498398e7-11b1-494b-9cd3-6d669dc3de33 (Accessed: 02 May 2023).

United Nations Environment Programme. (2020). *Emissions Gap Report 2020 – Executive Summary.* Nairobi. https://wedocs.unep.org/bitstream/handle/20.500.11822/34438/%20EGR20ESE.pdf?sequence=25 (Accessed: 02 May 2023).

CHAPTER 18

Health, environment, and disease

Eric D. Carter

Introduction

In March 2020, the city of Guayaquil was briefly the center of world media attention. As the Covid-19 virus swept through Ecuador's largest city, its public health system, hospitals, and mortuaries were overwhelmed. Before long, Ecuadorian officials imposed a strict "lockdown," lasting several months, to prevent the spread of the virus. Although many died, a public health catastrophe was avoided, and the following year Ecuador distributed the Covid vaccine at a record-setting pace: nine million people, about half the country's population, received the vaccine over the course of just 100 days in 2021. In the end, Ecuador fared the pandemic about as well as many much wealthier countries.

Many lessons can be drawn from Ecuador's experience of Covid-19. Latin America is highly connected to the global networks that accelerate the spread of novel infections, and yet, despite persistent poverty and inequality, most countries of the region have responsive and robust public health systems. Important to the aims of this chapter, the recent pandemic also serves to demonstrate the tight linkages among health, disease, and the environment. An airborne pathogen like Covid-19 is transmitted within an environment shared with human beings. Especially before a vaccine became available, controlling the pandemic relied on manipulating the environment to prevent the spread of the virus, mainly by separating people from one another (isolation), reducing densities of people in tight spaces (social distancing), and limiting the circulation of people between places (travel restrictions).

Effective public health strategies also recognize that social and economic factors might make some groups more vulnerable to infection than others. From time to time, novel diseases affect all sectors of society indiscriminately, irrespective of age, gender, race, occupation, or social position. However, as we saw with the Covid-19 pandemic, disease is nearly always a socially differentiated phenomenon, meaning demographic and social factors matter as much as environmental ones to understand the nature of disease spread. Particularly in open and democratic societies, successful public health

DOI: 10.4324/9781003430926-26

efforts also require a shared frame of reference – agreement about the causes of disease, the role of scientific expertise, and the efficacy of medical interventions.

This brief discussion of Covid-19 points us towards a geographical framework for exploring the relationship among health, disease, and the environment that incorporates ecological, structural, and epistemological dimensions. Some might call this a "human ecology," "political ecology," or "social medicine" approach, but following the lead of Paul Farmer (2001), the late physician-anthropologist, we may call this a "biosocial" perspective. With this lens, we can see that public health issues in Latin America and the Caribbean (LAC) are connected to larger questions about identity, development, equality, social inclusion, and environmental change – subjects that are often seen as outside of, or distant from, the health field.

It is important to point out, however, that many countries of LAC today have population health conditions on par with those of more affluent countries of the world (see Table 18.1), thanks to steady investments in health systems and a broad commitment to health care as a right (Carter, 2023). Thus, while this essay necessarily focuses on public health *problems*, we should not lose sight of the palpable public health *progress* across the region over the last several decades.

Table 18.1 Key health, demographic, and economic indicators, select Latin American countries (and comparison countries), 2019 (or most recent year).

Country	GNI per capita, Atlas method (current US$)	Life expectancy at birth (years)	Mortality, under-5 (per 1,000 live births)	Maternal mortality ratio (per 100,000)	Physicians per 1,000 people	Health spending, as percentage of GDP	Current health expenditure per capita (current US$)
Argentina	11,220	77.3	8.6	39	4.0	9.51	946
Bolivia	3,440	67.8	26.6	155	1.0	6.92	246
Brazil	9,220	75.3	14.9	60	2.2	9.59	853
Chile	14,830	80.3	7.1	13	4.9	9.33	1,376
Colombia	6,590	76.8	13.7	83	3.7	7.71	495
Costa Rica	12,040	79.4	8.2	27	1.3	7.27	922
Cuba	8,920	77.6	5.2	36	8.3	11.34	1,032
Ecuador	6,100	77.3	13.4	59	2.2	7.82	486
Guatemala	4,620	73.1	24.6	95	0.3	6.21	271
Mexico	9,660	74.2	14.2	33	4.8	5.43	540
Peru	6,770	76.2	14.9	88	0.8	5.22	370
Venezuela	13,010	72.2	24.2	125	1.7	5.37	339
Canada	46,540	82.1	5.1	10	2.4	10.84	5,048
Japan	42,010	84.4	2.5	5	2.5	10.74	4,360
United Kingdom	43,380	81.2	4.3	7	5.8	10.15	4,313
United States	66,130	78.8	6.4	19	2.6	16.77	10,921

Source: All data derived from the World Development Indicators and Sustainable Development Goals databases of the World Bank DataBank (https://data.worldbank.org/). Most data are from 2019; if 2019 data unavailable, the next most recent year's data were used. Data from 2020–21 avoided due to the Covid-19 pandemic.

In the following section, I discuss how outsiders have often viewed the LAC region as a "hotbed" of infectious tropical diseases and how urban environments were transformed in the name of public health. The second section centers on Mexico City and Buenos Aires, Argentina to examine environmental health issues that continue to impact urban residents, including air pollution, contamination of water, and exposure to dangerous chemicals. In the last section, I explore how Indigenous health beliefs and practices were suppressed historically, and how a new model of "intercultural health" has the potential to not only improve health equity but also to contemplate more integral connections between human health and environmental sustainability.

Tropical imaginaries

In early 2016, international news media issued frantic reports of a major outbreak of Zika virus, a mosquito-borne illness, in Brazil. As one news report put it, "The latest virus to break out of the tropics may be the most frightening" (Rosen, 2016). As the virus spread to nearby countries, many Americans cancelled their travel plans "south of the border" while the US intensified border surveillance in the name of health security.

Although the Zika virus had serious consequences, media coverage of the epidemic also resurrected old tropes about Latin America as a source of dangerous tropical diseases, always threatening to "break out" from warmer climates, transgress national borders, and impact well-off countries of the Global North. This geographical imaginary of LAC as a hotbed of exotic and deadly tropical diseases dates back to the 1700s, if not earlier, when regular bouts of tropical fevers threatened to foil the ambitions of French, British, and Dutch colonizers in the Caribbean. Indirectly, tropical disease intensified the trans-Atlantic slave trade, since Africans were viewed as more resistant to illnesses like malaria.

The nineteenth century, when most Latin American countries achieved their political independence, was also a period of frequent global epidemics. During the "first age of globalization" in the latter part of the century, as LAC was integrated more fully into the international economy, rising numbers of passengers aboard larger and faster steamships incidentally accelerated the movement of pathogens, producing frequent and dire epidemics. Recurrent waves of cholera and yellow fever swept through key ports in the Americas (New Orleans, Havana, Rio de Janeiro, Buenos Aires, Cartagena, Callao in Peru) and made their way inland, leaving devastation in their wake. An 1871 epidemic of yellow fever in Buenos Aires, for instance, killed over 10,000 people (about 8% of the city's population at the time), and the same disease took some 15,000 lives between 1890 and 1898 in Rio de Janeiro, then Brazil's capital. Meanwhile, endemic and persistent infectious diseases, such as malaria, hookworm, leprosy, and Chagas disease (which bears the name of the Brazilian scientist who discovered this vector-borne illness in the early 1900s), also flourished, often in rural areas where advances in hygiene and sanitation took a long time to arrive.

Concerns over hygiene and sanitation drew countries of LAC more tightly under the influence of the regional hegemon, the United States. During the building of the Panama Canal (see Chapter 6), the US government turned Panama into a virtual laboratory for scientific initiatives to control tropical diseases like malaria and yellow fever. In the 1910s, the US-based Rockefeller Foundation (RF) started medical research and public health programs in Brazil, Mexico, Costa Rica, and other Latin American countries, which helped reduce the burden of infectious diseases (hookworm, yellow fever, and malaria, particularly), promoted collaborations with Latin American medical experts, and, in some cases, built a nucleus of a public health system, especially in neglected rural areas. The RF also provided a durable template for international health efforts, founded on infusions of philanthropic funding for scientific research and technical interventions that mostly ignored broader social and political conditions. What became known as the PAHO, or Pan American Health Organization, has been based in Washington, DC since its inception, and for decades it steered regional health policies to serve US interests.

Such US-led public health efforts may seem like an imperial imposition, but they resonated with Latin American elites who viewed improvements in hygiene and sanitation as essential elements of modernization. Generally, these elites were liberal (in the sense of embracing free markets), positivist (believing that science and technology were key levers of social progress), and, not surprisingly, elitist. Steeped in the scientific racism of the era, modernizing elites pushed for European immigration as a means of racial improvement; investment in public health was meant to attract new immigrants and, over the long run, raise the "quality" of the national population. They made the control of diseases like malaria part of a larger "civilizing" mission and developed homegrown centers of expertise in tropical medicine (Carter, 2012).

The leaders of Latin America's scientific hygiene movement refashioned urban environments, inspired by developments in European capitals. The aforementioned 1871 yellow fever epidemic accelerated the transformation of Buenos Aires into the "Paris of South America." Wealthy *porteños* relocated to the city's northern zone, away from the epicenter of the epidemic and its suspected environmental causes (understood then as "miasmas," fever-causing emanations from cemeteries, slaughterhouses, and stagnant waters). Potable water and sanitary sewer systems began to reach the city's population in the 1880s. The overcrowded tenement buildings of Buenos Aires, the *conventillos*, were condemned as dank and suffocating spaces that spawned vice and disease. As historian Diego Armus (2011) has pointed out, a modern, healthy city meant well-ventilated and sunny spaces. Trees, viewed as "the lungs of the city," were planted everywhere: along widened boulevards, in plazas, and in new parks, like the Bosques de Palermo, a landscape of sand and scrublands that became the city's largest expanse of greenspace and recreation in the affluent northern district. Similarly, the hills of central Rio de Janeiro were levelled with dynamite and steam shovels, and miasmatic coastal marshes were filled to make way for a new business district and upper-class homes. Assisted by the father of Brazilian public health science, Oswaldo Cruz, mayor Francisco Pereira Passos styled himself as a "tropical Haussmann," seeking to recreate the orderly boulevards of fin-de-siècle Paris in what was then Brazil's capital city and primary port. Notably, many people displaced by urban sanitary reforms became residents of the city's earliest *favelas*, or shantytowns, on nearby hillslopes (Meade, 1997).

The incidence of infectious and vector-borne diseases has certainly diminished over time, and their causes are better understood, but the threat has never disappeared. One nagging public health problem is dengue fever, caused by a virus spread by the notorious *Aedes aegypti* mosquito (the same culprit in the transmission of yellow fever and Zika) (see Figure 18.1). Mild cases of dengue produce fever, aches, and fatigue,

Figure 18.1 Public health agent collecting samples from suspected Aedes aegypti breeding site, Tucumán, Argentina, 2016.

Source: Ramón Teves/Ministerio de Salud de la Provincia de Tucumán.

but a severe form called dengue hemorrhagic fever, a result of exposure to multiple strains of the virus, is often deadly. A PAHO campaign to eradicate *Aedes aegypti* in the Western hemisphere, using chemical insecticides like DDT in the 1950s and 1960s, was mostly successful; the mosquito still survived, however, in parts of the Caribbean and the southern US. From these embers the mosquito spread like wildfire in the 1980s and 1990s, as a massive economic crisis led to cuts in government disease surveillance budgets. Unfortunately, *Aedes aegypti* is almost perfectly adapted to tropical urban environmental conditions, especially in zones where residents live in close quarters, in substandard housing with irregular water supply – as in the informal settlements of urban peripheries in countries like Argentina, Brazil, Colombia, or Ecuador. Global climate change is likely to make conditions more favorable for the proliferation of Aedes mosquitoes, but only marginally; preventing dengue means addressing the living conditions of low-income neighborhoods, which presents vexing logistical and political challenges.

Urbanization and environmental crisis

The growth, improvement, and sanitation of Latin American cities has always been an uneven process. Healthful amenities, like parks and plazas, have never been spread evenly across urban space; similarly, the location of environmental hazards generally reflects the heterogeneous socioeconomic geography of the cities. Put more simply: to be poor in the city means being more vulnerable to environmental health risks – landslides, flooding, pollution, deadly traffic, unsafe water, and so on (**see Chapter 16**). The environmental crises of cities can be seen as a side effect of broader political-economic transitions, and perhaps even intrinsic to the workings of industrial capitalism.

The history of Mexico City serves to illustrate how environmental crisis accompanies economic development and population growth. Long Mexico's largest city, even before the arrival of the Spanish in the early 1500s, it experienced dramatic growth in the second half of the twentieth century, eventually becoming a "megacity" home to over 20 million residents. Its natural setting has exacerbated problems of air pollution, sewage treatment, and water supply. Smog is trapped and concentrated in a high-altitude, bowl-like valley, increasing the risk of chronic health problems like cardiovascular disease and lung cancer. The ecologically rich and attractive landscape of lakes and wetlands on the valley floor has been gradually drained and paved over. One of the last remnants of the valley's aquatic landscape, Lake Xochimilco, is home to the axolotl, a curious-looking salamander that has become the unlikely mascot of nature's tenacious hold in the city. Extraction of groundwater to supply homes and industry has led to subsidence, a gradual yet uneven sinking of the city at a rate of around 50 centimeters per year. This unstable geology intensifies the effect of earthquakes, like the catastrophic temblor of 1985 that killed over 10,000 people in Mexico City.

In 1992, the United Nations labeled Mexico City "the most polluted city in the world," due mainly to its hazardous air pollution (Hoffmann, 2022). By then, the city government had already taken some action, establishing the "Hoy No Circula" program in the 1980s, intended to take some automobiles off the street on designated days. While those restrictions were notoriously easy to evade, a series of comprehensive regulations (including the introduction of cleaner technologies and the control of emissions from power plants, factories, and cars) helped to reduce air pollution even as the population continued to grow, and the number of motor vehicles increased at an even faster rate. Despite having one of the largest public transit systems in the Americas, Mexico City continues to be car-dependent due to its sheer sprawl. And while the city's environmental health conditions have generally improved in many dimensions over the last few decades, the poor continue to experience the worst impacts from polluted environments, whether from long commutes, working conditions that expose them to hazardous outdoor air, or lack of access to adequate healthcare.

Until recently, Latin American academics and policymakers mostly ignored the environmental dimensions of poverty and marginalization. The story of the Matanza-Riachuelo watershed in Buenos Aires exemplifies how environmental justice analysis and activism can produce policy change with potential to improve living conditions in informal settlements. Before urbanization, the Matanza River was a typical watercourse of the Argentine Pampas, draining nearly flat terrain covered by native grasses, marshes, and riparian forests. By the late nineteenth century, the lower reaches of the river, known as the Riachuelo, had been straightened and dredged in order to facilitate navigation at the city's port and drain the effluents from burgeoning industries along its banks – first tanneries and slaughterhouses, and later industrial establishments of all sorts, including oil refineries and chemical plants. The 1871 yellow fever epidemic, centered on the neighborhoods of southern Buenos Aires nearest the Riachuelo, led affluent families to flee to the northern zone, thus cementing a lasting division between the affluent north and the working-class south, a pattern that extends into the suburbs of the metropolitan area. Unregulated use of the river, treated as an open sewer and garbage dump, made it one of the most polluted environments in all South America.

Nevertheless, along the Riachuelo informal settlements sprang up out of necessity: though polluted, foul-smelling, and flood-prone, the area also had land that was practically free for the taking. The introduction of neoliberal (free-market) reforms in the 1990s, followed by a cataclysmic economic crisis at the end of 2001, led to a spike in unemployment and worsening poverty: "From 2001 to 2006, the number of people living in slums, shantytowns, and squatter settlements in Greater Buenos Aires almost doubled, rising from fewer than 639,000 to more than 1.14 million. In the same time frame, the number of settlements rose from less than four hundred to a thousand" (Auyero & Swistun, 2009, p. 23). By 2009, more than four million people lived in the Matanza-Riachuelo watershed, around 500,000 of them in informal settlements. Among the latter, only about 10% had access to sanitary sewers and around half had access to clean running water. Residents were exposed, to different degrees, to arsenic, lead, mercury, and other hazardous chemicals.

Around this time, the plight of residents of the Riachuelo began to receive more attention from the Argentine media, sparking calls for government action. However, since the watershed crossed the boundaries of two provinces and 14 municipalities, coordinating sanitation efforts was a persistent challenge. Authorities tended to treat pollution and its health impacts "*as a problem whose solution is always someone else's responsibility*" (Auyero & Swistun, 2009, p. 100, emphasis in original). In 2004, a group of local residents and environmental activists from the neighborhood of Dock Sud, adjacent to a major petrochemical facility, sued for damages against various government authorities and private companies; a decision of the national supreme court compelled the authorities to clean up the area, under the authority of a new coordinating body, ACUMAR, established in 2008. There have already been tangible results: the removal of tons of garbage, stricter regulation of industrial effluents and emissions, and the construction of monumentally expensive engineering works to divert sewage away from the Riachuelo to a new treatment plant.

From one angle, the cleanup of the Riachuelo (see Figure 18.2) seems like an environmental justice success story, demonstrating the power of coalitions uniting academics, scientists, and residents. At the same time, the political resolution of the problem, with an unexpected high court decision, was unique, and the proximity of the Riachuelo to the center of Buenos Aires – the political, financial, and media capital of the country – made its problems hard to ignore. Concern over environmental inequalities has not yet taken hold more broadly across Argentina, and policy instruments to address the environmental problems of the urban poor continue to be weak.

Indigenous medicine and alternative visions of health and environment

To this point, this chapter has mostly taken for granted that medicine is synonymous with Western medicine, or biomedicine; we could also argue that disparities in health outcomes result from a failure to equitably distribute the full benefits of biomedicine to populations in need. However, there is another, critical perspective on this issue. For all of its advances and achievements, Western medicine can also be seen as another tool of European domination over Indigenous and Afro-descendant groups in the Americas, who have their own distinctive health beliefs and practices. These groups may well find Western medicine alienating or culturally unacceptable. A serious consideration of other kinds of medicine may lead not only to more socially just results, but also a

Figure 18.2 Cleanup of the Riachuelo by the Ministry of Environment and Public Space, City of Buenos Aires, 2014.

Source: Photo by Gabriela Seijo, licensed under Creative Commons Attribution License.

different approach to understanding connections between human health and the integrity of the natural environment (**see Chapter 13**).

From the colonial period onward, Latin America was a "dynamic frontier between Western and non-Western medicine" – a mix of European, Indigenous, and African influences, but within a sociocultural hierarchy that clearly placed Western perspectives above others (Cueto & Palmer, 2015, p. 59). Officially sanctioned medical education aspired to "Europeanize" national cultures while eradicating, or at least obscuring, alternative, non-Western, and "popular" forms of medicine. The suppression of Indigenous knowledge (particularly), in medicine and other domains, has been fittingly called "epistemicide," or "the killing of a knowledge system" (Breilh, 2021, p. 25).

However, such an erasure was never complete, especially in Meso-America, the Andean Countries, and Brazil, where Indigenous and/or African medicine was practiced clandestinely for centuries. Western biomedicine has often simply co-opted Indigenous practices when convenient. Notably, the bark of the Andean shrub known as *cinchona*, long known for its effectiveness against fevers, served as the raw material for quinine, whose power to prevent and cure malaria made it one of the most valuable drugs in the world, for a time. The leaves of another medicinal plant from the same region, coca, served as the raw material for cocaine, considered a legitimate stimulant and pain reliever in Western medical practice until the early 1900s. More recently, the use of ayahuasca, a medicinal herbal tea used in shamanic rituals in the Andean-Amazonian region, has become popular among Western tourists seeking relief from anxiety, PTSD, and other mental health issues.

However, Indigenous medicine offers more than medicinal plants and other "alternative" therapies; it also expresses a cosmovision, or worldview, that contrasts sharply with Western ideas and norms. Biomedicine is reductionist, assuming divisions between body, mind, spirit, and environment; the cause of disease must be traced back to some physiological or biochemical mechanism in human cells and tissues. It also centers on the health of individuals, and the trend towards genetic medicine, with the promise of therapies customized to each person's genetic code, only reinforces this individualistic tendency. By contrast, even as we allow for considerable diversity among Indigenous groups, Indigenous health beliefs are holistic and concerned with a social collective that extends into the non-human realm. The coca leaf, when masticated, has demonstrably effective pharmacological properties that make sense in a biomedical framework, but it can also be an "essential mediator between human and spiritual worlds," "essential for the maintenance of good social relations" (as a token gift that strengthens bonds of reciprocity), and a sacred symbol of Andean people and their cultural survival (Grisaffi, 2010). Similarly, ayahuasca is not simply a psychoactive chemical compound, but one element in a religious ritual rooted in a cosmovision that embraces the spirituality of nature (Irigaray et al., 2016). Within Indigenous cosmovisions, like that of *Kallawaya* healers of Bolivia, people, animals, and plants have a shared fate, and thus human health is dependent to a large degree on a harmonious and integral relationship with nature – a notion that is foreign to Western medicine (Harris, 2017).

After centuries of repression, Indigenous medicine has recently received wider recognition and legitimacy, an outgrowth of a broader political movement for Indigenous rights that flourished across the Americas starting in the early 1990s. In Bolivia, the practices of *Kallawayas* and other traditional healers were decriminalized starting in the 1980s (see Figure 18.3). After becoming president of Bolivia in 2006, Evo Morales, a coca farmer of Aymara descent, promulgated a model of "interculturality" in the medical field, embodied in a maternal and family health program known by its Spanish acronym SAFCI; this program attempted to make clinical spaces, like birthing rooms, more culturally acceptable to Indigenous women and to flatten professional hierarchies that were also racialized and gendered. SAFCI was also meant, of course, to reduce Bolivia's terrible maternal mortality rates, among the worst in Latin America (see Table 18.1), and even worse among the country's Indigenous women. SAFCI and other intercultural health policies were woven into a new governing philosophy of *sumak kawsay* (*buen vivir* in Spanish, or living well), a negation of the environmentally destructive model of neoliberalism (Hartmann, 2019) (see Chapter 10).

The weaving together of Western and Indigenous practices in medicine has not always gone smoothly. Formal medical education in LAC is still dominated by training in the biosciences, relatively few doctors identify as Indigenous, and the benefits of non-Western practices are not easily captured in the techniques and language of "evidence-based medicine." Nevertheless, there is now a more clear-eyed view of the reality of everyday therapeutic practices: people regularly alternate between mainstream and folk medicine – this is true broadly, even among people who do not identify as Indigenous per se. Just as importantly, there is wider recognition of the paternalism, elitism, and even racism of mainstream medical practice. Although there is still work to be done, intercultural programs like SAFCI have the potential to make medicine more effective, ethical, and accessible (Herrera et al., 2019).

Figure 18.3 *Kallawaya* healer conducting welcoming ritual at Latin American social medicine meeting, La Paz, Bolivia, 2018.

Source: Photo by author.

Conclusion

The promotion of public health cannot be easily disconnected from discussions of broader issues in Latin American geographies: undoing the legacies of colonialism, distributing the benefits of economic development, promoting social equity and inclusion, and addressing ongoing and intensifying environmental crises. Broadly speaking, population health conditions have improved dramatically across the region over the last two centuries, partly due to advances in medical care, but also to health-promoting environmental modifications. This long-term process exemplifies what Cueto and Palmer (2015) call "health in adversity," the tendency in Latin America towards health improvement in spite of high levels of poverty and inequality, and because of tireless efforts of "health professionals, activists, and popular leaders" who view health as a human right, prioritize equity, take a biosocial perspective on disease causation, and oppose efforts to commodify health care, while also promoting bioscientific achievement under challenging conditions.

A biosocial lens helps us see the complex relationships between health, disease, and environment, in ecological, structural, and epistemological dimensions. Whether in the case of infectious diseases like malaria or polluted waterways like those of the

Riachuelo, we can see that human health is influenced by the conditions of the environments where people live, work, and play. The risk of illness from environmental influences is not evenly distributed in space; in fact, there is a remarkable pattern of socio-spatial differentiation, meaning that the poor and marginalized are more vulnerable to the unhealthy side effects of modernization. Possibly the answer to such social and environmental crisis lies in an intensified commitment to biomedicine and other Western science and technology. However, this chapter highlights alternative and complementary ways of thinking about health – in particular, intercultural health, a bridge between Western and Indigenous medicine, which could give us new insights into more equitable, ethical, and environmentally conscious ways to promote human health and well-being.

Summary

- Disease is a spatially differentiated phenomenon, combining social and environmental factors.
- Latin America has been unjustly portrayed as a hotbed of dangerous tropical diseases.
- Urbanization sparked various environmental crises, with disproportionate impacts on poor and marginalized communities.
- Intercultural health bridges the divide between Western and Indigenous medicine, with the potential for more ethical, effective, and environmentally conscious health policies.

Review questions

1. How did concerns over public health lead to the social and physical modification of Latin American cities, starting in the nineteenth century?
2. Why is environmental health risk concentrated among poor and marginalized communities living in large Latin American cities?
3. What are some of the major differences between Western and Indigenous health beliefs and practices in Latin America?
4. What is intercultural health and what are some of the obstacles to its success?

Further reading

Auyero, J., & Swistun, D. A. (2009). *Flammable: Environmental suffering in an Argentine shantytown*. Oxford University Press.

Although published before the major Argentine supreme court decision that mandated the cleanup of the Riachuelo, this book explains how residents along its banks, adjacent to a major petrochemical facility, learn to live with environmental risk and navigate its legal system to gain some measure of justice.

Carter, E. D. (2012). *Enemy in the blood: Malaria, environment, and development in Argentina*. University of Alabama Press.
A close examination of how changes in scientific understanding and concern for regional economic development converged to control malaria in Argentina's impoverished Northwest.
Cueto, M., & Palmer, S. (2015). *Medicine and public health in Latin America: A history*. Cambridge University Press.
Authored by two leading historians, this indispensable book synthesizes decades of research on the history of public health and medicine in Latin America; most of the first section of this chapter is drawn from this book, or sources cited within it.

Keywords

Biomedicine: sometimes called Western, allopathic, or mainstream medicine, it hinges on the principles that all disease is caused by biological, chemical, or physical phenomena in the body, and that healing comes from the clinical application of lab science findings.

Biosocial framework: a framework that incorporates economic, political, and environmental factors to understand disease dynamics.

Epistemicide: the killing of a knowledge system, analogous to, and sometimes accompanying, cultural genocide.

Intercultural health: a model for integrating (Western) biomedical and Indigenous health beliefs and practices, based on principles of mutual respect, flexibility, and willingness to interact across cultural boundaries.

References

Armus, D. (2011). *The ailing city: Health, tuberculosis, and culture in Buenos Aires, 1870–1950*. Duke University Press.
Breilh, J. (2021). *Critical epidemiology and the people's health*. Oxford University Press.
Carter, E. D. (2023). *In pursuit of health equity: A history of Latin American social medicine*. University of North Carolina Press.
Farmer, P. (2001). *Infections and inequalities: The modern plagues*. University of California Press.
Grisaffi, T. (2010). We are originarios . . . 'we just aren't from here': Coca leaf and identity politics in the Chapare, Bolivia. *Bulletin of Latin American Research*, 29(4), 425–439.
Harris, D. M. (2017). Mountain-bodies, experiential wisdom: The Kallawaya cosmovisión and climate change adaptation. *Third World Thematics: A TWQ Journal*, 2(2–3), 376–390.
Hartmann, C. (2019). Bolivia's plurinational healthcare revolution will not be defeated. *NACLA Report on the Americas*. https://nacla.org/news/2019/12/19/bolivia-plurinational-healthcare-revolution-evo-morales
Herrera, D., Hutchins, F., Gaus, D., & Troya, C. (2019). Intercultural health in Ecuador: An asymmetrical and incomplete project. *Anthropology & Medicine*, 26(3), 328–344.
Hoffmann, B. (2022, March 23). The unequal effects of air pollution on health and income in Mexico City. *Inter-American Development Bank*. https://blogs.iadb.org/ideas-matter/en/the-unequal-effects-of-air-pollution-on-health-and-income-in-mexico-city

Irigaray, C. T. J., Girard, P., Irigaray, M., & Da Silva, C. J. (2016). Ayahuasca and Sumak Kaw-say: Challenges to the implementation of the principle of "Buen Vivir," religious freedom, and cultural heritage protection. *Anthropology of Consciousness*, 27(2), 204–225.

Meade, T. A. (1997). *Civilizing Rio: Reform and resistance in a Brazilian city, 1889–1930*. Penn State Press.

Rosen, M. (2016, January 22). Rapid spread of Zika virus in the Americas raises alarm. *Science News*. www.sciencenews.org/article/rapid-spread-zika-virus-americas-raises-alarm

Urbanization

CHAPTER 19

Urban peripheries

Matthew A. Richmond

Introduction

The term **urban periphery** is widely used in geography and urban studies to analyse the spatial organization and development of cities. As cities worldwide expand outwards at a rapid rate, many scholars argue that we must look to the periphery if we are to understand our increasingly urban world. Indeed, some propose that a "peripheral turn" is underway within the field (Ren, 2021), involving a complete rethinking of urban theory "from the outside in" (Keil, 2017). If that is the case, Latin American geographers have much to offer: researchers from the region have been analyzing and theorizing the periphery for decades. Indeed, the urban periphery can rightly be considered a fundamental – even foundational – category in Latin American urban geography. It is also dynamic one that has been reworked over time. This is both a product of intense theoretical debates about how to define and understand the social and spatial dynamics of urban peripheries, as well as a response to empirical transformations that the peripheries themselves have undergone.

However, despite the global relevance of this large and rich body of work on urban peripheries, there are also specificities that underpin Latin American approaches. Latin America's urban peripheries have traditionally been understood as spaces of poverty, informal land development, and **autoconstruction** (the self-building of homes) (Caldeira, 2017). This stands in stark contrast to traditional visions of suburbia in much of the Global North as spaces of large homes, picket fences, and proximity to nature that caters primarily to the middle classes. In fact, in recent decades, these starkly contrasting stereotypes have been somewhat eroded by complex socio-spatial urban transformations in both the Global North and South (Dear and Dahmann, 2008; Sposito and Sposito, 2020). Nonetheless, they do reflect important real-world differences in urban processes, especially between the Anglophone countries of the Global North on the one hand and Latin America on the other.

Attempts to understand how Latin America's urban peripheries originally expanded as spaces of poverty and state neglect were closely tied to dependency theory

DOI: 10.4324/9781003430926-28

(see Chapter 11). Under this framework, the urban peripheries came to be seen as a spatial manifestation of the region's peripheral and dependent position within the global economy. More recently, researchers have explored long-term transformations of the urban peripheries, emphasizing the agency of peripheral populations, the increasing presence of the state, and their growing socio-spatial heterogeneity and fragmentation.

This chapter contains three main sections. The first discusses the emergence of a dominant "centre-periphery" pattern of urban segregation in Latin American cities between the 1950s and 1980s, and of the importance of dependency theory in analyses of these changes. The second section offers an overview of the key transformations urban peripheries have experienced since the 1990s, including claims of a process of "socio-spatial fragmentation" in Latin American cities. The third section presents an empirical example of a peripheral urban neighborhood in São Paulo, identifying its historical development and recent transformations.

Defining and theorizing the urban periphery: 1950s–80s

As already indicated, geographers around the world use the term periphery in a variety of ways. It can be employed as a straightforward spatial reference, designating territories lying at the edge of (or even outside) an urban agglomeration, regardless of their particular social or urban characteristics. Used in this way, "periphery" might be considered more or less interchangeable with terms like "suburb", "exurb" or "edge city" (eg. Dear and Dahmann, 2008; Keil, 2017). Others have conceived of peripheries in terms of spatial hierarchies, to refer to spaces that are economically and politically marginalized. This might include rural hinterlands that service large metropolises or post-industrial regions that are "peripheralized" by processes of economic restructuring (Fischer-Tahir and Naumann, 2013). Finally, the term has been used in a more metaphorical sense to refer to marginalized groups living in various urban settings (including central areas), their collective practices of survival, and transgressive forms of political agency (Simone, 2010).

Latin American scholars, by contrast, have tended to define urban peripheries in more specific ways that combine spatial location with socio-economic and institutional criteria. Peripheries are thus usually understood as areas that are geographically distant from the city centre (though *within* urban agglomerations), but which are also poor and socially and politically marginalized. This more specific definition is a legacy of the historical development of a centre-periphery growth model that emerged in most large Latin American cities during the mid-twentieth century, dominant roughly from the 1950s to the 1980s. Under this model, urban infrastructure, services, employment opportunities and wealthy populations were concentrated in more central areas, while the poor tended to live in underserved areas at the edge of the city. Certainly not all the urban poor lived in the periphery, even at the height of the center-periphery model. For example, slum tenements, informal settlements and other kinds of low-income housing persisted in central areas. Nonetheless, at the urban scale, it could be seen that the majority of the urban poor lived in peripheral areas and the peripheries were almost entirely populated by low-income populations.

Empirical studies in the mid-twentieth century chronicled the ways in which poor populations, many of them Black, Indigenous and mestizo migrants from poor rural areas, came to settle at the edge of the city. This typically occurred in two different ways, which varied across different countries and cities within the region. The first was squatter settlement, whereby homeless groups would illegally occupy public or private land in the hope that these settlements would be retrospectively regularized (made legal) or at least tolerated by the authorities. The second was through the irregular sub-division and sale of undeveloped plots of land to individual buyers in ways that did not meet urban planning regulations and often lacked a valid transfer of ownership titles. In both cases, the neighbourhoods that emerged typically lacked basic infrastructure and services found elsewhere in the city and were populated by low-income populations living in autoconstructed homes.

The following passage by Peruvian anthropologist José Matos Mar, provides a description of the process of squatter settlement in Lima's peripheries in the 1950s:

> A peripheral barriada, or marginal quarter, may be defined as a social agglom-eration formed by a population which takes over waste land, usually owned by the state, the public welfare authorities or the municipalities or by private owners who do not make use of it. These areas are situated on the perimeter of the city, and a glance at a map of the city will clearly show how they form a ring around it. When a given area is thus invaded, the land is first divided into plots of various sizes and is allotted to the families which have applied for it. Each family, using any kind of building materials, immediately proceeds to build a dwelling on its plot in order to establish squatter's rights. A barriada is always set up in this organized manner.
>
> *(Matos Mar, 1958: 11)*

Early empirical accounts of this kind described in detail how such processes unfolded. However, over the 1960s and 1970s, Latin American scholars began to theorize this process by drawing on the emergent body of ideas that known as dependency theory. This approach identified Latin America as a "peripheral" region within the global capitalist economy, occupying a subordinate, or *dependent*, role in relation to "core" regions such as North America and Western Europe. The core/periphery dichotomy central to dependency theory was subsequently "transferred in an implicit way to the cities" (Hiernaux and Lindón 2004: 111). As Hiernaux and Lindón explain:

> The voice of 'periphery', with regard to cities, carries two inheritances: on the one hand, the geometric inheritance proper to the word periphery (the outer cir-cumference), on the other hand, it is an heir of the social theory of the sixties. . . . The conjunction of both inheritances gave a new meaning to the voice: the outer circumference of the city where the poor, the dominated, the dispossessed are located.
>
> *(Hiernaux and Lindón, 2004: 111)*

Dependency theory drew attention to the precarious position of the urban poor within the labor market. To varying degrees, Latin American countries experienced some industrialization during the mid-twentieth century, though this had led to a far lower levels of employment in modern industry than had occurred in the core countries (see **Chapter 6**). In an influential analysis, Peruvian sociologist Aníbal Quijano (1968) developed the notion of "social marginality" to argue that a significant portion of this large, unskilled population was essentially surplus to the requirements of the industrial economy. As such, he argued, the large number of unemployed people did not play the functional economic role envisioned by Marx's concept of the "reserve army of labor", because, beyond a certain point, they did not affect the bargaining power of employed laborers. While researchers would come to question the marginality theory over subsequent years (Caldeira, 2008), these insights about the implications of Latin America's position within the global division of labor for the socio-economic and spatial dynamics of its cities would persist.

However, Quijano was sceptical about the specificity of the urban periphery itself within this formulation. He argued that analyses of marginality were often used with "primitive physical connotations" to discuss issues of substandard housing and referring exclusively to precarious new neighborhoods at the edge of the city (Quijano, 1968: 8). He stressed the point (already made above), that such conditions were not only found in new neighborhoods at the urban periphery, but also in traditional central areas. Instead, he sought to develop into a more "rigorously delimited" notion of marginality, constructed in relation to the labor market under conditions of dependency, rather than living conditions or urban spatial location.

Notwithstanding Quijano's scepticism, the persistent expansion of low-income settlements in the urban outskirts over this period demanded a more specific account of the relationship between urban space and marginalization. Brazilian urbanist Lúcio Kowarick (1979) made a notable intervention, bringing the themes of labour and housing together by discussing the *social reproduction* of the workforce. He noted that housing costs, particularly rent or mortgage loans, represented the single largest cost for workers. Therefore, any "solution" to the housing problem faced by the urban poor would also indirectly benefit employers, because "by eliminating an important item such as housing from the workforce's survival costs, wages are limited to covering other essential expenses, such as transport and food" (Kowarick, 2024[1979]: 41). Peripheral modes of urbanization, which entailed lower housing costs largely shouldered by the poor themselves, could therefore be understood as a mechanism by which capitalism at the global periphery remained profitable. In this way, it also provided a perverse incentive for Latin American governments (which, during this period, were mostly military dictatorships) *not* to invest in urban infrastructure and housing in the peripheries.

Kowarick called this system "urban plunder" (*espoliação urbana*), to denote the extraction of value from workers not only through their labor, but also through various forms of extortion they faced while trying to achieve social reproduction in the city:

> The periphery as a formula for reproducing the workforce in cities is a direct consequence of the type of economic development that has taken place in Brazilian society in recent decades. It enabled, on the one hand, high rates of labor exploitation, and on the other hand, it forged forms of exploitation that took place at the level of the very urban condition of existence in which the working class was kept.
> *(Kowarick, 2024[1979]: 41)*

The "conjunction" (Hiernaux and Lindón, 2004) of the urban and global periphery in the work of dependency theorists thus provided the foundations for a distinctly

Latin American approach to urban theory that continues to underpin research to this day. However, some aspects of the approach would come to be questioned in the light of both societal and intellectual shifts over subsequent years. The dependency approaches of the 1960s and 1970s were strongly economistic, seeking to derive conclusions about diverse social, political, institutional and urban phenomena from economic factors alone. For similar reasons, they tended to emphasize structure over agency, and to be less attentive to the ways in which the residents of peripheries produced their own territories and contested the constraints imposed on them (see Chapter 12). These limitations would become more apparent as a wave of democratization spread across the region in the 1980s and, to varying degrees, became consolidated during the 1990s and 2000s (see Chapter 4). On the other hand, economic neoliberalization and "post-neoliberal" policies (see Chapter 8) would produce important impacts in urban peripheries, though not always in ways emphasized by dependency theory. This complex scenario has prompted new attempts to theorize the urban peripheries in recent decades.

Transformations and (re)theorization of the peripheries: 1990s–present

Since the 1990s, Latin America's urban peripheries have experienced significant socio-economic, institutional and physical transformations, which have prompted new forms of theorization. These are partly a result of broad shifts in Latin American societies and polities that have manifested in particular ways in urban peripheries. For example, since redemocratization in many countries, public policies in areas such as health and social assistance have expanded and consolidated, improving outcomes for peripheral populations. Meanwhile, economic growth and redistributive measures adopted by left-leaning governments during the so-called "Pink Tide" of the 2000s prompted a significant increase in consumption and commercial activity in peripheries, as well as some social mobility (even if many of these gains were reversed by economic crisis in the 2010s).

There have also been more direct transformations to the built and lived environments of urban peripheries. A highly visible change over recent decades has been the mass construction of state-subsidized social housing in several countries via policies delivered from different levels of government. While such policies may be targeted at different income groups, and thus vary in terms of size, location and building standards, in the peripheries they tend to be directed towards lower-income populations. This "Fordist production of social housing", as Mexican urbanists Emilio Duhau and Angela Giglia (2016: 74) have called it, now competes with irregular land development/squatting and autoconstruction as the primary means by which new housing is produced in the peripheries. Examples of such forms of development include the Conjunto San Buenaventura in Ixtapaluca, metropolitan Mexico City Duhau and Giglia, 2016) and those discussed by Salcedo (2010) in Peñalolén, metropolitan Santiago.

While these programmes have helped to reduce official housing deficits, they have also been criticized for furnishing construction companies with lucrative contracts and driving the financialization of low-income housing. Meanwhile, by operating within capitalist land markets, they often involve the production of large numbers of small housing units in highly segregated areas, far from labour markets and public services. As such, while representing a departure in terms of the role of the state in housing production, there is significant continuity in terms of their working-class character and relative marginalization in terms of access to public services and amenities.

Meanwhile, more subtly, older, autoconstructed peripheries have also continued to develop and diversify. Brazilian anthropologist Teresa Caldeira (2017) describes how homeowners in São Paulo have gradually rebuilt and developed their homes so that today they are often far larger, sturdier and better equipped than they were 20 or 30 years ago. In addition, through various "transversal" means, they have mobilized to demand the upgrading of their neighbourhoods by governments with urban infrastructure and services (see the case study of Fazenda da Juta, below). These processes have occurred unevenly, producing increasingly heterogeneous landscapes in peripheries. Duhau (2014), for example, distinguishes between "progressive" autoconstructed areas in Mexico City, which have experienced significant urban improvements and social mobility, and "nonprogressive" areas that remain physically precarious and socially marginalized.

However, despite these changes, consolidated autoconstructed neighbourhoods still broadly fit traditional definitions of urban peripheries. They are still overwhelmingly working-class, remain distant from centres of economic and political power, and still display the physical legacies of the peripheral urbanization process. As Brazilian architect Raquel Rolnik explains, even "consolidated areas, with comprehensive infrastructure, still bear the mark of a process of urbanization *ex post*, that is, in which the city arrived long after its inhabitants and their homes" (Rolnik, 2022: 88). In other words, neither the growth of social housing nor the diversification of autoconstructed peripheries fundamentally challenge the original formula of urban peripheries as defined by spatial distance and socio-economic and political marginalization.

By contrast, other processes point towards a potentially more dramatic rupture to this historic relationship. Most notable is the appearance of some elite gated condominiums in peripheral areas. Since the 1990s, such developments have been observed in cities across Latin America, from Buenos Aires (Prévôt-Schapira and Pineda, 2008) to Managua (Rodgers, 2004). Researchers attribute these processes to a number of factors, including growing perceptions of insecurity in Latin American cities and new security, transport, and communication technologies that facilitate connections between self-contained spaces. More broadly, globalization, financialization, and neoliberalization have altered the logics by which cities are produced and governed. Under these conditions, real estate actors have taken advantage of the availability of land beyond the traditional urban core to implant new spaces catering to the "globalized" lifestyles of urban elites and middle classes, such as the district of Barra da Tijuca in Rio de Janeiro (Herzog, 2013) or the municipality of Pilar, just north of Buenos Aires (Libertun de Duren, 2006). Beyond gated communities, researchers have identified the production of other spaces beyond the traditional urban core, such as shopping centres and office complexes, connected by infrastructures that "disembed" (Rodgers, 2004) these circuits from their lower-income surroundings.

Unlike other transformations, the presence of elite spaces in the peripheries seems to disrupt the historical link between the urban periphery and socio-economic and political marginalization. As a result, some scholars argue that Latin America is shifting from a "centre-periphery" model to one of socio-spatial fragmentation (Sposito and Sposito, 2020; Prévôt-Schapira and Pineda, 2008). Through this process, different urban territories become increasingly detached from one another and less dependent on particular, pre-determined locations within the overall geography of the city. On the other hand, it is important to note that, elite spaces in the periphery remain the exception rather than the rule. Indeed, many Latin American cities are also seeing processes of regeneration, gentrification, and displacement from central areas that tend to reinforce traditional centre-periphery dynamics (see Chapter 20). Meanwhile, many new residential and consumption spaces cater to an upwardly mobile peripheral working class rather than elites. Perhaps it would be most accurate to say that processes of socio-spatial fragmentation to date *overlay* and *complexify* the urban geography of Latin American cities, but without (yet) fundamentally disrupting the centre-periphery structure that these cities inherited from the twentieth century.

Case study: Fazenda da Juta

To illustrate some of the historical features and contemporary transformations of Latin America's urban peripheries, this section offers a brief case study of the neighbourhood of Fazenda da Juta in the East Zone of São Paulo.[1] Fazenda da Juta began to urbanize in the late 1970s, at the height of the centre-periphery model. At this time, it was a large area of farmland that had already been encircled by São Paulo's eastward urban expansion. Seeking to profit from the growing demand for housing in the area, the landowner decided to subdivide and sell plots to low-income buyers who signed up for long-term repayment plans. The buyers autoconstructed single-family homes in the expectation of being able to remain on the land. However, as land values in the area rose dramatically, the landlord sought to raise their monthly payments and, when they were unable to pay, to evict them. This conflict coincided with the beginning of Brazil's redemocratization process and the emergence of a strong housing movement. Housing lawyers came to the defence of the buyers, while activists lobbied the authorities, held marches and coordinated new squatter occupations in the area. Meanwhile, the landowner had accumulated a large land tax liability, which came to exceed the estimated value of the land. Eventually, the state government expropriated the property, paying compensation to the landowner.

Most of the original settlers who took part in these occupations autoconstructed their homes, initially with basic materials such as wood and corrugated iron and later with bricks and mortar. Over subsequent years, many would add additional floors and extensions to accommodate their growing families, as well as garages, security gates and other features. Today, there is significant diversity in the appearance of homes built during the original settlement, reflecting the varying needs and means of different families and the ways these have shifted over time (see Figure 19.1).

Meanwhile, during the 1990s, two other types of housing were established in the area on land ceded by the state government. Firstly, a series of "*mutirão*" (housing cooperative) projects were established through partnerships established between housing movements, the municipality (which, at the time, was under a radical left-wing

Figure 19.1 Consolidated autoconstructed homes in Fazenda da Juta.
Source: Author.

administration) and an architectural firm. Through this model, participants in the movement would build their future homes with materials provided by the city government and under the supervision of architects and engineers. Secondly, numerous social housing projects were built by construction companies contracted by São Paulo's state housing development company.

During the 1990s, the neighbourhood remained largely unurbanized. It not only lacked basic infrastructure like paving, street lighting, and drainage, but also services like schools and health centres, meaning residents had to leave the area to access these. However, via local mobilization through different channels, Fazenda da Juta was gradually transformed. In response to resident demands, first basic infrastructure was installed, and later various services were established. While many residents still complain about the quality of these services, clearly the state is very present in Fazenda da Juta. However, state performance remains inadequate in areas like waste collection, street cleaning and the enforcement of planning regulations, meaning that public spaces suffer from neglect and degradation. In this context, the inhabitants of the *mutirões* and social housing blocks have created residents' organizations that charge a small monthly fee to maintain and regulate their internal spaces. They have also installed security systems that to some extent mimic the conditions of elite gated communities, though without producing such a sharp rupture with their surroundings (see Richmond and Kopper, 2022).

In recent years, some further processes have impacted Fazenda da Juta. A new metro station has been inaugurated in the area, finally connecting it to the city's rail network. This has bolstered local housing markets in various ways. Although no major housing construction has taken place within Fazenda da Juta itself, nearby areas like São Mateus have seen significant real estate activity, with new condominiums catering to upwardly mobile (though not wealthy) groups (see Figure 19.2). Within Fazenda da Juta, real estate activity within the neighbourhood's diverse housing stock has intensified micro-scale socio-spatial differences. Meanwhile, new precarious squatter occupations have also cropped up across the area, occupying

Figure 19.2 New vertical condominiums under construction in nearby São Mateus.
Source: Author.

Figure 19.3 Wooden shacks in a precarious new squatter settlement within an environmental
protection zone in Fazenda da Juta.
Source: Author.

residual, environmentally precarious locations, such as hillsides, riverbanks, and
forested environmental protection zones (see Figure 19.3). While catering to an
extremely marginalized population, today such settlements also embody a mercantile logic. The criminal groups who usually coordinate them charge rents to homeless

families to live in makeshift huts, while waiting for their eventual consolidation at which point they are sold off for higher values.

Fazenda da Juta, then, displays many of the long-term transformations and contemporary dynamics present across Latin America's urban peripheries. Over time, the neighbourhood has seen a gradual consolidation of housing, infrastructure and various public policies both as the result of resident action and the expansion of state presence. However, conditions on the whole remain far below those found in wealthier areas of the city and the population is still almost entirely working class. Nonetheless, there is increasing stratification between those who have achieved a certain degree of financial and housing stability, and others living in extreme poverty and precarity. Internal differences also have a spatial dimension, with infrastructure and services present in more consolidated parts of the neighbourhood and largely absent from more precarious areas. Finally, stratified land and housing markets have become increasingly important in shaping socio-spatial organization. At one end of the spectrum, new vertical condominiums cater to a comparatively wealthy segment of the peripheral population, while at the other, criminal groups coordinate new squatter settlements to meet the housing needs of the very poor.

Conclusion

This chapter has argued that urban peripheries are a particular type of space found in most large and medium-sized Latin American cities. Unlike the Global North, where suburbs grew as desirable spaces for middle classes fleeing the inner city, in Latin America the peripheries were settled by working-class residents unable to access housing via formal channels in the regulated urban core. They did so instead by autoconstructing homes on land that was either squatted or irregularly subdivided in areas that lacked basic infrastructure and services. Influential analyses inspired by dependency theory identified these processes as the consequence of Latin America's peripheral and dependent position within the global economy, which had given the region a disproportionately large low-skilled workforce relative to employment opportunities.

However, over time, and especially since the 1990s, urban peripheries have transformed dramatically. Residents have gradually consolidated their homes and neighbourhoods and states have taken on an increasing role in the provision of housing, infrastructure and public services. In this context, researchers have retheorized urban peripheries, emphasizing the agency of their residents and their growing socio-spatial heterogeneity. Meanwhile, increasingly stratified land and housing markets cater to different segments of the peripheral population and, in some cases, even elite groups. This has led some researchers to posit a long-term transition from the traditional centre-periphery model of urban organization to a logic of socio-spatial fragmentation. However, if such a process is indeed emerging, it is in a very incipient stage. Peripheries continue to be overwhelmingly inhabited by working-class populations and, even in better-served areas, still have far lower levels of public service and infrastructure provision than in wealthier, central zones.

Summary

- Urban peripheries are a particular type of space in Latin American cities, defined by being distant from the urban core and by their low-income, socially and politically marginalized populations.
- Between the 1950s and 1980s, urban peripheries grew primarily via the squatting or irregular subdivision of land and autoconstruction of housing, which Latin American scholars theorized by drawing on the core insights of dependency theory.
- Since the 1990s, urban peripheries have been transformed via the consolidation of older neighbourhoods and the growing presence of the state in the provision of housing, infrastructure and services.
- In light of these transformations, scholars have retheorized peripheries emphasizing resident agency, growing socio-spatial heterogeneity, and fragmentation.
- Despite growing state presence and heterogeneity, on the whole, peripheries today can still be understood as marginalized spaces in comparison to more central areas.

Review questions

1. How are urban peripheries in Latin America distinct from peripheral urban areas in the Global North?
2. What are the respective roles played by residents, democratic governments, and land markets in the transformation of urban peripheries since the 1990s?
3. How are changes in urban peripheries linked to broader processes of socio-spatial organization of Latin American cities?

Further reading

Caldeira, T. (2017). 'Peripheral urbanization: Autoconstruction, transversal logics, and politics in cities of the global south', *Environment and Planning D: Society and Space*, 35(1), pp. 3–20.

An important theoretical contribution, identifying the distinctive features of autoconstruction as a form of peripheral urbanization and drawing on research from São Paulo and other global South cities.

Duhau, E. (2014). 'The informal city: An enduring slum or a progressive habitat', in Fischer, B., McCann, B., & Auyero, J. (Eds.), *Cities from Scratch*. Duke University Press.

An analysis of transformation of Mexico City, presenting a helpful typology of urban neighbourhoods across and identifying their changing conditions over time.

Hiernaux, D., & Lindón, A. (2004). 'La periferia: Voz y sentido en los estudios urbanos', *Papeles de Población*, 10(42), pp. 101–123.

A Spanish-language overview of the changing ways in which peripheral urban territories in Latin America have been understood, from the colonial period to the late twentieth century.

Richmond, M. A., & Kopper, M. (2023). 'Walling the peripheries: Porous condominiums at Brazil's urban margins', in Forte, G., & Hwa, K. (Eds.), *Embodying the Periphery*. Firenze University Press.

Explores an emergent type of residential space – the "peripheral condominium"– and its key features, through ethnographic cases drawn from three Brazilian cities.

Keywords

Autoconstruction: the self-construction of homes by residents, without contracting professional architects or construction workers. Autoconstruction is often initially carried out using basic materials, with homes gradually consolidated, expanded and embellished over time according to the means and needs of their residents.

Centre-periphery model: the dominant model of socio-spatial organization in most Latin American cities during the mid-to-late twentieth century, when infrastructure, services, employment opportunities were concentrated in central areas and low-income groups were forced to live in the periphery.

Socio-spatial fragmentation: a process whereby different urban territories become increasingly detached from one another and less dependent on particular, pre-determined locations within the overall geography of the city.

Urban periphery: a particular type of space in Latin American cities that is distant from the urban core and populated by a low-income and socially and politically marginalized population.

Note

1 This case study draws on previous work published by the author: Richmond (2020), Richmond (2022) and Richmond and Kopper (2022), as well as Ferreira (2021).

References

Caldeira, T. (2008). 'Marginality, again?!', *Internatinoal Journal of Urban and Regional Research*, 33(3), pp. 848–853.

Caldeira, T. (2017). 'Peripheral urbanization: Autoconstruction, transversal logics, and politics in cities of the global south', *Environment and Planning D: Society and Space*, 35(1), pp. 3–20.

Dear, M., & Dahmann, N. (2008). 'Urban politics and the Los Angeles school of urbanism', *Urban Affairs Review*, 44(2), pp. 266–279.

Duhau, E. (2014). 'The informal city: An enduring slum or a progressive habitat', in Fischer, B., McCann, B., & Auyero, J. (Eds.), *Cities from Scratch*. Duke University Press.

Duhau, E., & Giglia, A. (2016). *Metrópoli, Espacio Público y Consumo*. Fondo de Cultura Económica.

Ferreira, D. (2021). *Fazenda da Juta: Uma Trilha entre o Rural e o Urbano*. Editora CRV.

Fischer-Tahir, A. & Naumann, M. (Eds.). (2013). *Peripheralization: The making of spatial dependencies and social justice*. Springer.

Herzog, L. A. (2013). 'Barra da Tijuca the political economy of a global suburb in Rio de Janeiro, Brazil', *Latin American Perspectives*, 189(40:2), pp. 118–134.

Hiernaux, D., & Lindón, A. (2004). 'La periferia: Voz y sentido en lose studios urbanos', *Papeles de Población*, 10(42), pp. 101–123.

Keil, R. (2017). *Suburban planet: Making the world urban from the outside in*. Polity.

Kowarick, L. (2024) [1979]. *Urban spoliation* [English Edition]. ABCP & CEM.

Libertun de Duren, N. (2006). 'Planning à la Carte: The location patterns of gated communities around Buenos Aires in a decentralized planning context', *International Journal of Urban and Regional Research*, 30(2), pp. 308–327.

Matos Mar, J. (1958). *Migration and urbanization: The "Barriadas" of Lima: An example of integration into urban life*. United Nations Economic and Social Council.

Prévôt-Schapira, M., & Pineda, R. C. (2008). 'Buenos Aires: la fragmentación en los intersticios de una sociedad polarizada', *Revista Eure*, 34(103), pp. 73–92.

Quijano, A. (1968). *Notas Sobre el Concepto de Marginalidad social*. Comisión Económica Para América Latina (CEPAL).

Ren, X. (2021). 'The peripheral turn in global urban studies: Theory, evidence, sites', *South Asia Multidisciplinary Academic Journal*, 26, pp. 1–8.

Richmond, M. A. (2020). 'O devir-lugares das periferias urbanas: Transformações socioespaciais no bairro de Fazenda da Juta', in Richmond, M. A., Oliveira, V. C., Kopper, M., & Garza, J. (Eds.), *Espaços Periféricos: Política, Violência e Território nas Bordas da Cidade*. São Paulo: EdUfscar.

Richmond, M. A. (2022). 'The pacification of Brazil's urban margins: Peripheral urbanisation and dynamic order-making', *Contemporary Social Science*, 17(3), pp. 248–261.

Richmond, M. A., & Kopper, M. (2022). 'Walling the peripheries: Porous condominiums at Brazil's urban margins', in Forte, G., & Hwa, K. (Eds.), *Embodying the Periphery*. Firenze University Press.

Rodgers, D. (2004). '"Disembedding" the city: Crime, insecurity and spatial organization in Managua, Nicaragua', *Environment & Urbanization*, 16(2), pp. 113–123.

Rolnik, R. (2022). *São Paulo: O Planejamento da Desigualdade*. Fósforo.

Salcedo, R. (2010). 'The last slum: Moving from illegal settlements to subsidized home ownership in Chile', *Urban Affairs Review*, 46(1), pp. 90–118.

Simone, A. (2010). *City life from Jakarta to Dakar: Movements at the crossroads*. Routledge.

Sposito, E. S. S., & Sposito, M. E. B. (2020). 'Fragmentação socioespacial', *Mercator*, 19, pp. 1–13.

CHAPTER 20

Informality and public space

Veronica Crossa

Introduction

"¡Lleve de lo bueno!": the street vendors' chant, the musical sound they emit to announce their products, is an integral part of the urban experience when walking through certain streets and public spaces of urban Latin America. The chant refers to a song entitled "Callejero" by the Chilean group Juana Fé which, through the fusion of cumbia and ska, describes street vendors' musical mantra. On the one hand, the song mentions in a somewhat pejorative tone the adaptive nature of a street vendor who "sells anything that comes his way and will dress elegantly if needs be". On the other hand, street vendors are described as opportunists who adapt to any circumstance as long as they are not subsumed to the orders of a hierarchical labor structure: a street vendor does whatever it takes as long as nobody bosses him around. These representations of popular life are widespread in Latin American geographies. They reflect and feed into common imaginaries about street vendors' daily practices, their motivations, interests, behaviors, and ways of navigating urban life.

On the streets, in the metro, on the sidewalks, in plazas, parks, and hidden corners, it is inevitable to face the visual, auditory, tactile, and fragrant reality of street vending. As Tenorio-Trillo (2013, 293) argues when referring to Mexico City, "any walker develops ingratitude and blindness in order to live and survive the sidewalks of Mexico City". For a significant proportion of urban inhabitants, the presence of street vendors is often associated with the perceived chaos of a city governed by anarchy and disorder. Street vending is commonly viewed as a form of chaos arising from vendors' lack of understanding of basic notions of the public good, such as respect, civility, cleanliness, and adherence to legal regulations. Consequently, many urban residents perceive the presence of street vendors in public space as a sign of a declining public order, where the primacy of private interests takes precedence over a general sense of collective well-being. Under this logic, street vendors are seen as transforming urban public spaces, which ideally should be dedicated to promoting public life, into spaces driven by personal gain and private pursuits. These viewpoints, as Warner (2012, 17) highlights, are

DOI: 10.4324/9781003430926-29

rooted in the profound tension between the concepts of the public and the private. The presence of street vendors, who actively use and "appropriate" public spaces, embodies a complex subject that encapsulates the numerous political, economic, social, and cultural contradictions that are pervasive in contemporary Latin American cities and their geographies.

Men, women, children, disabled individuals, and the elderly traverse the diverse spaces of the city, offering a wider array of products including food, sweets, shoes, makeup, clothes, electronics, glasses, bags, gifts, stationary, souvenirs; the list is indeed endless. This is the urban street economy, understood as a series of practices sustained through a complex network of localized economic, political, cultural, and environmental processes that in most cases transcend the region. Understanding this complexity is not an easy task, and has been the endeavor of academics, policymakers, urban planners, NGOs, and politicians for more than half a century.

For many urban residents, the existence of the street economy provides access to thousands of products at lower costs, often products that would otherwise be unreachable. For others, however, the presence of street vending activities is an endemic problem that corrodes the fabric of a desirable urban life, indeed, a practice that embodies the weakness of the state. In this chapter, I argue that these valorizations which prevail in many Latin American cities are sustained through what I identify as two conceptual myths, that is, two ideas that have dominated the public sphere.

The first analytical myth is the division between the formal and the informal economy. This dichotomy has been used to classify economic activities into rigid categories which, despite more recent critiques, continue to dominate the general understanding of urban economic activities in public space. The second conceptual myth, more recently developed, is linked to the idea of urban public space. It is common to encounter urban policies, political discourses, and academic work that depict public spaces as inherently inclusive, fostering collective and participatory socialization, and representing the epitome of democratic contexts. However, when such abstract concepts are translated into public policy, the scenario is more menacing. To this day, and for the last two decades, street vendors in many cities across Latin America, are facing recurrent waves of policy interventions that seek to redefine the informal economy and its spaces (Bromley and Mackie, 2009; Crossa, 2018; Swanson, 2007). By breaking down these myths, the chapter provides important tools for rethinking Latin American geographies more broadly, and the geographies of informality more specifically.

The formal/informal divide

Research on the informal economy has traditionally viewed this activity as a hindrance to the development of a modern economy. Originally considered an activity confined to underdeveloped countries, the informal economy, also known as the pre-modern sector, was defined as a set of activities that did not contribute to national economic growth. Academic work following this line of thought were aligned with classical theories of modernization, particularly Rostow's linear modernization model, which outlined various stages of economic development for a country. The prevailing idea was that informal activities were a residual that needed to be reduced so that underdeveloped economies could reach their full potential. Moreover, workers in the informal economy were simply seen as poor individuals who worked or lived outside the bounds of the law (De Soto, 1986).

Within this general interest on informality, correlated with a linear view of modernity and development, a sub-sector within urban studies also showed interest in understanding the emergence of this form of economic subsistence, especially in cities across the Global South. This approach emphasized significant changes in the composition of cities, both in terms of their growth and their changing sociodemographic and economic composition. Following the culture of poverty perspective that drew heavily on the theoretical baggage of the Chicago School of urban ecology, this stance led to the development of a perspective that analyzed practices outside the margins of the state (primarily in relation to housing) as a result of rural migrants' inability to fully become urban dwellers. This situation tacitly gave rise to a system of behaviors that explained the relationship between urban growth and disorder (Lewis, 1961).

During the 1970s and 1980s, special attention was given to the contexts of Latin America, where cities experienced unprecedented levels of growth (see **Chapter 7**), driven, among other factors, by high levels of industrialization that resulted in large migratory flows from rural areas to cities (Tockman, 1991; Gilbert, 2004; Castells and Portes, 1989; Bromley, 2007). Research on the urban informal economy started from dichotomous frameworks based on definitions of informality always in contrast to those that were unquestionably considered formal (Moser, 1994). Contrasting self-employed workers with wage earners, workers with benefits versus those without such rights, or organized workers versus those who did not fit into an organized socio-political structure, the objective was to differentiate what was clearly a type of work that could not be explained by the formal political and economic structures of society (De Oliveira and Roberts, 1993). Therefore, informality came to be understood as anything that did not fit into the formal legal-economic order. It could be said that this dualistic approach was a product of the need to identify, specify, and measure a growing urban activity that was not previously defined in conceptual terms, in order to develop mitigation tools and public policies to address the phenomenon (Rakowsky, 1994).

However, the dualistic approach, which considered informality as a social residual, and where efforts concentrated on uncovering the ways in which the growing urban poor could (should) access formal public services, was largely an Anglophone debate. Within Latin America, particularly within the fields of urban anthropology and urban sociology, the analytical frameworks used to understand growing levels of urban poverty and informality provided more nuanced accounts of the relationship between urbanization, poverty, and (in)formality. The work, for example, of Larissa Lomnitz argued that to understand the constitutive relationship between the so-called formal and informal sector, it was essential to understand relations of intermediation and power underlying social relations in urban contexts (De la Peña, 1994, 7). As she states, informality is not a residue of traditionalism, but "an intrinsic element of "formality" insofar as it is a response to the inadequacies of formalization. It is an adaptive mechanism that simultaneously and in a vicious cycle, reinforces the shortcomings of the formal system" (Lomnitz, 1988, 41–43). Similarly, Milton Santos (2008) carefully described the ways in which urban economic systems are based on the mutually constitutive relationship between what he identified as two intricate circuits of an urban economy, that is, formal regulated systems of circulation, on the one hand, and its reliance on circuits of unemployment, poverty, and underemployment on the other (Santos, 2008).

In general, and through a detailed analysis of practices and circuits of exchange in urban contexts, both Lomnitz and Santos, together with other Latin American social scientists, sought to move beyond linear and dualistic notions of informality and focus on their interdependencies (Azuela, 1990; Connolly, 1990).

Recent Anglophone research on urban informality has perhaps unknowingly reprised debates which have circulated for more than four decades in the Latin American contexts (AlSayyad, 2003). Indeed, current research on the informal economy has been motivated by a concern to understand what has been termed the politics of informality, particularly the ways in which populations in the informal sector negotiate, transgress, or resist practices that arise from the changing relationship between the state-market-society in a context of the privatization of public life (McFarlane, 2012). This perspective raises "politics" in two analytical realms. On one hand, it emphasizes the need to think beyond static categories such as "formal and informal", which tend to place individuals in a set of almost unalterable temporal and spatial relationships. On the other hand, the politics of informality acknowledges the importance of looking at the heterogeneity within the practices that are typically reduced as "informal". However, and as Varley argues, recent "postcolonial urban scholarship has marginalised Latin America. This observation parallels concerns expressed by Latin American scholars about the neglect of the region by postcolonial theorists. In conclusion, we might ask, however, what postcolonial readings of informality from elsewhere could contribute to understanding the Latin American experience. Such a reading forms part of a recent manifesto for 'rethinking subaltern urbanism' by Roy" (Varley, 2013, 16).

However, despite the conceptual efforts made to provide a more nuanced account of the informal economy, the reality is that many individuals who engage in street vending activities continue to confront significant challenges in their daily lives. These challenges can vary depending on the specific context, but they often include issues such as limited access to formal financial services, lack of legal recognition and protection, inadequate infrastructure and facilities, inconsistent enforcement of regulations, and social stigmatization. Specifically, in the Latin American context, street vendors face continuous waves of exclusionary policies and forced displacements in many cities across the region, and many scholars within Latin America today seek to understand the socio-spatial implications of these policies and the problems posed to the informal street economy (Alba Vega and Braig, 2023; Crossa, 2018; Meneses, 2011).

In the last two decades, governments, urban authorities, and urban development planners sometimes view street vending as incompatible with their vision of a modern, orderly city (Delgadillo, 2016). As a result, policies aimed at beautifying urban public spaces and promoting commercial activities in designated areas have been at the expense of street vendors and other members of the so-called informal economy who depend on these spaces for their daily survival. Indeed, street vendors in Latin American cities now deal with the added burden of being defined not only as a residual to a formal political economy, but also as a hindrance to regeneration policies that rely on the beautification of urban public space for the future of more livable cities. Part of the challenge, as I argue in the next section, can be explained by what I call the conceptual myth of urban public space.

Urban public space

Public space is a difficult concept to define. Part of this difficulty lies in the fact that public space is intertwined with normative concepts such as democracy (**see Chapter 4**), citizenship, civil society, and rights (Ramírez Kuri, 2003). We are told that public space is synonymous with public life. For example, Jordi Borja, an urban planner with great influence in urban public policy in Latin America, explains: "The city is public space, and when it is not public space, we are in a city that would be diminished in its quality, in its citizenship. Public space is also the place where coexistence is expressed; it is the place that can build tolerance" (Borja and Muxi, 2001).

Indeed, the concept of public space is used to emphasize a normative idea of a desirable urban life. It is a sort of utopian language framed within the language of rights and refers to processes such as democracy, equality, participation, and inclusion. However, this abstraction is often detached from complex political, social, and economic realities where inequality, poverty, and underemployment prevail. When it takes concrete form and context in terms of public policy, public space is simply reduced to a physical place: a park, a square, an avenue. Moreover, by reducing public space to a confined physical space, in a Cartesian fashion, it becomes a technical problem, a design problem: where to place benches, tables, and lighting. In many respects, this critique of public space, indeed as a technical and not a social problem, coincides with academic concerns over the nature of urban neoliberalism, and what Anglophone geographers have coined as the "post-political" turn.

The use of the concept of public space in political discourses within Latin America is a relatively recent phenomena, and its emergence is not coincidental. It reflects a common understanding that has altered everyday language, reducing significant social issues to manageable inconveniences through technical and aesthetic strategies. This phenomenon represents the depoliticization of the political, which manifests in two notable ways: subtle yet influential changes in ordinary language, and the transfer of social problems to technical experts who employ a language that is technical, objective, scientific, and devoid of distinct political perspectives. The depoliticization of the political serves to mask the underlying power dynamics and structural inequalities that are intrinsic to social issues, including the roots of the street economy. By framing public space as a technical problem and relegating it to the realm of experts, the political nature of social challenges is obscured. This depoliticized discourse creates a sense of neutrality and impartiality, concealing the inherently political nature of decision-making processes and policy choices.

Drawing from the experience of recent policy interventions on urban public space and street vending in Mexico City, this chapter argues that informality today is not only considered a political and economic problem for cities, as past interventions highlighted, but also a technical and an aesthetic one.

The case of Mexico City

In the past two decades, numerous cities in Latin America, including Santiago, Quito, Sao Paulo, and Buenos Aires, to mention a few, have made significant efforts to transform specific areas within their urban landscapes. Streets and other public spaces have become primary targets for urban governments seeking to attract investments. Mexico City has been no stranger to these endeavors (Alba Vega, 2012). Under the rationale of

Figure 20.1 Historical center of Mexico City.
Source: Creative Commons.

"rescuing urban public space", both national and urban authorities developed a series of instruments with the goal of transforming public spaces of the city. The first was the recovery program implemented in Mexico City´s symbolic heart, the Historic Centre (see Figure 20.1). The "*Programa de Rescate*" (the Rescue Program) entailed the beautification of the historic area's streets, sidewalks, plazas, and other public spaces with the objective of attracting population and investment (Crossa, 2009). The *Programa de Rescate* was first announced in 2001 to address what the city government called the "crisis of the Historic Centre", a crisis defined by a so-called economic, demographic, and architectural deterioration of the area. Hence, the goal was to reactive the area's economy and stimulate new housing and commercial investment. A fundamental part of this involved the removal of thousands of street vendors from public spaces (see Figure 20.2). The *Programa's* underlying vision, then, was to reconstruct a Historic Centre devoid of street vending activities, with clean, tidy, and safe public spaces "mirroring the Soho, but of Mexico City" (Interview, July 2004).

The crisis of the historic center was essentially a crisis of its streets. The streets were discursively constructed as violent, insecure, dirty, and disorderly (Becker and Müller, 2013). Moreover, they were defined as spaces increasingly privatized by street vendors, who allegedly operate under a widespread culture of illegality. This message was repeated consistently in local newspapers and political discourses: "street vendors have taken over public space . . . leaving garbage and obstructing the free movement of pedestrians". Furthermore, vendors "extort authorities and assault citizens". Under these terms, the streets were a "battlefield".

So, while the streets are attributed descriptors like disorder, violence, deterioration, chaos, and danger, public space was conceived as its heroic solution, used without much questioning to describe an effort to improve collective urban life. The efforts to improve public urban life through the regeneration of public spaces was based on a very particular

Figure 20.2 Street vendors in downtown Mexico City.
Source: Creative Commons.

notion of the collective individuals. It was an individualized, abstract, apolitical collectivity that consumed aesthetics. Meanwhile, the collectivity of the streets was of a different nature: street vendors, parking attendants, and other members of the so-called informal economy were constructed as deceptive collectivities that operate according to private interests, mafia-like collectivities labeled as abusers of public order, lacking in civic virtue, and operating under illegal structures. Furthermore, they disrupt public order, operate through clientelist relationships, private interests, and anti-social practices. In contrast, families, cyclists, pedestrians, neighbors, the elderly, children, and the disabled are considered more amiable collectivities deserving of attention and public intervention: "A quality city must have public spaces where children, the elderly, the disabled, and the wealthy cannot resist going." They are the ones who should be given new sidewalks, lighting, bike lanes, and clean spaces. Those are the collectivities of public space.

In 2008, a few years after the launch of the *Programa*, Mexico City's authorities created the Public Space Authority (AEP) with the objective of coordinating, developing, and implementing policies for the improvement of urban public space. With its conception, public space projects migrated from the local borough to the scale of the city government. The AEP was comprised primarily of a team of technical experts (architects, urban planners, and designers), most of whom had no previous experience in the public sector. For the AEP, public space was a technical problem, and the discussions around these spaces focused on design issues, as mentioned by the director of the entity in 2014: "We take care of regulatory issues: finding the most suitable chair . . . universal accessibility." Urban problems, often manifested in urban public spaces, have been addressed through countless actions that reduce social complexity and class issues to a technical problem, belonging to the realm of designers, architects, engineers, and other "experts" who allegedly possess the ability to imagine safe, friendly, reliable, clean, and orderly spaces – spaces that generate a sense of predictability and certainty. All of this is framed within a vocabulary of democracy, equality, collective coexistence, and inclusion. These are attractive words, but they are fleeting and increasingly common in urban political discourses.

Conclusion

Although it may seem trivial, the words we use to describe spaces, actions, and people matter. Defining street vendors as "informal" means distancing oneself from their reality, their relations, their structures of constraints, indeed, from their everyday lives. Street vendors, together with other practices which have been identified as "informal" are the scapegoat of a public morality that is incompatible with an economic-political model based, among other things, on the individualization of inequality and poverty. Furthermore, street vendors have become the subject of social anxiety that translates into feelings of insecurity, disorder, and violence. The underlying structural background of inequality (**see Chapter 19**), which has often led to the increase in street vending, loses relevance, and the immediate problem becomes those who make illicit use of urban space. When constructing ideal city imaginaries, certain practices and people are included while others are excluded. The transformation of urban public spaces plays a central role in these representations and undoubtedly becomes a political issue. However, these efforts often hide behind technical and conceptual abstractions, as if the transformation of space were solely an infrastructure matter. Even if it were, the decision of which spaces deserve intervention, and the reasons behind it, are political decisions.

The depoliticization of the political is accompanied by a specific language and discursive framework. Public space is a term that carries a sense of objective fantasy. Unlike the street, public spaces are perceived as areas with the potential for socialization, often dominated by recreational activities, consumption or what Peck (2005, 760) calls "cappuccino urban policies". Discussions around these spaces and projects tend to focus on design issues because it is easier to talk about lighting, sidewalks, fountains, and playgrounds than to address deeper social problems such as poverty, homelessness, and unemployment, which are often visible phenomena in public spaces. It is simpler to concentrate on aesthetic or design problems rather than acknowledging the endemic issues rooted in different forms of structural inequality.

Summary

- The urban street economy is a series of practices sustained through a complex network of localized economic, political, cultural, and environmental processes that in most cases transcend the street.
- For many urban residents in Latin America, the presence of street vending activities is an endemic problem that corrodes the fabric of a desirable urban life, a practice that embodies the weakness of the state.
- Valorizations which prevail in many Latin American cities are sustained through two conceptual myths: the division between the formal and the informal economy, and the idea of urban public space.

Review questions

1. How has informality been understood in traditional Anglophone academic literature? Is there a distinction between these definitions and those proposed by Latin American scholars?
2. In what way does the notion of public space help reproduce processes of exclusion and displacement on the streets of Latin American cities?
3. What underlying assumptions about street vending and informality are mobilized in policies like *El Programa de Rescate*?

Further reading

Chen, Martha and Carré, Françoise (eds) (2020) *The informal economy revisited: Examining the past, envisioning the future*. Routledge.

A text which brings together more than three dozen authors who have studied the informal economy in different contexts. The book provides an extensive overview of theoretical, research, and policy responses to the phenomenon of informality. The contributors reflect on the past, present, and future of the informal economy drawing from examples from across the world.

Dürr, Eveline and Müller, Julian (eds) (2019) *The popular economy in urban Latin America: Informality, materiality, and gender in commerce*. Lexington Books.

An interesting perspective on the relationship between informality, materiality and gender based on detailed qualitative and ethnographic accounts of cities in Latin America. The texts compiled in this book provide a nuanced account of the relationship between informality and the state.

Salazar, Clara (ed) (2020) *Informality revisited: Latin American perspectives on housing, the state, and the market*. Wiley-Blackwell.

This edited volume provides an overview of recent debates around informality in Latin America, specifically with regards to land rights and housing. It explores the effects that neoliberal policies in Mexico, Peru, Brazil, and Colombia have had on the production of informal relations to land and urban space. Many contributors review and challenge past theoretical understandings on informality and provide more nuanced approaches, recognizing the emergence of a "new informality" in Latin America.

Keywords

Informality: usually understood as anything that falls outside the formal legal-economic order.

Public space: used to emphasize a normative idea of a desirable urban life.

Street vendors: anyone who sells goods in the street; they are a key feature of Latin American cities.

References

Alba Vega, Carlos (2012) La calle para quien la ocupa: las condiciones sociopolíticas de la globalización no hegemónica en México D. F. *Nueva Sociedad* 241: 79–92.

Alba Vega, Carlos and Braig, Marianne (2023) *Las voces del Centro Histrórico: la lucha por el espacio en la Ciudad de México*. El Colegio de México.

AlSayyad, Nezar (2003) Urban informality as a "new" way of life. In Roy, A. and AlSayyad, N. (coord.), *Urban informality: Transnational perspectives from the Middle East, Latin America, and South Asia*. Lexington Books, 7–30.

Azuela, Antonio (1990) Fuera del huacal, aún en la calle. El comercio y el espacio público en el centro de la Ciudad. *Trace* 17: 20–24.

Becker, A. and Müller, M. (2013) The securitization of urban space and the "rescue" of downtown Mexico city, vision, and practice. *Latin American Perspectives* 40(2): 77–79.

Borja, Jordi and Muxi, Saida (2001) *El espacio público, ciudad y ciudadanía*. Electa.

Bromley, Ray (2007) Foreword. In Cross, J. and Morales, A. (eds.), *Street entrepreneurs: People, place and politics in local and global perspective*. Routledge, xv–xvii.

Bromley, Rosemary and Mackie, Peter (2009) Displacement and the new spaces for informal trade in the Latin American city center. *Urban Studies* 46(7): 1485–1506.

Castells, M. and Portes, A. (1989) World underneath: The origins, dynamics, and effects of the informal economy. In Portes, A., Castells, M. and Benton, L. (eds.), *The informal economy: Studies in advanced and less developed countries*. John Hopkins Press, 11–37.

Connolly, Priscilla (1990) Dos décadas de 'sector informal'. *Sociológica* 5(12): 75–94.

Crossa, Veronica (2009) Resisting the entrepreneurial city: Street vendors' struggle in Mexico city's historic center. *International Journal of Urban and Regional Research* 33(1).

Crossa, Veronica (2018) *Luchando por un espacio en la ciudad*. El Colegio de México.

De la Peña, Guillermo (1994) Presentación. In Lomnitz, Larissa (ed.), *Redes sociales, cultura y poder: ensayos de antropología latinoamericana*. Flacso.

De Oliveira, Orlandina and Roberts, Bryan (1993) La informalidad urbana en los años de expansión, crisis y reestructuración económica. *Estudios Sociológicos* 11(31): 33–58.

De Soto, Hernando (1986) *El otro sendero: la revolución informal*. Editorial Sudamericana.

Delgadillo, Victor (2016) Selective modernization of Mexico city and its historic center. Gentrification without displacement. *Urban Geography* 37(8): 1154–1174.

Gilbert, Alan (2004) Love in the time of enhanced capital flows: Reflections on the links between liberalization and informality. In Roy, A. and AlSayyad, N. (eds.), *Urban informality: Transnational perspectives from the Middle East, Latin America, and South Asia*. Lexington Books, 33–65.

Lewis, Oscar (1961) *Antropología de la pobreza: Cinco familias*. Fondo de Cultura Económica.

Lomnitz, Larissa (1988) Informal exchange networks in formal systems: A theoretical model. *American Antrhopologist* 90: 42–55.

McFarlane, Colin (2012) Rethinking informality: Politics, crisis, and the city. *Planning Theory & Practice* 13(1): 89–108.

Meneses, Rodrigo (2011) *Legalidades públicas: el derecho, el ambulantaje y las calles en el centro de la Ciudad de México (1930–2010)*. CIDE/UNAM.

Moser, Caroline (1994) The informal sector debate, Part I: 1970–1983. In Rakowsky, C. (ed.), *Contrapunto: The informal sector debate in Latin America*. State University of New York Press, 11–30.

Peck, Jamie (2005) Struggling with the creative class. *International Journal of Urban and Regional Research* 29(4): 740–770.

Rakowsky, Cathy (1994) *Contrapunto: The informal sector debate in Latin America*. State University of New York Press.

Ramírez Kuri, Patricia (2003) *Espacio público y reconstrucción de ciudadanía*. Miguel Ángel Porrúa.

Santos, Milton (2008) O espaço dividido: os dois dircuitos da economia urbana. 2a ed. Universidade de São Paulo [first published in 1979].

Swanson, Kate (2007) Revanchist urbanism heads south: The regulation of indigenous beggars and street vendors in Ecuador. *Antipode* 39(4): 708–728.

Tenorio-Trillo, M., 2013. *I Speak of the City: Mexico City at the Turn of the Century*. University of Chicago Press

Tockman, Victor (1991) The informal sector in Latin America: From underground to legality. In Standing, G. and Tockman, V. (eds.), *Towards social adjustment: Labor market issues in structural adjustment*. Geneva: International Labor Organization, 141–157.

Varley, Anne (2013) Postcolonialising informality? *Environment and Planning D: Society and Space* 31: 4–22.

Warner, Michael (2012) *Público, públicos, contrapublicos*. Fondo de Cultura Económica, Umbrales.

Resistances

CHAPTER 21

Social movements

Renato Emerson dos Santos

Introduction

The term "social movements" is used to define wide-ranging forms of collective action. In Latin America, although profoundly unequal models of society have been maintained, social movements have been the protagonists of significant experiences involving struggles, demands, and fighting for rights, cultural and behavioural changes, a broadening of the political sphere, the resistances of groups under threat, as well as maintaining the forms and knowledge on the relationship with nature. Social movements are also spaces for the creation of transformative knowledge and actions, such as educational methodologies (e.g., the pedagogy of Paulo Freire), productive systems (such as the movement for agroecology) and mechanisms of collective decision-making which, through disputes (contentious politics) led by movements, influence private agents and states, in the form of public policies.

In the experience of Latin America, from Patagonia to Mexico, social movements have emerged as political subjects in the countryside, in the city, in the jungles and their rivers, and on the peripheries, in the form of grassroots organizations or regional and transnational articulations, disputing societies based on different themes and with different constitutions: urban social movements (those living in favelas or city residents in general), peasants, women, Indigenous peoples, Blacks, youth, environmentalists, human rights, cultural groups, and so on. There is often a mixture of this cross-section, constituting a particularly singular wealth of mobilizing experiences in Latin America. Over time, some of these struggles have engendered different forms of organization, which throughout the centuries have constituted traditions, such as the Black, Indigenous and peasant movements.

Recently, growing attention has been focused on the geographies of these social movements. Struggles that have sought to sustain their own territory, whereby they seek to constitute emancipatory forms of social relations, productive activities, and transformative projects of education as human formation and forms of political socialization reveal spatial strategies. Experiences such as the Movimento dos Trabalhadores

DOI: 10.4324/9781003430926-31

Rurais Sem Terra [Landless Rural Workers Movement – commonly referred to as "MST"] in Brazil, and of struggles that use "space as a trump card", called "socio-territorial movements" (Fernandes, 2022) since they seek to implement specific forms of sociability and economy in their settlements, have revealed Latin American Social Movements (LASM) as subjects that seek to build their own geographies, a condition for the reproduction of their struggles. The Zapatista Movement, in the region of Chiapas, Mexico, is another example of a struggle that, by taking land from farmers, operates based on the autonomous and democratic management of the territory. This involves all aspects, from the production and reproduction of life (food, economy, health), culture and identity, collective decision-making (the political sphere, by creating their own institutions, such as assemblies and committees) to even military control of their territories. Engendered in the political work of LASMs there is a broad diversity of spatial strategies and geographies.

This chapter contains three sections. First, the emergence of social movements is discussed as an object of concern for social thought and as political subjects. The second section addresses the various forms of organization adopted by the struggles. The final section examines some of the spatialities of LASMs. All the discussions are illustrated by referring to examples of these struggles and their geographies, through which the reader may then conduct further research.

Social movements in the historical and social formation of Latin America

Social movements (SM) became a significant topic for the social sciences worldwide between the 1960s and 1970s. On the one hand, this period was impacted by several significant emerging struggles, such as the Black civil rights movement in the USA, the protests of May 1968 in France, the emergence of the pacifist and environmentalist movements, as well as the struggles against dictatorships in Latin America. On the other hand, perceiving that the experiences of struggle and Marxist-inspired political regimes were weakening and generating disappointment, the social sciences begun to consider SM in political and analytical terms. Thus, SMs appeared as forms of collective social action that emerged from the social conflicts they revealed, although without the commitment of needing to embody a subject capable of leading a macrostructural transformation in the historical system, as advocated by some Marxist trends. Instead of the revolutionary seizure of power, SMs were viewed as complex, contradictory forms of collective organization, more committed to promoting cultural and political transformations centred on the daily lives of the people involved. While they also appeared as subjects that could have some kind of bond, they were autonomous in relation to the subjects and organisational nuclei of the transformative struggle traditionally valorised by Marxism, such as trade unions and political parties.

This intellectual turn was strongly influenced by the political struggles of SMs in Latin America. Key authors in the North Atlantic scenario, such as Manuel Castells and Alain Touraine, studied SMs in Latin America, and maintained an intense dialogue with Latin American authors. In Latin America, which during the 1970s was experiencing a "wave" of military dictatorships, there was a need to re-discuss politics, given that the control (of institutions and actors) exercised by these regimes brought visibility to LASM. With the suspension of oppositional political parties under military regimes (**see Chapter 4**), the strict control of trade unions and leaders from both these areas either in exile, disappeared or murdered, political resistance from other forms of collective action gained focus. Within this context, SMs appeared as (i) important political subjects in constructing resistance to dictatorships – such as the movement of the "Mothers of the Plaza de Mayo", in Argentina, who in 1977 began to condemn the arrest, torture and disappearance of their children by the dictatorship, or in Brazil, where urban popular movements, against food prices, for public transport, periphery and favela residents and the struggle for housing, among many others, articulated by the Base Ecclesial Communities of the Catholic Church, became important allies of the opposition to the military regime (Gohn, 2012); (ii) potential vehicles for building national-popular revolutions – as in Ecuador, where Indigenous people began to mobilize ethnicity as the cement of political action, and from the confluence of histories and mobilizations on a local scale, they began to organize regional and national articulations (and, thus, creating new political geographies), such as ECUARUNARI (Ecuador Runacunapac Riccharimi, "Dawn of the Indigenous People") in 1972, which intended to bring together the "runas", Indigenous people of the Ecuadorian Andean highlands and CONAIE (Confederation of Indigenous Nationalities of Ecuador) in 1986 (**see** Figure 21.1) (Cazar and Peralta, 2003); and even (iii) in a contextualised form, the embodiment of the class subject against the forces of capital, as in the March for Life in Bolivia in 1986, during which tens of thousands of miners, housewives, students and peasants made their way from different cities in the countryside toward the capital La Paz, contrary to the guidelines of union leaders committed to a social pact similar to "state capitalism" (Linera, 2015), and which led to a national state of siege being decreed and to the military surrounding the demonstrators. While this collective protest did not win, it did however, create the conditions for a new political culture that would return years later.

Figure 21.1 CONAIE assembly in Quito, Ecuador, 2011.
Source: www.flickr.com/photos/asambleanacional/6355512365/

Viewed as emerging actors, these struggles were called "new social movements". However, from their action, deeper criticisms of Latin American societies emerged, and later, the traditions that had existed before this context of the social movements and their struggles became recognised. In its formation and colonial heritage, Latin America has been marked by an "oligarchic conception of politics", which conceives "politics as a private business of the elites . . . resulting in an immense distance between civil society and politics" (Alvarez et al., 1998: 8).[1] In this region, the coloniality of power (Quijano, 2000) has instituted a tangled package of hierarchies of social domination, articulating race, class, gender, nationality, language, culture, and spirituality as technologies of power, linked by the inter-state system of institutions that impose their modern forms of organization and social coordination, with the use of their bellicose-military apparatus. Creating a political culture of exclusion, these are states built "against the people", invested in by elites with the mission of building nations through anti-Indigenous and anti-Black projects so as to whiten the population, which is also the "whitening of the territory" (Santos, 2009) with the historical, symbolic and cultural erasure of non-white groups – a biopolitical guideline that has been fought against by the geographies engendered by LASMs. Even political projects that set out to incorporate the masses, such as populism (see **Chapter 4**), nationalism, and economic developmentalism operate within the institutional confinement of public life in order to historically restrict popular participation, in the same movement that reproduces discourses indicating a "weak civil society".

Thus, also in the so-called democratic contexts, social authoritarianism of colonial origin rooted in Latin American societies is combined with an excluding social organization, in which state action becomes selective, and the economic gains of social production become concentrated. Even with the end of the cycle of dictatorships, this model has continued, and in the face of new political, economic and social scenarios, other collective struggles have emerged. This was the keynote, for example, between the 1990s and 2000s, with the wave of governments implementing neoliberal projects, against which a heterogeneous set of social mobilizations was organized in different countries. Faced with a deepening social crisis and unemployment, and with the closure of production units, initiatives, such as the Unemployed Workers Movement and the Association of Family Producers, flourished in Argentina, which rejected the government's income transfer programs and fought for the creation of "dignified jobs" through the organization of self-managed cooperatives. From the articulation of locally based initiatives such as these, in the early 2000s, the National Movement of Recovered Factories and the National Movement of Recovered Enterprises were created, parts of a struggle that included more than 400 recovered factories across the country (Zibechi, 2017), engendering new social and economic geographies in the territory. Also taking place in other countries in the region, a milestone in this movement's internationalization was the First Latin American Meeting of Worker-Recovered Enterprises, in Venezuela in 2005 (http://www.marxist.com/encuentro-latinamericano-trabajadores.htm). The self-management of productive activities spread strongly due to the response of the Latin American social movements to the neoliberal wave, reaching more than two thousand enterprises in Mexico (in the countryside and in the city), and more than thirty thousand enterprises in the solidarity economy in Brazil, where, according to Zibechi (2017), more than three million people were mobilized. Along the same lines, there were many other initiatives, also of self-managed infrastructures: in Colombia, there are more than twelve thousand community aqueducts. All these initiatives have constructed and disputed geographies, thereby constituting disputes over models of social and territorial organization.

In addition to self-management, the most striking SM response to the privatization of services and infrastructure implemented by neoliberal governments during the 1990s was the organization of large-scale street protests, some of which changed the political landscape of their countries. The Bolivian case deserves to be highlighted, in which, during the early 2000s, protests and roadblocks were organized that reversed the privatization process of the water services (including community systems, which had not been built by the state). This became known as the "Bolivian Water War", in which a strong role was played by FEDECOR (Cochabambina Departmental Federation of Irrigation Organisations). These protests and their consequent victory influenced others to take place over the following years, which put pressure on the Bolivian elites and brought new political actors onto the national scene, creating the conditions for the coming election of a progressive government led by the first Indigenous president in the country's history.

From grassroots organizations to movement networks

As in Bolivia, across several other countries across the region, from the 2000s onwards, social movements played a fundamental role in establishing a cycle of electoral victories for progressive governments (**see Chapter 8**). However, even within the contexts of these governments, modernizing projects were also confronted by grassroots mobilizations. In Brazil, the 2002 presidential elections were won by the Workers' Party, founded in 1980 with the strong participation of grassroots social movements, both in the countryside and in the city, under the strong influence of the Base Ecclesial Communities of the Catholic Church. Once in power. However, the party adopted a neo-developmentalist economic growth program, strongly anchored in expanding the export of primary goods and in energy and natural resource-intensive activities, a project that was "exported" to neighbouring countries under the Brazilian geopolitical leadership in the continent (**see Chapter 17**). Thus, in several countries across the region, large infrastructure works (many of them binational), such as the construction of roads, ports, and hydroelectric plants, in addition to an increase in agricultural monocultures and mining for export, provoked strong reactions from social movements.

One exemplary case of the discursive entanglements and struggles that arose within this context was the construction of large hydroelectric plants along the Madeira River, one of the main tributaries of the Amazon basin, which encompasses parts of Bolivia and Brazil. Two dams, the Jirau and the Santo Antônio, located on the Brazilian section of this river, although close to the border, and with strong social and environmental impacts in Bolivia, generated a myriad of reactions. First, the precariousness of the work (including accusations that the work was carried out under slave-like conditions) engendered strong uprisings and strikes by contracted construction workers in 2011 and 2012. In addition to the workers being exploited by the hydro-energy business, the affected communities also mobilized, organized mainly by MAB (Movement of People Affected by Dams), an entity created in Brazil during the 1980s to fight against the impacts of the large hydro-electric plants adopted as a model for energy production. On the Bolivian side, mobilizations were already taking place even before the work began, with FOBOMADE (Bolivian Forum on Environment and Development), made up of non-governmental organizations, fishermen's associations, Indigenous leaders, professors from the Autonomous University of Beni, and politicians from the Department of Beni. From 2006, FOBOMADE was pressuring the government of the president, also progressive, and of Indigenous origin, Evo Morales, to intercede with the Brazilian government against the construction of the plants.

Thus, the affected, exploited groups – workers, and Indigenous, riverside, and urban peoples, with different identity formations – became united by belonging to the "sacrifice zones" created by the enterprises, and initiated processes to create political and solidarity networks against these plants. In Brazil, MAB, in addition to territorialized local resistance, also played a leading role in dialogue with a key actor in disseminating this development model, BNDES (National Bank for Economic and Social Development). At the end of the 2000s, this state-owned bank, which financed undertakings both at home and abroad, became one of the largest development banks in the world, with a budget even greater than the combined capital of the World Bank, the Inter-American Development Bank (BID), and the US Eximbank. Then, in 2009, the First South American Meeting of Populations Affected by BNDES-Financed Projects was held, which launched the "Charter of Those Affected by BNDES". By provoking dialogue with the BNDES, the Movement for People Affected by Dams not only problematized the local impacts of the dams, but also articulated their impacts on the daily lives of affected groups with structural issues, criticizing the development model. Thus, this not only brought about the convergence of other groups with similar problems in different locations, but also groups with apparently different problems, but whose origins were similar. Subsequently, when these plants went into operation and the impacts became even greater (such as severe flooding in Bolivia in 2015), and with the continuity of this model and the projected construction of new plants, the Pan-Amazon Rivers Alliance was created, evidencing new geographies of political alliances of affected groups.

The case of the struggle against hydroelectric dams in the Amazon basin exemplifies a fundamental aspect of Latin American social movements: the articulation between the creation of locally based organizations and the formation of solidarities, entities and national and transnational alliances, in the form of social movement networks, forums, events and a wide range of spaces for the convergence of militancy. For example, in a region where one of the fundamental pillars of its historical colonial formation was the expropriation of land (see Chapter 3), in different historical contexts the struggles for the democratization of access to land and agrarian reform have engendered the creation of national entities in almost all countries, such as the CCP (Peasant Confederation of Peru, of 1947), CONTAG (National Confederation of Agricultural Workers in Brazil, of 1963), MUCECH (Unitary Peasant and Ethnic Movement of Chile, of 1987) or UNAG (National Union of Agriculturists and Cattlemen of Nicaragua, 1981). International solidarities have also flourished from these national articulations: it was at a UNAG meeting in 1992 in Managua, to which leaders of various American and European social movements were invited, that the idea of creating the Via Campesina [The Peasants' Way] was born, a planetary articulation of social movements entitled "a movement of movements and the global voice of peasants that feed the world".

In addition to taking on this leading role in various articulations on a global scale (such as the World Social Forum and the People's Summit), LASMs have also built a tradition for the convergence of supranational struggles across the region, such as the Hemispheric Meetings to Oppose the FTAA (Free Trade Area of the Americas) and entities such as ASOCODE (Association of Central American Peasant Organizations for Cooperation and Development). The constitution of both grassroots struggles and supranational entities, articulations and movement networks on a continental scale enable us to observe that, in addition to their plurality, the LASM also operate multi-scale political strategies. The case of the Brazilian Black Movement would be a good example, which in the process of preparing the UN Third World Conference against Racism, Racial Discrimination, Xenophobia and Related Intolerance (held in Durban, South Africa, in 2001), became mobilized during preparatory conferences on local, regional, and national levels, as well as at a Regional Conference of the Americas, to push through the commitments of the American states with their agendas. In this process, discussions and decisions agreed upon during these conferences at the suggestion of Black Movement leaders were used as political pressure for subsequently gaining rights and public policies on topics such as education, university access (the quota policy), ethnic recognition of the right to land and territory, and health, among many others, thereby constituting a process of the "boomerang effect" of the anti-racism struggle in Brazil (Santos and Soeterik, 2015), a situation in which local actors mobilized institutions, actors and power games on a global scale to pressure its antagonists and interlocutors on the local scale. Another example of the LASMs' "politics of scales" (Swyngedouw, 1997) in the Brazilian case, would be the strategy of the Movimento dos Trabalhadores Rurais Sem Terra (MST), which during the presidential term of Fernando Henrique Cardoso (from 1995 to 2002) occupied the president's own farm on three occasions, thus transforming a local act into a fact of national visibility, and thereby mobilized public opinion for changes in the government's policies for agrarian reform.

Geographies of LASMs

The abovementioned examples of the "politics of scales" lead, in this last section, to the examination of other LASM geographies. Manipulating the scales of political games is a spatial strategy of social action (Santos, 2011). Another manner of mobilizing space when constructing LASM's strategies of struggle is through holding demonstrations in public spaces and roadways, moments in which decisions such as the location of the act, the way of distributing individuals within the spaces, etc., are repertoires of action that produce specific geographies, which provide certain singularities and contribute to the identities of the protests. The location of a protest, for example, may be a tool to jog the memories of political and ideological traditions that have marked certain places. This was the case, for example, in protests in Rio de Janeiro during the early 2000s. It had been notorious that left-wing demonstrations were held in certain squares and major thoroughfares in the city centre, traditional locations for working-class acts and acts against the dictatorship, political memories that the protests sought to trigger. On the other hand, demonstrations by right-wing groups (including acts led by groups from evangelical churches) tended to take place in the South Zone, the richest part of the city, such as Copacabana beach (in addition to occurring predominantly on weekends, while left-wing demonstrations were more predominant on weekdays in the late afternoon). Such spatial decisions express political strategies for activating the sense of place and symbolic geographies as a social mobilisation tool.

Spatial strategies also characterized, for example, the protests of the so-called "piqueteros", one of the main strands of the Unemployed Workers' Movement in Argentina during the late 1990s and early 2000s. The strategy of blocking streets, roadways or bridges, an action to interrupt and intercept the flows in space (and, by this way, influencing geographies of material flows, above all, economic ones), was characterized by the act of people sitting on the roads as a form of protest against job vacancy resulting from neoliberal policies. A distinct form of protest along the public roadways was the huge marches, developed by Indigenous peoples in countries such as Bolivia, Ecuador, Peru, starting in the Andean region, but also spreading to groups in the lowlands of these countries. Indigenous marches towards the capitals, such as La Paz and Quito, or large cities such as Cuzco, brought together thousands of people who, in some cases, such as the various marches held in the 2010s against the construction of a road crossing the Indigenous Territory Parque Nacional Isiboro Sécure (TIPNIS) in the Bolivian Amazon, travel more than 600km and even walk for up to two months, paralyzing roads and mobilizing efforts for the support of local residents along the way (see **Chapter 10**). The marches, with their strategy of arranging individuals in well-disciplined, spaced lines in order to occupy as much space as possible along the roads, often wearing traditional costumes from their native Indigenous cultures, have become strong symbolic images of Indigenous struggles in the region, which have emerged since the 1970s.

These spatial strategies, with the flow of Indigenous bodies and cultures filling the main roadways, in addition to the direct clash against specific political projects and undertakings, have also functioned as acts that strongly question the symbolic and racial political ordering of such countries, historically marked by the domination of whites and mestizos over Indigenous peoples. Emerging popular projects have centred on a sense of ethnic belonging, expanding the notion of citizenship by linking it to their struggle through recognition. Such struggles redefine the spatialities and temporalities of historical understandings that have organized and defined the rules for the insertion of peoples into their societies. They claim historicities from the anti-colonial resistance and even other pre-Hispanic resistances, such as resistance against the domination of the Inca Empire in the case of the Aymara peoples, as part of an effort for the decolonization of thought (see **Chapter 2**). Thus, Indigenous movements reconstruct historical narratives and redefine their memories of struggles, thereby opposing the idea, initiated in the 1970s, that they are "new social movements".

The struggles of Indigenous peoples in recent decades have mobilized, re-signified and transformed geographies across the continent. The struggles of the Aymaras were structured based on the organizational forms of their community, the Ayllus. They were the basis for neighbourhood structures of self-government, mainly in the city of El Alto, next to the national capital La Paz, which in the early 2000s engendered what became known as the "Gas War", which led to the resignation of President Sanchéz de Lozada. They also fought for territorial autonomy, calling for the retaking of Qullasuyo, an Andean region of Aymara predominance that covers parts of Bolivia,

Chile, and Peru, in discursive and political articulations that configured "Indigenous geopolitics" (Ramírez, 2005), triggering territorial, economic-productive, cultural, and socio-political frameworks. Thus, the space-time forms of organizing the social relations that constitute the cultures of Indigenous peoples are geographies that construct them as political subjects, while representing, at the same time, a structuring factor and object with which to claim for their struggles. In Bolivia, as a result of these struggles, the 2009 Constitution re-founded the state, now called plurinational, which involves recognising the pre-colonial existence of Indigenous nations in their territories, and their self-determination, autonomy, and right to be consulted through their own institutions on any issue that involves them (see Chapter 4). There is also, since the 1990s, the case involving CONAIE in Ecuador, that has called for the recognition of a "plurinational and intercultural State", respecting the Indigenous ways of existence, and has challenged the universalist, developmentalist senses of space and territory, under which historically the national states have operated.

This paradigm of the differential understanding of territory in a plurinational state results from the constitutive geographies of the mobilization processes of these movements. A similar situation may be observed in the struggles of other traditional groups across the region, such as the Black communities that have remained from the struggle against slavery (in Brazil, called quilombos), who claim "we don't want land, we want territory" so that they may defend their ways of life, their specific forms of relationship with nature and cultural ties, something that goes beyond the right to land titles (see Chapter 15). Quilombos in Brazil, Palenques in Colombia, Cumbes in Venezuela, Garífunas in Honduras, Guatemala and Belize, Saramakanos in French Guiana, or Cimarrones in several countries: these Black communities have been fighting for their ethnic and territorial rights, associating racial and cultural guidelines with debates on agrarian reform in the region. Thus, they fight for the right to remain in territories with the geographies that constitute them as collective subjects. They are opposed to alliances with economic groups interested in the destructive exploitation of nature and biosociodiversity and, as has happened with the Black Communities Process, the articulation of Colombian Black movements, they face violent offensives against their leaders and territories.

A brief look at the Brazilian Black Movement (see Figure 21.2) demonstrates the diversity of geographies that the anti-racism struggle has built and taken on. In response to the complex forms of oppression that racism engenders, anti-racism activism has many levels, involving a range of agendas for their claims and organizational forms, and as a result, the Black movement appears as an "action system" (Melucci, 1994). Thus, in addition to the struggles for (and through) the territory of the more than three thousand quilombola communities recognized by the Brazilian State, for example, Black groups have disputed the senses of place as a repertoire in their actions. The group called Frente 3 de Fevereiro [The February 3rd Front], created in São Paulo in 2004, protesting after the police had murdered a Black youth who had completed his university training as a dentist, began a set of interventions in the urban space, comprising artistic presentations, political acts and the production of urban graphics. One of these was the construction of "horizontal monuments", which were paintings and sculptures of bodies lying on the ground next to dates, in places where young Black men had been murdered. This act of "graphing the land", therefore, "geo-graphing", aimed to bring awareness into the dispute over the genocide of Black youths through

Figure 21.2 March of Black women in downtown São Paulo, 2022.
Source: Daniel Arroyo/Ponte Jornalismo.

writing an urban grammar by activating a memory of and in places. This is similar to the disputes for the construction and state recognition of monuments and toponyms of streets, squares and other places named after Black historical and cultural personalities, as groups have done in the old port area of Rio de Janeiro, which due to its strong cultural and Black political presence was called "Little Africa" in the early twenty-first century. Graphing in points of space, which become references for rereading history, mobilized by contents attributed to the past of these places, reframes the relationship between individuals and their ancestry, causing changes in identity and belonging. It is the dispute over the meanings of places that is a tool of social struggle.

Conclusions

Social movements are an important aspect of the political, social and cultural landscape of Latin America. They produce and are produced by its geographies. In this chapter, some of these geographies have been revealed: spatial strategies (politics of scale, the demonstrations and occupations of major roadways to block the flow of traffic and performances in public spaces), geographies that build subjects (or spaces as the basis of organization, from cultural forms and territorialized social relations), in addition to subjects that build their geographies (through various forms of territorial control and the imposition of forms of cultural, political, and economic sociability, constituting processes of territorialization). The engendered spatial forms that have emerged from the action of organized struggles are innumerable. LASMs are not "new social movements"; many of them are the traditions of struggles, the continuity of resistance and liberation and emancipation projects which have occurred for centuries, thus constituting re-existences (Porto-Gonçalves, 2002), as a reinvention of the ways that subaltern and oppressed groups in the region have found to exist.

Summary

■ While in European and North American social thought, social movements gained attention between the 1960s and 1970s as alternative forms of collective mobilization to trade unions and political parties, in the Latin American experience they constitute traditions of struggle that go back much further.

■ In different political and economic contexts in Latin America, including populism, military dictatorships, neoliberal democracies, and the recent neo-developmentalist cycle, various subaltern social groups have established mobilizations in the form of social movements.

■ As re-existences (i.e., ways of resisting hegemonic projects of society and territory, but also, in this same action, ways of reinventing their existences), the LASM have produced social experiences that have redefined territories, territorialities, and senses of place and thus engendered new geographies.

■ LASMs also use various spatial strategies in their forms of collective action, such as the occupation of roads and public spaces, disputes and manipulations of symbolic meanings of spaces, and politics of scale, amongst other geographies that they produce and of which they are composed.

Review questions

1. What roles (economic, political, social, cultural) have social movements played in the constitution of Latin American societies?
2. In a region with a colonial, racist historical formation dominated by elites that do not identify with the people, what are the potentialities of social movements in constructing alternative projects for society?
3. What geographies do social movements produce in Latin America? In what ways do they oppose the hegemonic forms of spatial control imposed by states and large corporations?

Further reading

Oslender, U. (2016). *The Geographies of Social Movements: Afro-Colombian Mobilization and the Aquatic Space*. Duke University Press.
This presents the constitutive geographies and territorial struggles of Black Colombian communities.
Rossi, F. (Ed.). (2023). *The Oxford Handbook of Latin American Social Movements*. Oxford University Press.
A collection of texts which addresses Latin American Social Movements within different theoretical, political and empirical approaches, including a wide range of cases.

Zibechi, R. (2012). *Territories in Resistance: A Cartography of Latin American Social Movements*. AK Press.
A panoramic look at the political, economic and territorial dimensions of social movements in Latin America, based on a wide range of different struggles, stretching from Patagonia to Mexico.

Keywords

Politics of scale: ability of a social actor to mobilize other actors, arenas and political games in different scales, as local, national, regional, or global, to reinforce his struggle. When, for example, a local actor tries to directly access global arenas or actors, without the national mediation, it is called "jumping scale".

Scale: here it is understood as the spatial reach of games and political arrangements, which organize the actions of agents. Thus, scales such as local, regional, national, or global are not just metric measures, but a form of space-time organization of political action, defining possibilities and limits to the subjects' action.

Social movements: a wide set of forms of collective action, involving activisms that can be organized and institutionalized or not. Social movements cannot be confused with organizations – some of them are, but others can mix entities and individual activists (who can also take part of some organization) and develop as traditions around themes (e.g., environment, development model and their impacts), identities (Indigenous, Blacks, Youths, workers), or both combined, creating action systems.

Spatial strategies: any strategic decision in which groups mobilize material, political, or imagined geographies to produce political impacts or reactions of other social actors.

Note

1 This and all other non-English citations hereafter have been translated by the author.

References

Alvarez, S., Dagnino, E. & Escobar, A. (1998). Introduction: The Cultural and the Political in Latin American Social Movements. In: Alvarez, S., Dagnino, E. & Escobar, A. (Eds.), *Cultures of Politics, Politics of Cultures: Re-Visioning Latin American Social Movements*. Westview Press.

Cazar, F. G. & Peralta, P. (2003). *El poder de la comunidad: ajuste estructural y movimiento indígena em los Andes ecuatorianos*. Clacso.

Fernandes, B. M. (2022). Territories of Hope: A Human Geography of Agrarian Politics in Brazil. *Environment and Planning E-Nature and Space*, 6, 1–16.

Gohn, M. G. (2012). *Teorias dos movimentos sociais: Paradigmas clássicos e contemporâneos*. Edições Loyola.

Linera, A. G. (2015). *Plebeian Power: Collective Action and Indigenous, Working-Class and Popular Identities in Bolivia*. Haymarket.

Melucci, A. (1994). Qué hay de nuevo en los nuevos movimientos sociales? In: Laraña, E. & Gusfield, J. (Eds.), *Los nuevos movimientos sociales: de la ideología a la identidad*. Madrid: Centro de Investigaciones Sociológicas.

Porto-Gonçalves, C. W. (2002). Da geografia às geo-grafias: um mundo em busca de novas territorialidades. In: Ceceña, A. E. & Sader, E. (Eds.), *La guerra infinita: hegemonia y terror mundial*. Clacso.

Quijano, A. (2000). Coloniality of Power, Eurocentrism, and Latin America. *Nepantla*, 1(3), 533–580.

Ramírez, P. M. (2005). *Geopolíticas indígenas*. El Alto: CADES (Centro Andino de Estudios Estratégicos).

Santos, R. E. N. (2009). *Rediscutindo o ensino de Geografia: Temas da Lei 10.639*. CEAP.

Santos, R. E. N. (2011). *Movimentos Sociais e Geografia: Sobre a(s) espacialidade(s) da ação social*. Ed. Consequência.

Santos, R. E. N., & Soeterik, I. M. (2015). Scales of Political Action and Social Movements in Education: The Case of the Brazilian Black Movement and Law 10.639. *Globalisation, Societies and Education*, 13, 1–19.

Swyngedouw, E. (1997). Neither Global nor Local: 'Glocalization' and the Politics of Scale. In: Cox, K. (Org.), *Spaces of Globalization: Reasserting the Power of the Local*. The Guilford Press, 137–166.

Zibechi, R. (2017). *Movimientos Sociales en América Latina: El "mundo otro" em movimiento*. México: Bajo Tierra A.C., Comunidad Autonomía y Libertad (Comunal), El Rebozo.

CHAPTER 22

Agrarian inequalities and conflicts

Bernardo Mançano Fernandes

Introduction

The development of agrarian inequalities is a deep-rooted process in Latin America and has critical implications for the region's geographies. They respond to an accumulation of times and spaces, of processes of domination, tied to historical movements of colonization – appropriating the territories of original peoples and exploiting natural resources – and the production of the plantationocene. Against a predatory model that produces exclusion and dependency, there has always been resistance and conflictualities.

The debate on the agrarian question is also a key paradigm for understanding the inequalities of global systems. This chapter highlights the Amazon biome as one of the areas of resistance to the predatory model. It further discusses the relevance of agrarian reform as a condition for recovering the territorial rights of Latin American peoples, with implications for other regions. Decolonizing material and immaterial territories, such as the land and the mind, has been a long road in the history of resistance by Latin American peoples, overcoming the plantationocene and implementing a sustainable model of development is necessary for Latin America and for the Earth.

The agrarian question is a territorial question, because the main issue is the use and control of land. There are many territorial disputes between corporations and communities that determine the different uses of territories. The corporations are only interested in the territories, and not people. For the communities, territories and people matter. This chapter analyzes the process of territorial domination and the process of territorial liberation based on different commodities, from trees, minerals, agriculture, and livestock to the recent production of renewable energy.

The main argument of this chapter revolves around debates over territories and development models (**see Chapters 10 and 17**). For over five centuries, Indigenous peoples and peasants have been advocating for sustainable development, which involves including people in their territories, as opposed to the predatory model that takes people away from their territories. This issue is particularly significant in the Amazon

DOI: 10.4324/9781003430926-32

region, where indigenous peoples are fighting for the sustainable development of their territories and are in conflict with corporations that aim to displace Amazonian populations to take control of the territories.

This chapter is organized into two parts. The first explores the historical process of production of agrarian inequalities. It introduces key concepts to understand this process, beginning with colonization in the 16th century and concluding with the exploitation of natural resources in the Amazon in the 21st century. In the second part, the chapter analyzes the prospects for overcoming agrarian inequalities, presenting different readings on the types of predatory development and sustainable development, and relating it to hunger and food sovereignty.

The historical process of producing agrarian inequalities and conflictualities

By its very nature, colonization is a model based around the development of inequalities. This model appropriates the natural resources of colonized territories, keeping the colonies dependent on the metropolises. Even today, countries in Latin America, colonized by Spain and Portugal, have enormous inequalities that are legacies of permanent coloniality (see Chapter 2). This model impacted all dimensions, creating dependent economies, geopolitical control, cultural erasure, social unfeasibility, and environmental and social destruction. The creation of colonies facilitated the exploitation of natural resources and people, yet also generated resistance.

The beginning of the colonization of Latin America, in 1492, initiated a series of conflicts with the original (Indigenous) peoples, mainly due to the expropriation of their territories, enslavement, or servitude of the Indigenous peoples for the exploitation of natural resources. Both in America colonized by the Spanish and in America colonized by the Portuguese, there were wars of extermination of Indigenous peoples such as the Tupinambás in Brazil, Incas in Peru, and Aztecs in Mexico. The processes of extermination of Indigenous peoples continue to the present day, such as the Yanomami peoples in the Brazilian Amazon (Kopenawa and Albert, 2023). The territorial formation of the Latin American peoples was constituted by the original peoples (Indigenous), the peasantry (mixtures of all peoples), the quilombolas (African-Americans), and other traditional populations. Agrarian inequalities and conflicts in Latin America are marked by the continued deterritorialization of these peoples by landowners, and national and multinational corporations, especially at the beginning of the 21st century with the consensus of commodities (Svampa, 2015). The hegemony of commodity production produces sets of conflicts, territorial disputes, and predatory and sustainable development models, which mark the agrarian question of the 21st century.

For centuries, Indigenous peoples were exploited, deterritorialized, or exterminated. From the beginning of colonization in the 15th to 20th centuries until the independence of countries on several continents in the 18th to 20th centuries, territorial exploitation took place through the establishment of plantations for the production of agricultural and mineral commodities. The independence of the colonized countries did not put an end to the predatory model of plantations for the production of commodities. On the contrary, the exploitation of natural resources intensified, including the monoculture of trees and the production of renewable energy. These social relations produced the so-called "agrarian question", one of the main dimensions of inequality, understood as the concentration of wealth and land. The neocolonial predatory model has been structured for more than five centuries and has become hegemonic in the production of inequality in all its dimensions, but mainly the environmental dimension with the production of climate change, which are the result of mega projects of extraction of natural resources throughout time (**see Chapter 17**). At the beginning of the 21st century, several countries began to think about how to mitigate climate change, but their proposals do not attack the structure of the predatory development model.

Plantationocene in the 21st century

The Anthropocene is an unofficial time for humanity, in which we have become more than a force for transforming the planet's landscape: we can also destroy it if we pursue a model of the development of inequalities. The Anthropocene is a concept debated in several sciences and its period has not yet been defined. The destructive force of humanity also has other concepts, such as Plantationocene or Capitalocene (Haraway, 2015). The Plantationocene is a concept that helps us better understand the production of commodities in the present, thinking about the past and the future. The plantation, formed by extensive fractions of territory for the exploitation of natural resources, was implanted during colonization and continues today as the basis of the predatory model. Plantations used slave labor, wage labor, and high-tech equipment to produce commodities. Plantationocene is an alternative name for the Anthropocene. In the processes of transforming nature, in addition to climate change, there is also an intensive use of pesticides in the pollution of land, water, and people. This concept makes it possible to understand that the modernization of relations and technologies for the production of agricultural commodities, minerals, and renewable energy does not modify the conservative structures of the predatory development model.

In the first decade of the 21st century, the process of appropriating the territories of Latin American peoples intensified, adding another element to the agrarian question: the foreignization of the land. This element is part of a global system that involves dozens of high-income countries, whose corporations invest in the exploitation of natural resources in Latin America. It was the result of global changes, mainly with the end of the Soviet Union and the fall of socialism. The Washington Consensus, in 1989, formulated a set of structural adjustment policies with a neoliberal orientation. This adjustment made it possible to make labor and environmental policies more flexible, expanding possibilities for exploration investments in new territories. In the last decade of the 20th century, several studies on territorial development began to be prepared to support governments, multilateral agencies, and national and multinational corporations linked to agribusiness. The concept of territory started to be used by several sciences to justify the territorialization of the neoliberal model. Territory was seen as a possibility for expanding production with the increased exploitation of natural resources. Neoliberalism reinforced the predatory model. Latin American peoples reinterpreted the concept of territory as a living space, opposing the sense of exploring land, water, and forests (**see Chapter 12**). Via Campesina, a world coalition of peasant movements, at its 2004 international meeting in Brazil, claimed the defense their territories as a condition of existence, and have continued to mobilize through a global network (**see Figure 22.1**). Foreign investments in Latin America and other continents

Figure 22.1 Via Campesina forum for social and environmental justice, Cancun, Mexico, 2010.

with the acquisition of large tracts of land created a new agricultural frontier (Messerli et al., 2014).

The Amazon

The Amazon biome comprises parts of the territories of Brazil, Bolivia, Peru, Ecuador, Colombia, Venezuela, Guyana, Suriname, and French Guiana. Most of the Amazon is in Brazilian territory. In this immense biome live more than two hundred Indigenous peoples and other traditional peoples. But in all these countries their territories are threatened, characterizing the Amazon agrarian issue as one of the most violent in the world. The advance of agricultural frontiers and mineral exploration have impacted the region with deforestation and pollution of land and water. In recent decades, the territorialization of livestock has increased deforestation, whose lands are used for planting soy monoculture or for palm oil. The deforestation process begins with landowners who threaten peoples and their territories, opening the way for the entry of national and multinational corporations. This landholding-agribusiness alliance has been the main front for the territorialization of commodities. The images below give a sense of the scale of deforestation in the Amazon biome in Brazil (see Figure 22.2). This deforestation is the result of the Brazilian government's policy of investment in the Amazon. Another front is the exploration of ores such as petroleum and artisanal mining. These have been the causes of threats and deaths in Amazonian communities. Deforestation is at the heart of the problems in the Amazon; it is directly associated with deterritorialization and climate impacts on a continental and global scale (Nobre, 2014). The existence of the Amazon biome defies predatory development because it can only exist under conditions of sustainability.

The Amazon has been a forest inhabited for at least 14,000 years. There are over two million people who form a rich diversity of peoples that live in and produce the Amazon

Figure 22.2 Industry and deforestation in the Amazon.
Source: www.flickr.com/photos/72482589@N07/45567621401/

biome. To give an idea of this diversity, let us consider only a few peoples. In the Brazilian Amazon, more than 180 Indigenous peoples live, such as the Yanomami, Matis, Marubo, Ashaninka, and Kayapó. These peoples have developed various knowledges that have enabled the creation of methods and techniques for food production and the use of spaces in the humid tropical forest. In the Venezuelan Amazon, among many peoples, live the E'ñepa; part of this population is also present in Brazil. In the Colombian Amazon, for example, live the Miraña, Yucuna, Letuama, Makuna, and Tanimuka peoples. In the Peruvian Amazon, the Jíbaro, Alamas, Yagua, Bora, and Kukama Kukamiria peoples live. In the Ecuadorian Amazon, the Kichwas, Siekopai, Achuar, Waorani, Andoas, Cofanes, Záparas, Shiwiar, Sionas, and Shuar peoples live. In the Bolivian Amazon, the Cavineños, Cayubabas, Canichanas, Chacobos, Chiquitanos, Guarayos, Guarasugwe, Itonamas, Lecos, Machineris, Maropas, Moré, Mosetenes, Movimas, Moxeños, Nahuas, Pacahuaras, Sirionós, Tacanas, Toromonas, Tsimane, Yaminahuas, Yukis, Yuracares, Araonas, Ayoreos, and Baures peoples live. On the borders between Brazil, Suriname, and French Guiana, the Tiriyó and Waiãpi peoples live.

Since the arrival of Spanish and Portuguese colonizers in the 16th century up to the present times of the 21st century, the territories of these peoples continue to be threatened and destroyed by large corporations for extractive exploitation of minerals, trees, agriculture, and in this early 21st century, renewable energies such as solar and wind power. The Amazon biome and the peoples living within it can only exist inseparably. The destruction of the people means the destruction of the territories and vice versa. Therefore, the Amazon is a territory that challenges predatory development because it can only exist through sustainable development.

Perspectives for overcoming agrarian inequalities

The Latin American **agrarian question** changed during the transition from the 20th to the 21st century. In the last century, a tendency of the agrarian question paradigm explained that most of the peasantry would be destroyed and a small part would be "integrated" into the capitalist system. This system is responsible for monoculture production and the extraction of other natural resources, such as mining, a model that became known as extractive, denominated here as a model that produces inequalities or a predatory model. This "integration" was believed by those promoting it to result in a metamorphosis of the peasantry into capitalists. There was also a widespread policy of cultural erasure of Indigenous, peasant, and Afro-descendant identity. However, these strategies were not able to succeed due to the resistance of these peoples and because the production of commodities on a large scale is not part of the history, territories, or culture of these peoples (**see Chapter 13**). In this century, these peoples began to build their own territorial development model based on **food sovereignty** and agroecology. The advance of neoliberal policies in the countryside with the financialization of agriculture and the **deterritorialization** of peoples generated a global agrarian issue. The agrarian question of the 21st century has among its components the territorial struggles for land and food, such as agrarian reform, the resumption of Indigenous territories and the defense of quilombola territories, the production of healthy food in a perspective of sustainable development (Fernandes, 2016).

Sustainable agriculture

Agrarian reform is a public policy for the deconcentration of rural property to democratize access to land. This policy is used to promote sustainable development. More than a territorial redistribution policy, agrarian reform also contributes to food production, industrialization, commercialization, job creation, housing, education, and health, among many other conditions to improve the quality of life. Land reform is associated with the land rights of peasants, but other peoples are also claiming their land rights, such as Indigenous peoples and quilombola peoples (**see Chapter 15**), who were deterritorialized by landowners and agribusiness corporations. In Latin America, there are countries with the highest land concentrations in the world, as in the case of Paraguay and Brazil. Struggles for land and territories are collective actions by people to pressure governments to carry out agrarian reform. However, the landowner-agribusiness alliance has prevented the implementation of this policy, promoting the predatory model as if it were a sustainable model, thus preventing the expropriation of large properties. Peasant and Indigenous movements have been fighting for their territorial rights in the countryside and forests. Progressive governments have implemented timid land distribution policies (**see Chapter 8**), while conservative governments have destroyed these policies, weakening communities to sell their land to capitalist corporations (Fernandes and Welch, 2019).

Agroecology is a model of sustainable development in the making. From a multidimensional perspective, it comprises planting, production, commercialization, industrialization techniques, and technologies, but also includes social relations in all its systems. The origins of agroecology are in Indigenous and peasant agriculture, whose knowledge of nature is associated with the treatment of pests and diseases from the ecosystem itself. Agroecology is cultivated in small biological productive units in resistant and resilient systems. Since the beginning of plantations, many of these systems have collapsed with the territorialization of monoculture systems. Currently, agroecology and agribusiness compete for territories, but aerial spraying of pesticides is fatal for small agroecological units. The principle of sustainability in agroecology requires that healthy food be produced with the participation of people in the community without causing inequality. That is, for agroecological production, it is necessary to respect nature in all its dimensions, the land, the plants, the water, and the human and non-human animals. Agroecological experiences take place in all Latin American countries, not only in fields and forests but also in cities. In several countries, there are agroecological transition policies for peasant farmers who abandon monoculture systems and opt for diversified systems (Rosset and Altieri, 2017). Yet agroecology faces a central threat of agribusiness.

Agribusiness

It is essential to understand the vision of the world that transformed agriculture into agribusiness. This involved the replacement of culture by business, creating a development model that increasingly artificializes food through the intensive use of pesticides and genetic changes. This model intensified disputes between agribusiness corporations that became hegemonic. In Latin American countries, these corporations dominate rural development policies, that is, they control the production and sale of seeds and chemical inputs used in agriculture that pollute the land, water, and people. Among Latin American countries, the most impacted by the use of pesticides are Mexico, Guatemala, Haiti, Honduras, Ecuador, Colombia, Brazil, and Paraguay. The increase in the use of pesticides in these countries is associated with the production of the following commodities: soy, sugar cane, corn, palm, and cotton. The growing use of poisons in agriculture, in addition to pollution and disease, creates other problems, such as the emergence of plants and insects that are increasingly resistant to poisons, increased production costs, reduced ecosystem biodiversity, and the deterritorialization of communities around the agribusiness territories, due to health problems (González, 2020).

The concept of agribusiness was created by John H. Davis and Ray A. Goldberg from Harvard University in the middle of the last century (Davis and Goldberg, 1957). Their readings were limited to understanding the integrated functioning of the set of systems: agricultural, livestock, industrial, mercantile, and financial. This reductionist definition of agribusiness from the paradigm of agrarian capitalism avoided incorporating predatory impacts because it sought to serve the interests of corporations in defending the idea that agribusiness is good for everyone. This is how the "Harvardian" concept of a set of systems contributed to the creation of a hegemonic global development model (Davis and Goldberg, 1957). Patricia Flynn and Professor Roger Burbach's concept of agribusiness includes predatory impacts based on the logic of the agrarian question paradigm. This concept was created when they were researching predatory impacts on the American continent in the 1960s and 1970s. They studied the processes of deterritorialization of traditional populations by US corporations such as Del Monte and Cargill. The formation of the set of systems and the predatory impacts was uneven in the Americas because only in the United States were the systems more developed, while Latin America and the Caribbean had only the agricultural and industrial systems and in conditions of dependence on North American imperialism (Burbach and Flynn, 1980).

Hunger

In 2021, around 10% of the population of Latin America was starving, while the territories of all countries were occupied by the production of commodities. Agribusiness has stated several times that it can put an end to hunger, but this has never happened. Moore (2016) argues that modes of thought are more difficult to overcome than the modes of production that they interpret. This can help us understand why agribusiness has become hegemonic, even though it is a predatory development model. The ideology of the superiority of capitalist agriculture makes the looting it practices through extractivism invisible, being one of the most impactful factors of environmental changes on a planetary scale. From the seed to the production process, on land, to the formulation of ultra-processed foods, in industry, there are impacts on territorial health: people, cultures, policies, social relations, and environments (**see Chapter 18**).

Ultra-processing is at the heart of globalized agribusiness food systems. These systems created the paradox of agribusiness, combining high productivity with hunger and obesity. Food sovereignty and agroecology have driven socio-territorial movements (Halvorsen et al., 2019) toward healthy agriculture. In several Latin American countries, there are initiatives to build public policies to strengthen experiences in food production, popular and institutional markets, and the solidarity economy. These experiences confront the power of corporations that determine agricultural policies. Land and food are current disputes between agribusiness and agroecology. Peasant movements are food movements in the construction of emancipatory public policies from the local to the national scale for the construction of education projects and appropriate and open technologies. Brazil was removed from the UN hunger map in 2014, because of a set of policies for the production of healthy foods led by peasant movements, for schools, and distribution to the starving population. The destruction of these policies by an ultra-neoliberal government under Bolsonaro made Brazil return to the hunger map in 2020.

Conclusion

Predatory development models have reached their limits across the entire planet. New paradigmatic propositions are being born in Latin America and Europe with post-extractivism and degrowth as a way of thinking about the sustainability of life (Acosta and Brand, 2017) (**see Chpater 17**). These proposals could lead to the overcoming of inequalities and agrarian conflicts, with the territorial organization of agricultural production that respects nature and people. Latin America has generated experiences that allow us to think about this overcoming. The persistence of the hegemonic and predatory model may increase the planetary problems of the Plantationocene, putting our existence at ever greater risk. After centuries of destruction, we have experiences, practices, and policies that allow us to live sustainably. We have to make the right decision.

Summary

- The development of inequalities provides a counter narrative to the linear readings of development as progress.
- Although many corporations defend sustainable development, they rarely create sustainable results.
- The agrarian question in the 21st century is global. In every country in the world, there are investments for the exploitation of natural resources in fields and forests with the participation of multinational corporations.
- Local communities affected by the impacts of megaprojects are not aware of the origin of the companies and investment funds that are deterritorializing them.
- Latin American peoples are organized in socio-territorial movements and permanently fight in defense of their territories.
- Hunger is a political problem present in several countries of the world, mainly in Latin America. The agribusiness development model was never able to end hunger; however, public policy experiments for the production of healthy food by the peasantry have been successful.

Review questions

1. Think, using references, about why inequalities generate conflicts and resistance.
2. Development does not always lead to progress. Discuss this idea from the concepts of sustainable development and predatory development.
3. Food production is essential for health, but the use of pesticides contradicts this. What are the differences between agribusiness and agroecology?
4. The Amazon biome is attacked every day by corporations that exploit this territory for the production of agricultural and mineral commodities. Discuss the limits of this process by reading the book *The Falling Sky: Words of a Yanomami Shaman*.
5. Is agrarian reform a necessary policy to end inequality? Discuss this theme by creating a dialogue between this chapter and other chapters in this book of your choice.

Further reading

Foreignization in Latin America
www.fdcl.org/wp-content/uploads/2014/03/factsheet_landgrabbing_latinamerica_engl_web.pdf
Foreignization of the land or land grabbing are words that help to interpret how colonizers yesterday and corporations today expropriate Latin American peoples.

Land grabbing and human rights: The involvement of European corporate and financial entities in land grabbing outside the European Union (for the European Parliament, along with the International Institute of Social Studies)

https://handsontheland.net/wp-content/uploads/2016/06/EXPO_STU2016578007_EN.pdf

Many institutions are allies of socio-territorial movements and support their resistance so that they can continue to exist in their territories.

Land Reform in Latin America: Past, Present, and Future

https://drive.google.com/file/d/1ujWTU28MJloeotrdNHWcNN5nfC15ewhV/view

Agrarian reform is one of the most important policies for sustainable development, as it guarantees land and food for people all over the world.

McDonald's Linked to Amazon Deforestation in New Report

https://pulitzercenter.org/stories/mcdonalds-linked-amazon-deforestation-new-report

Some multinational corporations take advantage of predatory development to increase their economic power. The Amazon is one of the biomes that suffer most from this type of attitude.

Resistance

www.nationalgeographic.com/history/article/meet-survivors-taino-tribe-paper-genocide

The Taíno people say that it was they who discovered Christopher Columbus and the Spaniards. An essential indicator of the resistance of indigenous peoples in the Americas is their continued existence, despite the plantation.

Keywords

Conflictualities: a concept that makes it possible to understand that conflicts are not isolated acts, but are part of a complex set of relationships built over time and in different spaces. Which are contradictory and represent territorial disputes and development models.

Deterritorialization: a process of destruction of territories and their peoples. This geographic process is very present in Latin America in the countryside and forests, where people are dispossessed by mega commodity production projects with the support of national and multinational corporations.

Food sovereignty: in building a sustainable development model, Via Campesina defends food sovereignty as a food production policy that recovers the biodiversity of the territories. This policy aims to produce the greatest possible diversity in each territory and that these foods are consumed by local populations.

Territorial disputes: agribusiness and agroecology are distinct development models that compete for territories and produce different landscapes and types of food. The struggle for land and deterritorialization are hallmarks of territorial disputes.

Territory: is formed in the appropriation of geographic space. For a critical understanding of this concept, it is necessary to consider that the territory is composed of different dimensions and scales. This chapter deals with the territory as the geographic space of Latin American countries as national territory, the space of communities as local territories at the same time that they are fractions of the national territory and that are analyzed in their political, cultural, environmental, and socioeconomic dimensions.

References

Acosta, A.; Brand, U. (2017) *Salidas de laberinto capitalista. Decrescimiento y postextractivismo*. Icaria Editorial.

Burbach, R.; Flynn, P. (1980) *Agribusiness in the Americas*. Monthly Review Press.

Davis, J. H.; Goldberg, R. A. (1957) *A Concept of Agribusiness*. Harvard University Press.

Fernandes, B. M. (2016, March) Development Models for the Brazilian Countryside Paradigmatic and Territorial Disputes. *Latin American Perspectives*, 43(207), 48–59.

Fernandes, B. M.; Welch, C. A. (2019) Contested Landscapes: Territorial Conflicts and the Production of Different Ruralities in Brazil. *Landscape Research*, 44, 1–16.

González, J. C. M. (2020) *Agrotóxicos em América Latina*. FIAN Brasil.

Halvorsen, S.; Fernandes, B. M.; Torres, F. V. (2019) Mobilizing Territory: Socioterritorial Movements in Comparative Perspective. *Annals of the American Association of Geographers*, 109(5), 1454–1470.

Haraway, D. (2015) Anthropocene, Capitalocene, Plantationocene, Chthulucene: Making Kin. *Environmental Humanities*, 6(1), 159–165. https://doi.org/10.1215/22011919-3615934.

Kopenawa, D.; Albert, B. (2023) *The Falling Sky: Words of a Yanomami Shaman*. Harvard University Press.

Messerli, P.; Giger, M.; Dwyer, M. B.; Breu, T.; Eckert, S. (2014, September) The Geography of Large-Scale Land Acquisitions: Analysing Socio-Ecological Patterns of Target Contexts in the Global South. *Applied Geography*, 53, 449–459.

Moore, J. W. (Org.). (2016) *Anthropocene or Capitalocene? Nature, History, and the Crisis of Capitalism*. PM Press.

Nobre, A. D. (2014) *The Future Climate of Amazonia – Scientific Assessment Report*. ARA, CCST-INPE, INPA.

Rosset, P.; Altieri, M. (2017) *Agroecology: Science and Politics*. Practical Action Publishing. http://dx.doi.org/10.3362/9781780449944

Svampa. M. (2015) Commodities Consensus: Neoextractivism and Enclosure of the Commons. In *Latin America South Atlantic Quarterly*, Vol. 114. Duke, 65.

CHAPTER 23

The politicization of life

Mariana Arzeno and Mónica Farías

Introduction

Since the 1990s, inequality has been on the rise in Latin America. Neoliberal measures of all kinds have resulted in increased socio-spatial exclusion (**see Chapter 19**), forced migration (**see Chapter 9**), health problems (**see Chapter 18**), the expansion of police and military control, physical violence, and the stigmatization of certain, often racialized, groups and communities (**see Chapter 16**). This has led to a wave of ascendant contentiousness, a new cycle of protest that has put social movements and collective forms of grassroots organizing at the forefront of struggles against neoliberalism (**see Chapter 21**). One important aspect of these movements is that they focus on life – human and non-human – as a central category of political struggles and resistances. Debates raised by social movements around the "defence of life", the "land to live in", and against "dispossession" and "death projects" expose the violence that characterizes socio-economic neoliberal processes in the region. This chapter introduces the idea of politicization of life to describe the political strategies, both material and discursive, by which Latin American movements coalesce several interrelated claims for the preservation of life in the face of neoliberal politics of death.

We will discuss these things through the lens of two case studies: the Lenca people in Honduras and the *Ni Una Menos*/Not One Women Less movement in Argentina. The first one refers to a movement that exposes the interrelated forms of violence that come with mega-projects, such as the construction of dams, and the work of profit-seeking corporations. The Lenca people and their late leader Berta Cáceres fight against this with an anti-capitalist, anti-patriarchal, and anti-racist resistance. The second example refers to a feminist movement that originated in Argentina in 2015 that coalesces a number of struggles for women's rights and that has spread around the continent making visible patriarchal and neoliberal violence against women. Both movements put the defence of life at the center of the debate.

DOI: 10.4324/9781003430926-33

Neoliberalism and the politics of death

The neoliberal turn in Latin America has always taken a violent form. In numerous countries, neoliberalism arrived in the 1970s alongside illegal and antidemocratic military governments that sought to eradicate all traces of leftist thinking and organizing. The military regimes worked to destroy the social fabric and to stop all expressions of solidarity. Neoliberalism's goal was to discipline society in order to create the conditions to advance economic reforms. This was done by installing a regime of terror that included direct repression, torture, and death in actions that sometimes demanded coordination across states. The rollback of the state on matters of social reproduction and the economic measures taken resulted in the impoverishment of the working classes and the increase of social inequality. The harsh deterioration of the quality of life for many and the lack of opportunities for entire sectors of society have had long-lasting effects.

The depoliticization of society and the politics of death that accompanied these socioeconomic reforms were continued by democratic governments through structural adjustment policies (SAPs) in the 1990s. SAPs were imposed by financial agencies as a condition for the rescheduling of sovereign debt, at the expense of welfare programs, labor laws, and public health and education. These measures have taken distinctive forms depending on the country, but all in all, they have had pervasive effects on most of the population, as they led to the privatization of private assets, increased inequality, and the expansion of poverty.

Since the early 2000s, violence and dispossession have come, in great part, by a neo-extractivist turn that took place during the so-called "Pink Tide" (see Chapter 8) when many governments – most of them with a progressive and redistributive agenda – advanced a politics based on the reprimarization of the economy (see Chapter 17). This implies the extraction at a large scale of minerals using controversial techniques such as fracking, the mass production of transgenic crops, particularly soy, and the building of mega projects like dams for energy production. This politics, known as the "consensus of commodities" (Svampa, 2013), further aggravated previous sociospatial injustices, creating new ones, and led to all sorts of human and environmental rights abuses by companies and governments. In fact, some authors describe "sacrifice areas", "socially voidable" towns or rural areas (for instance Svampa, 2014), or "territories of exception" (Malheiro y Cruz, 2019) to refer to those places where the lives of certain populations (e.g., Indigenous peoples) are seen as less important than the need for so-called development.

This neo-developmental turn reveals a machinery of death, a resurgence of violence and dispossession, which led to the struggles that focus on the defence of life, as we will see in the following section.

The politicization of life

Faced with the processes of dispossession described above, social movements are constructing a narrative that situates the struggle in terms of "defence of life". Latin American academia has taken this into account and, in close dialogue with the movements, has created new concepts to describe this particular aspect of the struggle. For instance, Mina Navarro Trujillo (2020) defines the struggles for the defence of life as "the actualization of a way of managing the political to organize one's own existence in interdependence with others, placing at the center the reproduction of human and non-human life" (2020: 94). Some geographers propose to use the concept of "r-existence" or "re-existence" to refer to this dimension of struggles that implies the "power of restarting, of regeneration, of giving new senses or renewing the senses of existence" (Hurtado and Porto-Gonçalves, 2022: 5; see also **Chapter 12**). Other concepts used by geographers and other scholars that are important to current struggles for the defence of life in the region are "body-territory" (for instance, Cruz Hernández, 2016) and territory-body (of the earth) discussed by Haesbaert (2020) (see **Chapter 14**). Both speak to the inseparability of body and territory as it has been established by many Indigenous movements.

Our proposal speaks to this literature, but we focus on the very ideas of life-death and the "defence of life" as categories of political practice built by movements to give an account of the severity and violence of the extractivist neoliberal wave in Latin America. We follow here the distinction proposed by Haesbaert (2014) between categories according to their uses and contexts of enunciation. In this sense, the categories of political practice are those that arise and are activated in the context of struggle with political purposes and meanings. We consider life in these terms: life as an object of demand and defence by social movements, that is understood in a broad sense (human and non-human, life as existence, enjoyment) and that it is used to expresses a situation of violence as a result of socioeconomic processes that impose "death" on them, whether it is a real or concrete idea of death (such as the assassination of community referents and members, femicides, etc.) or a figurative one (the deterioration of the conditions for existence, such as the depredation of natural resources and consequently of certain ways of life, or the poverty and indebtedness that limits the possibility of women to free from violent relationships).

Space is at the center of the struggles and the defence of life – not only in terms of territory or body-territory, as has already been highlighted by different authors, but also in terms of public space (see **Chapter 20**). The appropriation of public space through marches, performances, and occupations is one of the most powerful instruments used by movements to make their struggles visible. In doing so, they affect the production of public space, creating new geographies, even if they are ephemeral. Urban public space, especially, becomes at the same time an object and a means of struggle, as we will discuss through these cases.

The struggles of the Lenca people in Honduras

The Lenca people's struggle in Honduras constitutes a paradigmatic example to analyze how popular mobilizations have multiplied and reorganized around the idea of "defence of life" since the 1990s. At that time, neoliberal economic reforms prompted the rollback of the agrarian reform and processes of land grabbing (see **Chapter 22**), which resulted in land concentration. Also, the opening of the economy to international competition and foreign investment was facilitated by free trade agreements such as the Puebla Panama Plan (see **Chapter 6**). All of this gave way to an intense cycle of protest and social conflict led by workers, peasants, teachers, and students, but also by new social actors: Indigenous and Afro-descendant people, women's and feminist organizations, environmentalists,

Figure 23.1 Image of Berta Cáceres against backdrop of protest with signs that read: "Water will NOT be sold, it is cared and defended!" and "Waters for life".

Source: Copinh Honduras (April 6, 2021) #JusticiaParaBerta. Facebook. www.facebook.com/copinh. intibuca/photos/a.1609398112667295/2853575811582846/?type=3

and regional territorial movements (Sosa, 2015; Posas, 2019), particularly in rural areas, although they also had strong repercussions and supporters in urban areas. These mobilizations denounced the different forms of violence and dispossession that accompanied the neoliberal escalation, with particular attention to that which resulted from resource exploitation – such as mining, forestry, agricultural, energy – and touristic activities. The 2009 coup that ousted President Zelaya sought greater state and corporate control over the areas of the country that were of interest to national and transnational companies for resource exploitation. In addition, these also were areas with greatest mobilization and resistance. As resistance grew strong, so did persecution, threats and even murder, particularly of the leaders of the organizations.

Of all these movements, we focus on the Civic Council of Popular and Indigenous Organizations of Honduras (COPINH), an organization led by Berta Cáceres and a symbol of popular resistance in the country (see Figure 23.1). COPINH is an Indigenous-based organization that articulates different types of struggle defined as anti-capitalist, anti-racist, and anti-patriarchal. In COPINH's discourse and actions it is possible to identify three axes of violence that hinder Honduran Indigenous, peasant, and Garífuna's lives around which struggles organize: a) the exploitation of resources and the construction of infrastructure; b) the militarization of the territory and criminalization of struggles; and c) the assassination of leaders and members of the organizations.

The defence of life is organized against the "projects of death", an idea used by COPINH to refer to the violence that comes with "development" projects, in particular forestry and timber activities – which have received a great impulse since the beginning of the 1990s – and the construction of dams for the generation of hydroelectric power in the following decade. It is a way to describe the magnitude of the impact that these projects have on the lives of the populations, particularly those living in the exploitation zones.

In relation with forestry and timber activities, for example, they refer to the idea of "plundering of life" (Berta Cáceres, in Korol, 2018: 27) to express the impact of indiscriminate logging, not only in terms of biodiversity but also in terms of the alteration of the hydrological cycle and the drying up of natural water sources. The idea of looting also seeks to focus on the illegality and corruption associated with logging in Honduras. That is why mobilizations around the defence of forest around Tegucigalpa (the main city of the country) were named as "Marches for Life".

Public space is also appropriated through performances. Symbolic "crucifixion" (Indigenous Lenca people tied themselves to crosses) in front of the House of Government, were used to represent both the death and the suffering experienced by the populations directly affected by neoliberal policies.

Another dimension of this violence around which struggles are organized is the construction of dams. After the 2009 coup d'état, the government opened the concession of water resources to companies for periods of between 20 and 30 years, which actually meant the privatization of the country's water resources and a serious harm to the entire Honduran population, particularly the communities living in the areas where the dams were to be built. A paradigmatic case against the construction of dams is the struggle organized by COPINH to protest for the Agua Zarca project on the Gualcarque river, considered sacred by the Lenca community. Since April 2013, the community of Río Blanco are fighting for the definitive cancellation of Agua Zarca under the slogan: "Freedom for the rivers!" showing the meaning that rivers have for the communities: rivers as spiritual entities that are being "captured" by dams.

The defence of life against "projects of death" also alludes to what surrounds the deployment of these projects, such as the militarization of the territory and the constant threat to the physical integrity of residents and activists. Even though this is neither new nor is exclusive to Honduras, it has deepened in the last decade as resistance against neoliberal projects grew. The criminalization of protest and the persecution of leaders and members of popular organizations increased after the coup with the passing of anti-terrorist laws that enabled the intervention of the army.

A key event of violence against organizations was the assassination of Berta Cáceres in March 2016 by hired assassins sent by a company involved in Agua Zarca's project. Even though Berta had suffered persecution, death threats, arrests, and prosecution for years, she never gave up on the fight. In fact, the "defence of life" contemplated, to some extent, her own death. In relation to the struggles they faced against the construction of dams, Berta expressed:

> It is not understood why there are such fierce and profound struggles in defence of the common goods, which are the basis of the unique balance of life. It seems to me that the issue of autonomy, of self-determination, allows us not to lose our identities, because if there is something the gringos and imperialism in general are betting on, it is to destroy our diverse identities. They know that it is one of our weapons, and they know that if they finish with this, they have free reign over everything else. That is why an Indigenous person clings so much to the land, to stay and give his life for that, for the water, for the land.
>
> *(Berta Cáceres, in Korol, 2018: 32–33, authors' translation)*

Several COPINH leaders were also killed after Berta, as well as so many others in Honduras. The "defence of life" thus emerges in the face of a politics of death encompassing not only the destruction of the means of existence of several communities, but also the actual murder of Indigenous leaders and activists of Honduran movements of resistance. It is interesting to see how the deaths of the leaders is resignified around the idea of life that feeds back the struggle. For example, COPINH refers to the death of Berta as "the sowing of Berta", a way of resignifying the deaths of those who have fought for the defence of life, through an idea that symbolically accounts for the reproduction of the struggle, a death that was not in vain, because it leaves a seed germinating.

The "Ni Una Menos" movement

Another case that provides a window to look at the centrality of the defence of life in popular struggles in Latin America is the movement Ni Una Menos (NUM)/Not One Less (see Figure 23.2). NUM emerged in Argentina in 2015 after the murder of Chiara Perez on May 9th, a 14-year-old girl who was two months pregnant. Despite the existence of the law "Comprehensive Protection to Prevent, Punish and Eradicate Violence Against Women" (Law 26.485) passed in 2009, violence against women had continued to grow in Argentina, providing the horrific number of one woman killed every 30 hours and 277 women killed only in 2014.

NUM recognizes itself as part of a historical movement made by previous – and ongoing – struggles led by women for the protection and expansion of rights, who might or might not have been organized around a gender identity, like the Mothers and Grandmothers of Plaza de Mayo and the National Campaign for the Right to Legal,

Figure 23.2 "My body, my rules", a Ni Una Menos protest in Argentina.
Source: Vanina Dolce (March 8, 2018).

Safe and Free Abortion, among others (see NUM's Organic Chart). Even though femi-nist activism had accumulated numerous achievements through the years, NUM made the long-dated claims and struggles visible in a way in which they could no longer be ignored in the public and political agendas.

NUM/Ni Una Menos refers to a poem by Mexican poetess Susana Chavez and it means that not one more woman's (and gender non-conforming people's) life must be lost to gender violence. The initial agenda of NUM included the demand for the effec-tive implementation of Law 26.485 and the compilation and publication of official sta-tistics on violence against women, among other things. These claims were broadened as NUM spread to other countries in the region and LGTBQ+, Afro-descendant, and Indigenous movements joined it.

Signs with phrases like "*vivas nos queremos*/we want ourselves alive" are frequent in NUM's marches. Besides claiming for the protection of life itself (Rose, 2001), this assertion subverts the potentially passive role of women victims of violence. It bestows agency on women who, as a collective, want to and can protect their lives and their right to "*vivir las vidas que queremos vivir*/live the lives we want to live". That is, to live a life without mandates that if not followed, puts life at risk.

Through the documents and actions of NUM we identify three types of interrelated violence (which do not exhaust other possible forms): physical/emotional violence, institutional/state violence, and economic violence.

Physical violence and emotional/psychological violence (harassment and verbal abuse) are the most evident forms of violence against women and gender non-conforming peo-ple, with murders being the most extreme manifestation. The gendered nature of these deaths is represented in the term *femicide*, the conjunction of female and homicide. This term denaturalizes women's deaths and allows us to understand that these are not random cases or coincidences, but the violence that results from a patriarchal system that works towards controlling women's bodies and lives, a system that exerts its vio-lence when women rebel against an unwanted relationship or an expected way of life. Similarly, the terms *travesticide/transfemicide* signal to the punishment of those who rebel against the patriarchal and heterosexual pact.

NUM denounces this violence as a disciplining tool to remind women that they cannot decide for themselves, be independent and exert their autonomy as subjects with their own desires and wishes. Thus, the defence of life implies the right to refuse, choose, and decide for oneself as an autonomous and free subject. To refuse and not be killed, at the hands of a man or in a clandestine abortion facility, asserts a life that is dignified, free from poverty, and ruled by one's desire.

Institutional/state violence manifests in different ways, including the absence – up to 2015 – of official statistics about gender violence, the defunding of programs that assist victims of patriarchal violence, and importantly, the functioning of a patri-archal judicial system built over gender prejudices and stereotypes that re-victimize the victim of gender violence and leaves her/them unprotected – not to mention the participation of the judicial system (and the public forces) in networks of human and sexual trafficking.

The penalization of abortion is another manifestation of institutional violence. It forces pregnant bodies who do not want to continue with the pregnancy to resort to clan-destine practitioners, which, in the case of poor women, happens under very precarious and risky conditions. The idea of "Legal abortion is life. Without legal abortion there is not Ni Una Menos" shows that clandestine abortions and femicides are in fact related as part of a same system that controls women's lives and bodies with the same outcome. Clandestine abortions are seen as femicides in the hands of the state that opposes legal, safe, and free abortion. Even though Argentina passed law 27.610 that de-criminalized abortion in 2020, difficulties in its implementation continue to affect women's lives.

Another form of violence happens when economic hardships affect women's ability to refuse. NUM has pointed to the relationship between neoliberalism and gender violence and the need to create a "social commitment to build a new never again". The phrase "never again" – Nunca Más – was coined by Human Rights Organizations to mean that there could never be the conditions for the installation of an illegal and criminal regime, like the one that ruled the country between 1976 and 1983. The use of the Nunca Más connects the violence of the past, that drove the national economy into debt, with economic reforms and indebtedness expressed today in the feminization of poverty and the shrinking of the safety net that burdens women with more reproductive/care work, and the impossibility for women to get free from violent partners.

In the face of all these violences, most of which women suffer in their private spaces hidden from view, NUM calls for the creation of spaces to share women's experiences and collectively make sense of them. Coming together in public space is a key and powerful feature of the movement as a scenario where solidarity and sorority take place against violence.

The relevance of encountering others is expressed in Manifesto #10 when it says that "[i]n the street, ideas always come out, forces accumulate, plots are woven". Acuerparse – roughly translated as "to make body" – in the streets emerges as a method and as a tool of resistance but also as an act of (re)creation. It means bringing together bodies that oppose patriarchal, capitalist, colonialist, and racist oppressions, but in a happy, even cheerful way – a way that subverts the potentially passive role of women victims of violence and bestows agency and power on women who together, as a collective, want to, and can protect their lives to live them as they wish. During NUM manifestations, bodies come in ways that disrupt and contest hegemonic modes of feminine embodiment (Sutton, 2010) – naked, painted, screaming, marching, playing drums – constituting themselves as a source of power that defies violence and death.

Conclusion

Neoliberal projects have historically had a violent manifestation in Latin America. Ever since they were implemented in the 1970s, at the hand of illegal military regimes, they have been opposed by numerous movements of resistance. In the last few years, life emerged as a concept around which struggles and resistance movements are organized in the region. These claims are no longer just about access to land and housing, better salaries, or the protection of the environment, but about a life that is dignified and that must be lived autonomously and freely. The politicization of life puts on a common plane the diverse struggles that are becoming more acute in the neoliberal context.

The politicization of life in Latin America has an important spatial dimension. In effect, the spatial dimension in terms of territory, appropriated and defended space, becomes central. The defence of life implies the struggle for the space that contains it, whether it be the river, the forest, or even the body itself. Also, public space and its appropriation becomes central to the struggle. The marches, the performances, the road blockades in places full of symbolism (such as the House of Government, the Parliament, the road leading to the dam) are constitutive of these movements: without this appropriation of public space that allows them to make their demands, visible movements would not exist. Importantly, by meeting in public space and gathering strength from being together (acuerparse), the movements build a collective body with which to oppose death projects and defy them by holding to life, a dignified life even in the face of violence.

In sum, the defence of life not only means opposing a neoliberal and patriarchal "system of death", but more importantly, having a "life worth living".

Summary

- In Latin America, neoliberalism has produced different and interrelated forms of violence triggering the response of resistance movements.
- Numerous social movements in the region have politicized life by situating its defence, "the defence of life", at the centre of their struggles against neoliberalism and patriarchy.
- Movements mobilize "life" as a category of political practice through different and interconnected ideas, expressions, and performances.
- Space, its defence, and its appropriation are central to the strategies of movements. Space is not only important for subsistence, but also as a means through which to encounter one another in the struggle and gather strength.

Review questions

1. Can you think of other examples, from chapters in the book or elsewhere, of the politicization of life?
2. What are the geographies of life and how are they turned into the basis for political action?
3. Why is gender such an important dimension to the politicization of life?

Further reading

Littler, J. & Gago, V. (2022). "We want ourselves alive and debt free!," *Soundings*, 80(80), 9–21.
Interview with Verónica Gago, feminist scholar and member of NUM, that dwells on the meaning of women's strike and the relation between debt and gender violence.
Navarro Trujillo, M. L. (2020). "Struggles in defense of life within the context of dispossession and capitalist violence in Mexico: A closer look through the lens of the production of the common," *Latin American and Caribbean Ethnic Studies*, 1–20.
Overview of the recent proliferation of Indigenous and peasant struggles focused on the defence of life in Mexico.

Keywords

Defence of life: this term refers to the struggles that focus on defence of common goods, the environment, the right to choose, the right to live according to different identities, in sum, life itself.

Neoliberal violence: violence that is intrinsic to neoliberal doctrines, policies, and institutional arrangements in their prosecution of profit and capital accumulation.

Politicization of life: this term refers to the central place that life occupies in Latin American movements' claims and political strategies. The politization of life implies opposing the politics of death brought about by neoliberal economic, social, and political measures.

Politics of death: this term refers to effects of and violences implied in the "development projects" that are promoted by all Latin American countries, whether they are under progressive or conservative governments.

References

Cruz Hernández, T. (2016). "Una mirada muy otra a los territorios-cuerpos femeninos," *Solar*, 12(1), 45–46.

Haesbaert, R. (2014). *Viver no limite. Território e multi/transterritorialidade em tempos de insegurança e contenção*. Bertrand Brasil.

Haesbaert, R. (2020). "Do corpo-território ao território-corpo (da terra). Contribuições decoloniais," *GEOgraphia*, 22(48).

Hurtado, L. & Porto-Gonçalves, C. (2022). "Resistir y re-existir," *GEOgraphia*, 24(53), w/p.

Korol, C. (2018). *Las revoluciones de Berta*. América Libre.

Malheiro, B. & Cruz, V. C. (2019). "Geo-graphias dos grandes projetos de des-envolvimento: territorialização de exceção e governo bio/necropolitico do território," *GEOgraphia*, 21(46).

Navarro Trujillo, M. (2020). "El antagonismo de las luchas en defensa de la vida como proceso de repolitización de lo social en América Latina. Un diálogo con Juan Pablo Pérez Sáinz," *Encartes*, 2(4), 88–97.

Posas, M. (2019). "Movimientos sociales en Honduras", in R. Romero (comp.), *Antología del pensamiento hondureño Contemporáneo*. CLACSO, pp. 259–292.

Rose, N. (2001). "The politics of life itself," *Theory Culture Society*, 18(1), 1–30.

Sosa, E. (2015). *Democracia, procesos electorales y movimientos sociales en Honduras: de la transición política al golpe de Estado*. Documento de Trabajo, CLACSO.

Sutton, B. (2010). *Bodies in crisis: Culture, violence, and women's resistance in neoliberal Argentina*. Rutgers University Press.

Svampa, M. (2013). "Consenso de los Commodities" y lenguajes de valoración en América Latina," *Nueva Sociedad*, 244, 30–46.

Svampa, M. (2014). "¿Territorios vacíos o Territorios en disputa? Las sociedades locales, ¿entre las promesas incumplidas del desarrollo regional y el establecimiento de zonas de sacrificio? ¿Compite el avance de la explotación de yacimientos no convencionales con economías regionales preexistentes?" in *20 mitos y realidades del fracking*. El Colectivo, pp. 147–159.

Glossary of technical terms

Abya Yala: this name means in the language of the Kuna people (Panama and Colombia) 'land of vital blood,' and has recently been used by activists of Indigenous movements and by people committed to an opposition to cultural colonialism to designate the Americas as a whole, but especially so-called 'Latin' America. This latter expression has been heavily criticized, not only for its colonial origins (more specifically, nineteenth-century French imperialism), but also and above all for the fact that most of the continent's population is not simply or exclusively 'Latin,' but rather, in greater or lesser degree, descendant and culturally tributary of native and African peoples.

Agency: the capacity of taking autonomous initiatives, especially referring here to the action of subalterns performing revolts without waiting for instructions from above.

Agrarian reform: processes oriented to re-distribute land concentrated in large estates or *latifundios*. Some agrarian reform processes in Latin American countries were promoted by revolutionary movements (such as Mexican or Cuban reform), but others originated by top-down policies to prevent and/or contain emerging revolutionary processes.

Anglophone feminist geography: defined as feminist geography that is written in English and addressed to a Global North audience.

Balance of payments: the balance of payments monitors a nation's economic interactions worldwide, covering trade and financial transactions.

Body-territory: our first territory, defended mainly by (Afro-descendant and Indigenous) women, and which extends relationally to other scales in which these bodies transit and fight, in short, expand and become effective and affective spaces.

Boomerang effect: when a political actor accesses global arenas or actors to generate a feedback to their local or national scale as a political pressure or command.

Capital intensive goods: goods that require a substantial investment in capital, such as machinery and equipment, relative to labor costs, in their production processes.

Civilization: during the eighteenth century, influenced by the Enlightenment, the concept of civilization combined the idea of a process and a condition of organized social life. The idea of process involved the potential for moving away from an 'original barbarism' through a 'collective and uninterrupted' refinement. This was represented by European societies, which epitomized refinement and order. Thus, under these ideas and values, Europe was given the mission to civilize colonial populations. The idea of civilization as a condition celebrates modernity values like social order or order knowledge.

Clientelism: the exchange of political support for goods or services.

Coloniality of power: a hegemonic system of power based in racial and post-colonial difference. It is not a 'theory,' but a set of theoretical and political agendas brought by Latin American thinkers (led by Peruvian sociologist Anibal Quijano) from the 1990s. They argued that colonization was not only a process of territorial occupation, administration, and exploitation by the Europeans, but the establishing of an entangled package of power relations involving race, class, gender, culture, spirituality, and knowledge, among others. Moreover, those power relations remained after the end of colonization.

Communitarian feminisms: Indigenous women's political action and resistance practices against patriarchy, colonialism, and capitalism.

Creole: member of a cultural group resulting from colonial (voluntary or forced) settlement. In decolonial scholarship this term often indicates white Latin American settlers, but there are also, for instance, 'Black creole' identities in the Caribbean.

Delegative democracy: democratically elected governments, with a majority base of support (commonly appealing to the ballotage), but without the exercise of horizontal accountability, that is, governing without interference from the legislative, judicial and control bodies.

Ethnogeodiversity: while the term 'ethnodiversity' is already well known (as is the concept of biodiversity), 'geodiversity' is a more recent concept, and which has been explored in a very limited way, with an excessive emphasis on the geological and geomorphological diversity of the Earth. 'Ethnogeodiversity,' by uniting the two concepts, draws attention to the interrelations between ethno-cultural and geographic diversity, in a broad sense.

Favela: marginalized neighborhood in Brazilian cities. The fact that these neighborhoods are self-constructed and can be to some extent self-governed allowed historical geographers to consider early *favelas* as spaces of resistance.

Femicide: the killing of women or gender diverse peoples because of their gender.

Feminicide: hate crimes against women forged by social and state tolerance to systemic gender-based violence.

Fordism: mass production system with assembly line methods and high capital investment.

Garífuna: ethnic group originating from the encounter of African slaves, Arawaks, and Caribs in the seventeenth century. They are settled mostly in Guatemala and Honduras.

Global North: developed countries with advanced economies and higher living standards.

Gross domestic product (GDP):   measure of total value of goods and services produced within a country.

Import substitution industrialization: strategy to promote domestic industries by substituting imports with local production.

Labor intensive goods: goods requiring substantial labor input for production.

Living territories: the territory as a living being implies that it has agency and rights in the face of the damages and violence that affect it.

Marronage: all over the Americas and beyond, and all along colonial and postcolonial history, the phenomenon of (mostly) African slaves (Maroons) revolting and escaping in the hinterlands to build their own communities.

Military juntas: supreme organ of the civic-military dictatorship composed of the heads of the three Armed Forces (Army, Navy, and Air Force).

Neoliberalism: an ideology and an economic doctrine based on the supremacy of the market, a vision that has its roots in an idealized conception of competitive individualism and the rejection of social solidarity.

Neo-Malthusian: This term derives from the name of Thomas Robert Malthus (1766–1834), English clergyman and economist, who asserted that while the growth of food supply is linear, population growth is exponential, which inevitably leads to hunger and misery. To face this supposed problem (greatly exaggerated by him, who underestimated, among other factors, the technological capacity to raise food production to previously unimaginable levels), Malthus advocated measures typical of his time and of a clergyman like him, such as sexual abstinence and celibacy. Later, in the twentieth century, neo-Malthusianism, 'updating' Malthus, began to advocate measures such as population control through less or more authoritarian means (from the distribution of contraceptives by the state to mass sterilization campaigns).

Plan Puebla Panamá: project of economic and infrastructure integration of Mesoamerican countries, oriented to facilitate the extraction and exportation of natural resources.

Political ontology: an approach that acknowledges there are multiple ways of approaching reality and the relation between them is a political one that is itself an element of decolonial struggle.

Politics of everyday: women's everyday practices become political defences of their livelihoods and territories.

Post-development: the critique of the very notions of 'development' and 'underdevelopment' as discourses elaborated in the North to serve economically, politically, and culturally the agendas of neocolonialism.

Post-Fordism: shift from a model of standardized production to flexible, customized production methods.

Quilombo/Palenque: respectively, the Portuguese and Spanish words to indicate Maroons' settlements, likewise an important object of study for (historical) geography.

Quilombola: Brazilian quilombo settler.

Race: a social construction that produces mixedness as well as discrimination and violence against Indigenous and Black people, where Europeanness and whiteness are crucial aspects in which Latin American societies are invested.

Reprimarization of the economy: the reorientation of the economy towards productive activities with little added value, predominantly primary-extractive and export oriented (e.g. agricultural, mineral, forest and energetic commodities).

Semi-peripheral (capitalist) countries: While the *core countries* are the leading countries of world capitalism (headquarters of the main transnational companies and, in some cases, large military and geopolitical players, typically being former colonial powers) and the *peripheral countries* are those that most clearly contrast with them (countries whose economies are basically agrarian), *semi-peripheral countries* are those that play a kind of 'intermediate role' in the dynamics of the capitalist world system, presenting expressive industrialization and great economic complexity as well as possessing significant geopolitical relevance on a hemispheric or international level, but at the same time showing economic weaknesses and, above all, huge social inequalities.

Structural adjustment policies (SAPs): neoliberal reforms imposed by the International Monetary Fund and the World Bank to national economies as a condition for new loans and the rescheduling of sovereign debt. National governments must adopt a program of economic stabilization and economic structure reforms in accordance with the requirements of these international organizations.

Terricide: the most violent form of deterritorialization, in which the loss of territory means the loss of one's own life (economic, political and in the culture-nature bond), as occurs with many Indigenous or original peoples.

Territory: space built by the exercise of power in its multiple dimensions (legal-political, economic, symbolic-affective), scales (starting with the body), and relationships (class, ethnic, gender, and with forces of nature), always involving relations of domination and resistance.

Trade liberalization: substantially reducing or eliminating barriers to trade.

Transterritoriality: a relational and multi-scalar strategy that engages within and beyond the state, providing an important dimension to struggles to decolonize territory, as experienced by many migrants, Afro-descendant, and Indigenous groups. When the movement of transit between different territories/territorialities is emphasised, we can also use the term multi-territoriality.

Urbanocentrism: analytical bias that consists of analyzing the world and its problems through exclusively urban lenses (and especially from the viewpoint of social life in large cities and metropolises), ignoring or enormously underestimating the richness of existing socio-spatial reality.

Zoonosis: infectious disease transmitted between animals and humans.

Index

Page numbers in *italics* indicate figures; page numbers in **bold** indicate tables.